SCHAUM'S OUTLINE OF

THEORY AND PROBLEMS

OF

MATHEMATICS
of FINANCE

Second Edition

•

PETR ZIMA
Adjunct Professor, Department of Statistics and Actuarial Science
University of Waterloo

and

ROBERT L. BROWN
Professor, Department of Statistics and Actuarial Science
University of Waterloo

•

SCHAUM'S OUTLINE SERIES
McGraw-Hill

New York   St. Louis   San Francisco   Auckland   Bogotá   Caracas
Lisbon   London   Madrid   Mexico City   Milan   Montreal   New Delhi
San Juan   Singapore   Sydney   Tokyo   Toronto

PETR ZIMA was awarded the M.Sc. by Charles University of Prague and the M.A. (Quantitative Methods) by the University of Economics in Prague. He presently serves on the faculties of three Canadian institutions - Conestoga College, Wilfrid Laurier University, and University of Waterloo—and is the coauthor of several works in financial mathematics and operations research.

ROBERT L. BROWN, F.S.A., F.C.I.A., A.C.A.S., holds the B. Math and M. Arts from the University of Waterloo, where he is currently Professor in the Department of Statistics and Actuarial Science. He is author of three books and coauthor of a recent textbook in the mathematics of finance.

Schaum's Outline of Theory and Problems of
MATHEMATICS OF FINANCE

6 7 8 9 10 11 12 13 14 15 16 17 18 19 20 VFM VFM 07 06 05 04

ISBN   0-07-008203-0

Sponsoring Editors: John Aliano and Arthur Biderman
Production Supervisor: Pamela Pelton
Editing Supervisor: Maureen Walker

Library of Congress Cataloging in Publication Data

Zima, Petr.
    Schaum's outline of theory and problems of mathematics of finance
    Petr Zima and Robert L. Brown. – 2nd ed.
        p. cm. – (Schaum's outline series)
    Rev. ed. of: Schaum's outline of theory and problems of
contemporary mathematics of finance.
    Includes index.
    ISBN 0-07-008203-0
    I. Business mathematics.   2. Business mathematics – Problems,
exercises, etc. I. Brown, Robert L., date . II. Zima, Petr.
Schaum's outline of theory and problems of contemporary mathematics
of finance. III. Title.   IV. Title: Mathematics of Finance
V. Series.
HF5691.Z54 1996   96-113
332.8′ 01′51–dc20   CIP

McGraw-Hill
A Division of The McGraw·Hill Companies

Preface

Knowledge in the field of the Mathematics of Finance and Actuarial Science has increased in importance because of the variability and high level of interest rates. The widespread adoption of inexpensive electronic calculators has also changed the level of problems that can be solved by any student.

This book supposes that the student will be using an electronic calculator that has a full range of functions, including a logarithmic function and exponential function. Because of this, interest tables are dispensed with.

The book is designed for use either as a supplement to any current textbook or as an independent text for a formal course in the Mathematics of Finance and Actuarial Science. It should also prove useful as a reference book and as a self-study guide.

Each chapter is broken into sections containing statements of principles and formulas, together with solved problems pertinent to each subtopic. These sections are followed by a set of supplementary problems with answers. There are also 225 review problems at the end of the book for extra review or testing. In all, there are 1224 problems, of which 320 are solved in full detail. These serve as a complete review of the material presented.

We would like to thank Lynda Clarke for the fine work she performed in typing the manuscript.

PETR ZIMA
ROBERT L. BROWN

Contents

        10.1 Introduction   201
        10.2 Probability   201
        10.3 Mathematical Expectation   204
        10.4 Contingent Payments with Time Value   206

Chapter 11   LIFE ANNUITIES AND LIFE   **213**
              INSURANCE

        11.1 Introduction   213
        11.2 Mortality Tables   213
        11.3 Pure Endowments   214
        11.4 Life Annuities   216
        11.5 Life Insurance   218
        11.6 Annual-Premium Policies   220

        REVIEW PROBLEMS   227

        APPENDIX A   245
        APPENDIX B   246

        INDEX   248

Chapter 1

Exponents and Logarithms

1.1 EXPONENTS

The product $a \cdot a \cdot a \cdot a \cdot a$ may be abbreviated to a^5, which is known as the *fifth power* of a. The symbol a is called the *base*; the number 5, which indicates the number of times the base a is to appear as a factor, is the *exponent*.

For a first power, we have $a^1 = a$; a second power is called a *square*; a third power is a *cube*.

SOLVED PROBLEMS

1.1
(a) $x^4 = x \cdot x \cdot x \cdot x$

(b) $625 = 5 \cdot 5 \cdot 5 \cdot 5 = 5^4$

(c) $1\,000 = 2 \cdot 2 \cdot 2 \cdot 5 \cdot 5 \cdot 5 = 2^3 \cdot 5^3$

(d) $(49)^3 = 49 \cdot 49 \cdot 49 = 7 \cdot 7 \cdot 7 \cdot 7 \cdot 7 \cdot 7 = 7^6$

(e) $(1 + i)^5 = (1 + i)(1 + i)(1 + i)(1 + i)(1 + i)$

(f) $(1 - d)^n = (1 - d)(1 - d) \cdots$ to n factors

1.2 LAWS OF EXPONENTS

If m and n are positive integers and $a \neq 0$, $b \neq 0$, we have:

$$a^m \cdot a^n = a^{m+n} \tag{1.1}$$

$$\frac{a^m}{a^n} = a^{m-n} \quad (\text{when } m > n) \tag{1.2}$$

$$(a^m)^n = a^{mn} \tag{1.3}$$

$$(ab)^n = a^n b^n \tag{1.4}$$

$$\left(\frac{a}{b}\right)^n = \frac{a^n}{b^n} \tag{1.5}$$

SOLVED PROBLEMS

1.2 Prove (a) (1.1), (b) (1.2), (c) (1.3).

(a) $a^m \cdot a^n = (a \cdot a \cdots$ to m factors$)(a \cdot a \cdots$ to n factors$)$
$= (a \cdot a \cdots$ to $m + n$ factors$) = a^{m+n}$

(b) $\dfrac{a^m}{a^n} = \dfrac{a \cdot a \cdots \text{ to } m \text{ factors}}{a \cdot a \cdots \text{ to } n \text{ factors}} = a \cdot a \cdots$ to $m - n$ factors $= a^{m-n}$ $(m > n)$

(c) $(a^m)^n = a^m \cdot a^m \cdots$ to n factors $= a^{m+n+\cdots}$ to n terms $= a^{mn}$
in which the second equality follows from (a).

1.3 (a) $2^3 \cdot 2^2 = 2^{3+2} = 2^5 = 32$ (f) $(2^2)^3 = 2^{2\cdot3} = 2^6 = 64$

(b) $x^4 \cdot x^2 = x^{4+2} = x^6$ (g) $(x \cdot y)^3 = x^3 y^3$

(c) $\dfrac{a^8}{a^4} = a^{8-4} = a^4$ (h) $(2a^2)^3 = 2^3 a^6 = 8a^6$

(d) $\dfrac{3^6}{3^2} = 3^{6-2} = 3^4 = 81$ (i) $\left(\dfrac{x}{y}\right)^4 = \dfrac{x^4}{y^4}$

(e) $(a^3)^3 = a^{3\cdot3} = a^9$ (j) $\left(\dfrac{2a^2}{b^3}\right)^2 = \dfrac{2^2 a^4}{b^6} = \dfrac{4a^4}{b^6}$

1.3   ZERO, NEGATIVE, AND FRACTIONAL EXPONENTS

We extend the notion of an exponent to include zero, negative integers, and common fractions by the following definitions:

$$a^0 \equiv 1 \quad (a \neq 0)$$
$$a^{-n} \equiv \frac{1}{a^n} \quad (n \text{ a positive integer})$$
$$a^{m/n} \equiv \sqrt[n]{a^m} \quad (m \text{ and } n \text{ positive integers})$$

It can be shown that the laws of exponents, (1.1) through (1.5), hold when m and n are rational numbers (that is, positive and negative integers and common fractions, and zero), and even when m and n are arbitrary real numbers.

SOLVED PROBLEMS

1.4 (a) $1 = \dfrac{5^3}{5^3} = 5^{3-3} = 5^0$ (e) $(27)^{-2/3} = (3^3)^{-2/3} = 3^{-2} = \dfrac{1}{3^2} = \dfrac{1}{9}$

(b) $2^{-2} + 3^{-1} = \dfrac{1}{2^2} + \dfrac{1}{3} = \dfrac{1}{4} + \dfrac{1}{3} = \dfrac{3+4}{12} = \dfrac{7}{12}$ (f) $\left(\dfrac{a^3}{a^{-3}}\right)^{1/2} = (a^3 a^3)^{1/2} = (a^6)^{1/2} = a^3$

(c) $(49)^{1/2} + (8)^{1/3} = \sqrt{49} + \sqrt[3]{8} = 7 + 2 = 9$ (g) $\sqrt{\dfrac{x^{-2}}{y^6}} = \left(\dfrac{x^{-2}}{y^6}\right)^{1/2} = \dfrac{x^{-1}}{y^3} = \dfrac{1}{xy^3}$

(d) $(8)^{2/3} = (8^{1/3})^2 = (\sqrt[3]{8})^2 = 2^2 = 4$ (h) $(1.12)^8 (1.12)^{-2/3} = (1.12)^{8-(2/3)}$
$= (1.12)^{22/3}$

1.5 The following evaluations are made with a pocket calculator:

(a) $\sqrt[4]{5} = 5^{1/4} = 1.495348781$ (d) $\dfrac{5}{\sqrt[6]{809}} = \dfrac{5}{809^{1/6}} = 5(809^{-1/6})$
$= 1.637992532$

(b) $\sqrt[3]{\dfrac{2015 \times 0.2389}{21.87}} = \left(\dfrac{2015 \times 0.2389}{21.87}\right)^{1/3} = 2.802512$ (e) $\dfrac{(1.07)^{15} - 1}{0.07} = 25.12902201$

(c) $8500(1.083)^{-5} = 5705.275787$ (f) $\dfrac{1 - (1.07)^{-15}}{0.07} = 9.107914005$

1.6 Applying the laws of rational exponents, we use a pocket calculator to solve the following exponential equations with the unknown in the base:

(a)

$$700(1+i)^{12} = 1400$$
$$(1+i)^{12} = 2$$
$$1+i = 2^{1/12}$$
$$i = 2^{1/12} - 1$$
$$i = 0.059463094$$

(b)

$$15\,000(1-d)^{11} = 800$$
$$(1-d)^{11} = \tfrac{800}{15\,000}$$
$$1-d = \left(\tfrac{800}{15\,000}\right)^{1/11}$$
$$d = 1 - \left(\tfrac{800}{15\,000}\right)^{1/11}$$
$$d = 0.233922668$$

(c)

$$(1+i)^{12} = (1.08)^2$$
$$1+i = (1.08)^{2/12}$$
$$i = (1.08)^{1/6} - 1$$
$$i = 0.012909457$$

(d)

$$\frac{(1+i)^6 - 1}{.03} = 10$$
$$(1+i)^6 = 1.3$$
$$1+i = (1.3)^{1/6}$$
$$i = (1.3)^{1/6} - 1$$
$$i = 0.044697508$$

1.4 LOGARITHMS

Let N be a positive number, and let b be a positive number different from 1. Then the *logarithm, base-b*, of the number N is the exponent L on base-b such that $b^L = N$. The statement that L is the logarithm, base-b, of N is written briefly as $L = \log_b N$. For example,

$$\log_2 16 = 4, \text{ since } 2^4 = 16 \quad \text{and} \quad \log_5 125 = 3, \text{ since } 5^3 = 125$$

Logarithms, base-10, are called *common logarithms* and are used as aids in computation. We shall write simply $\log N$ instead of $\log_{10} N$ for the common logarithm of a number N. Hereafter, the word "logarithm" will refer to a common logarithm. By definition,

$$\log 1000 = 3, \text{ since } 10^3 = 1000 \qquad \log 0.1 = -1, \text{ since } 10^{-1} = 0.1$$
$$\log 100 = 2, \text{ since } 10^2 = 100 \qquad \log 0.01 = -2, \text{ since } 10^{-2} = 0.01$$
$$\log 10 = 1, \text{ since } 10^1 = 10 \qquad \log 0.001 = -3, \text{ since } 10^{-3} = 0.001$$
$$\log 1 = 0, \text{ since } 10^0 = 1$$

It should be noted that, while N is required to be a positive number, $\log N$ may be any real number — positive, negative, or zero. Since logarithms are (real) exponents, laws (1.1), (1.2), and (1.3) for exponents immediately give us the three fundamental laws of logarithms:

1. The logarithm of the product of two positive numbers is the sum of the logarithms of the numbers.

$$\log A \cdot B = \log A + \log B \tag{1.6}$$

2. The logarithm of the quotient of two positive numbers is the logarithm of the numerator minus the logarithm of the denominator.

$$\log \frac{A}{B} = \log A - \log B \tag{1.7}$$

3. The logarithm of the kth power of a positive number is k times the logarithm of the number.

$$\log A^k = k \log A \qquad (1.8)$$

In law 3, k may be any real number.

SOLVED PROBLEMS

1.7 A pocket calculator or a table of logarithms gives: $\log 2 = 0.301030$, $\log 3 = 0.477121$.

(a) $\quad \log 6 = \log(2 \times 3) = \log 2 + \log 3 = 0.301030 + 0.477121 = 0.778151$

(b) $\quad \log 1.5 = \log \frac{3}{2} = \log 3 - \log 2 = 0.477121 - 0.301030 = 0.176091$

(c) $\quad \log 8 = \log 2^3 = 3 \log 2 = 3(0.301030) = 0.903090$

(d) $\quad \log 200 = \log(2 \times 10^2) = \log 2 + \log 10^2 = 0.301030 + 2 = 2.301030$

(e) $\quad \log 0.003 = \log(3 \times 10^{-3}) = \log 3 + \log 10^{-3} = 0.477121 + (-3) = 0.477121 - 3$

(f) $\quad \log \sqrt[3]{2} = \log 2^{1/3} = \frac{1}{3} \log 2 = \frac{1}{3}(0.301030) = 0.100343$

1.5   CHARACTERISTIC AND MANTISSA; ANTILOGARITHMS

Any positive number can be written as a so-called *basic number* B $(1 \le B < 10)$ multiplied by an integral power of 10. For example:

$$5836 = 5.836 \times 10^3 \qquad 0.0032 = 3.2 \times 10^{-3}$$

Taking logarithms of these numbers, we obtain

$$\log 5836 = \log 5.836 + 3 \qquad \log 0.0032 = \log 3.2 - 3$$

The logarithm of a basic number is a nonnegative decimal fraction (since $\log 1 = 0$ and $\log 10 = 1$), and the logarithm of an integral power of 10 is, by definition, an integer. Thus, the logarithm of a positive number may be thought of as consisting of:

(i) An integral part, called the *characteristic*. The characteristic is the logarithm of the integral power of 10 and is determined solely by the position of the decimal point in the number. The characteristic may be any whole number (integer), positive, negative, or zero.

(ii) A decimal part, called the *mantissa*. The mantissa is the logarithm of the basic number and is determined by the sequence of digits in the number without regard to the position of the decimal point. The mantissa is a positive decimal (or zero, if the number happens to be an integral power of 10).

Standard textbooks in the mathematics of finance usually contain a table of mantissas (rounded off to six decimal places) of all numbers of four or fewer digits, that is, all numbers from 1.000 to 9.999. No such table is included in this Outline, as it is assumed that each student has a pocket calculator with a built-in logarithmic function. It should be noted that the logarithm of a number N, such that $0 < N < 1$, will be shown on the display of the calculator as a single negative number, the sum of the negative characteristic and the positive mantissa. In this case, the decimal part of the displayed negative number *does not* represent the mantissa. For example, if

$$N = 0.002 = 2 \times 10^{-3}$$

the characteristic is -3 and the mantissa is $\log 2 = 0.301030$. The calculator will display their sum,

$$\log N = -2.698970$$

and it is seen that 0.698970 is not the mantissa of $\log N$, but 1 minus the mantissa.

So far we have discussed the problem of finding the logarithm L of a given positive number N. Another problem when using logarithms in computation is: given the logarithm L of a number N, to find N. The number that corresponds to a given logarithm is called the *antilogarithm*; we write $N = \text{antilog } L$ when $\log N = L$. For example:

$$\text{antilog } 1.301030 = 20 \quad \text{since} \quad \log 20 = 1.301030$$

$$\text{antilog } (0.845098) = 0.7 \quad \text{since} \quad \log 0.7 = 0.845098 - 1$$

To calculate antilog L on a pocket calculator, use the inverse logarithmic function (INV LOG or 10^x), as explained in the owner's manual.

SOLVED PROBLEMS

1.8 Using a pocket calculator, find the characteristics and the mantissas of the logarithms of the numbers in the table below.

Number	Scientific Notation	Characteristic	Mantissa
248.3	2.483×10^2	2	0.39497672
3.81	3.81×10^0	0	0.580924976
0.08972	8.972×10^{-2}	-2	0.952889265
0.000006	6.0×10^{-6}	-6	0.77815125
303 891	3.03891×10^5	5	0.482717838

1.9 Given $\log 528 = 2.722633923$.

(*a*) antilog $0.722633923 = 5.28$ (*c*) antilog $(0.722633923 - 2) = 0.0528$

(*b*) antilog $4.722633923 = 52\ 800$ (*d*) antilog $(0.722633923 - 5) = 0.0000528$

1.10 Using a pocket calculator, find the antilogarithm in the table below.

L = log N	N = antilog L
1.3	19.95262315
2.8181726	657.9192597
-2.32587	0.004722044
-0.12356	0.752384777

1.6   COMPUTING WITH LOGARITHMS

The importance of logarithms in computations has diminished with the development of electronic calculators, which directly perform long multiplications or divisions, and take powers or roots. However, to solve exponential equations for an unknown exponent (see Problems 1.13, 1.14, 1.15), logarithms must still be used.

Problems 1.11 and 1.12 illustrate the use of logarithms in situations where they provide an alternate method of solution.

SOLVED PROBLEMS

1.11 Determine i (which represents the interest rate per interest conversion period) if

$$800(1+i)^{20} = 5000$$

Using logarithms, we have

$$
\begin{aligned}
\log 800 + 20\log(1+i) &= \log 5000 \\
20\log(1+i) &= \log 5000 - \log 800 \\
\log(1+i) &= \frac{\log 5000 - \log 800}{20} \\
\log(1+i) &= 0.039794001 \\
1+i &= 1.095958226 \\
i &= 0.095958226 \approx 9.60\%
\end{aligned}
$$

Direct solution without the aid of logarithms:

$$
\begin{aligned}
800(1+i)^{20} &= 5000 \\
(1+i)^{20} &= \frac{5000}{800} \\
1+i &= \left(\frac{5000}{800}\right)^{1/20} \\
i &= \left(\frac{5000}{800}\right)^{1/20} - 1 \\
i &= 0.095958226 \approx 9.60\%
\end{aligned}
$$

1.12 Determine d (the annual compound rate of depreciation) if $55\,000(1-d)^9 = 7000$.

Using logarithms, we have

$$
\begin{aligned}
\log 55\,000 + 9\log(1-d) &= \log 7000 \\
9\log(1-d) &= \log 7000 - \log 55\,000 \\
\log(1-d) &= \frac{\log 7000 - \log 55\,000}{9} \\
\log(1-d) &= -0.09947385 \\
1-d &= 0.795291151 \\
d &= 0.204708849 \approx 20.47\%
\end{aligned}
$$

Direct solution without the aid of logarithms:

$$55\,000(1-d)^9 = 7000$$
$$(1-d)^9 = \frac{7000}{55\,000}$$
$$1-d = \left(\frac{7000}{55\,000}\right)^{1/9}$$
$$d = 1-\left(\frac{7000}{55\,000}\right)^{1/9}$$
$$d = 0.204708849 \approx 20.47\%$$

1.13 Find x, given (a) $(2^x - 1)/3 = 12$, (b) $5^x = 3.28(2^x)$.

(a)
$$\frac{2^3-1}{3} = 12$$
$$2^x - 1 = 36$$
$$2^x = 37$$
$$x\log 2 = \log 37$$
$$x = \frac{\log 37}{\log 2}$$
$$x = 5.209453366$$

(b)
$$5^x = 3.28(2^x)$$
$$x\log 5 = \log 3.28 + x\log 2$$
$$x(\log 5 - \log 2) = \log 3.28$$
$$x = \frac{\log 3.28}{\log 5 - \log 2}$$
$$x = 1.29636084$$

1.14 Find n (the number of incorrect conversion periods) if (a) $800(1.0125)^n = 1200$, (b) $808(1.092)^{-n} = 90$.

(a)
$$800(1.0125)^n = 1200$$
$$(1.0125)^n = 1.5$$
$$n\log 1.0125 = \log 1.5$$
$$n = \frac{\log 1.5}{\log 1.0125}$$
$$n = 32.63952146$$

(b)
$$808(1.092)^{-n} = 90$$
$$(1.092)^{-n} = \frac{90}{808}$$
$$-n\log 1.092 = \log 90 - \log 808$$
$$n = \frac{\log 808 - \log 90}{\log 1.092}$$
$$n = 24.93728565$$

1.15 Determine n (which represents the number of payments) if:

(a) $\dfrac{(1.02)^n - 1}{0.02} = 15$ (b) $\dfrac{1-(1.11)^{-n}}{0.11} = 6$

(a)
$$\frac{(1.02)^n - 1}{0.02} = 15$$
$$(1.02)^n - 1 = 0.30$$
$$(1.02)^n = 1.30$$
$$n\log 1.02 = \log 1.30$$
$$n = \frac{\log 1.30}{\log 1.02}$$
$$n = 13.2489624$$

(b)
$$\frac{1-(1.11)^{-n}}{0.11} = 6$$
$$1-(1.11)^{-n} = 0.66$$
$$(1.11)^{-n} = 0.34$$
$$-n\log 1.11 = \log 0.34$$
$$n = -\frac{\log 0.34}{\log 1.11}$$
$$n = 10.33738504$$

1.7 THE ELECTRONIC POCKET CALCULATOR

A typical "electronic slide-rule," or "scientific," calculator is shown in Fig. 1-1. This calculator employs the Algebraic Operating System (AOS), which allows one to enter mathematical expressions in the same order as they are algebraically stated. The TI-36X SOLAR was used for all calculations in this book.

Fig. 1-1

SOLVED PROBLEMS

Carry out the following calculations using an electronic calculator.

1.16

$$x = \sqrt{5}\sqrt[3]{7} = \sqrt{5}(7)^{1/3}$$

Enter	Press		Display
5	$\boxed{\sqrt{x}}$	$\boxed{\times}$	2.236067978
7	$\boxed{y^x}$		7.
3	$\boxed{1/x}$		0.333333333
	$\boxed{=}$		4.277444161

The answer, to 10 significant digits, is $x = 4.277444161$. Note the use of the $\boxed{1/x}$ key to divide 3 into 1.

1.17

$$d = 1 - \left(\frac{700}{8000}\right)^{1/11}$$

Enter	Press		Display
700	$\boxed{\div}$		700.
8000	$\boxed{=}$ $\boxed{y^x}$		0.0875
11	$\boxed{1/x}$		0.090909091
	$\boxed{=}$		0.80134386
	$\boxed{+/-}$ $\boxed{+}$		−0.80134386
1	$\boxed{=}$		0.19865614

The answer is $d = 0.19865614$. In the calculation, the exponential function is evaluated first, then this value has its sign changed by the $\boxed{+/-}$ key, and finally, 1 is added.

1.18

$$n = -\frac{\log 120 - \log 809}{\log 1.09}$$

Enter	Press	Display
120	log −	2.079181246
809	log ÷	2.907948522
1.09	log	0.037426498
	=	−75.6184042
	+/−	75.6184042

The answer is $n = 75.618404$.

1.8 ROUNDING

In this book we shall observe the following rules for rounding the answers obtained from our calculator.

1. If the least significant digit is less than 5, then it is simply dropped. The remaining digits are unchanged. (This is known as *rounding down*.) For example, 0.83563 becomes 0.8356 when rounded to 4 significant digits.

2. If the least significant digit is greater than 5, then the next significant digit is increased by 1. (This is known as *rounding up*.) For example, 2.367 becomes 2.37 when rounded to 3 significant digits.

3. If the least significant digit is exactly 5, then the next significant digit is increased by 1 if it is odd. If the next significant digit is even, however, it remains unchanged. (This is known as the *odd-add rule*.) For example, 21.055 becomes 21.06 when rounded to 4 significant digits. However, 21.045 becomes 21.04 when rounded to 4 significant digits.

Multiple-digit rounding is carried out in the same manner. For example, if three significant figures are to be dropped, then we round down if the last three digits are less than 500, round up if they are greater than 500, and apply the odd-add rule if they are equal to 500.

Supplementary Problems

LAWS OF EXPONENTS

1.19 Prove:

$$(a) \quad \frac{a^m}{a^n} = \frac{1}{a^{n-m}} \quad (m < n) \qquad (b) \ (1.4) \qquad (c) \ (1.5)$$

1.20 Simplify:

$$
\begin{array}{ll}
(a)\ a^3 \cdot a^6 & (e)\ \dfrac{(a^3)^2 a}{a^5} \\[2mm]
(b)\ (2a) \cdot (4a^2)(8a^3) & (f)\ (a^2 b^3)^2 \\[2mm]
(c)\ \dfrac{x^8}{x^4} & (g)\ \left(\dfrac{2aa^2}{b^2 b}\right)^3 \\[2mm]
(d)\ (a^2)^3 & (h)\ (1.05)^3 (1.05)^{12} (1.05)^4
\end{array}
$$

Ans. (a) a^9; (b) $64a^6$; (c) x^4; (d) a^6; (e) a^2; (f) $a^4 b^6$; (g) $8a^9/b^9$; (h) $(1.05)^{19}$

ZERO, NEGATIVE, AND FRACTIONAL EXPONENTS

1.21 Simplify

$$
\begin{array}{ll}
(a)\ a^{1/2} a^{2/3} & (e)\ 25^{-1/2} + 81^{-1/4} \\[2mm]
(b)\ a^{1/2} \div a^{2/3} & (f)\ (1.01)^{-3}(1.01)^{1/4} \\[2mm]
(c)\ \dfrac{aa^{-2}}{a^{-1}} & (g)\ \left(\dfrac{a^{-2}}{b^{-3}}\right)^{-1/2} \\[2mm]
(d)\ (a^{3/4})^{1/3} \div a^{-1} & (h)\ \left(\dfrac{x^2}{y^3}\right)^{-2/3} \left(\dfrac{x^{5/3}}{y}\right)^2
\end{array}
$$

Ans. (a) $a^{7/6}$; (b) $a^{-1/6}$; (c) 1; (d) $a^{5/4}$; (e) 8/15; (f) $(1.01)^{-11/4}$; (g) $ab^{-3/2}$; (h) x^2

1.22 Simplify, using rational exponents:

$$
\begin{array}{ll}
(a)\ \sqrt{x}\,\sqrt[3]{x^2} & (c)\ \dfrac{\sqrt{5}\,\sqrt[3]{25}}{\sqrt[6]{5}} \\[3mm]
(b)\ \dfrac{a \cdot \sqrt[3]{a}}{\sqrt{a^3}} & (d)\ \left(\dfrac{\sqrt{a^4}\,\sqrt[3]{b^2}}{ab^3}\right)^{-3}
\end{array}
$$

Ans. (a) $x^{7/6} a^{2/3}$; (b) $a^{-1/6}$; (c) 5; (d) $a^{-3} b^7$

1.23 Evaluate, using a pocket calculator:

$$
\begin{array}{ll}
(a)\ \sqrt[3]{0.0468} & (e)\ 375(1.03)^{-2/3} \\[3mm]
(b)\ \sqrt[15]{\dfrac{24.60}{396}} & (f)\ \sqrt[4]{\dfrac{21.2}{(0.082)^2}} \\[3mm]
(c)\ \dfrac{37(23.3)^2}{\sqrt[3]{111.3}} & (g)\ \sqrt{3}\,\sqrt[3]{5}\,\sqrt[4]{7} \\[3mm]
(d)\ \dfrac{(1.065)^{15} - 1}{0.065} & (h)\ \dfrac{1 - (1.11)^{-13}}{0.11}
\end{array}
$$

Ans. (a) 0.36036999; (b) 0.830901089; (c) 4175.88482; (d) 24.18216933; (e) 367.6826338;
(f) 7.493367615; (g) 4.817537887; (h) 6.749870404

1.24 Solve the following exponential equations using a pocket calculator:

$$
\begin{array}{ll}
(a)\ 3500(1 + i)^8 = 5000 & (f)\ (1 + i)^{-10} = 0.9490 \\[2mm]
(b)\ 823.21(1 + i)^{60} = 15\,000 & (g)\ (1 + i)^{1/4} = 1.0113 \\[2mm]
(c)\ 17\,800(1 - d)^{20} = 500 & (h)\ (1 + i)^{20} - 1 = 80 \\[2mm]
(d)\ 8000(1 - d)^{11} = 800 & (i)\ (1 + i)^4 = (1.01)^{12} \\[2mm]
(e)\ 1000(1 + i)^{-20} = 35 & (j)\ (1 + i)^{12} = (1.05)^2
\end{array}
$$

Ans. (a) 0.045593188; (b) 0.049565815; (c) 0.163574046; (d) 0.188869169; (e) 0.1824876; (f) 0.005248373; (g) 0.045971928; (h) 0.24573094; (i) 0.030301; (j) 0.008164846

LOGARITHMS

1.25 Find the logarithm L.

$$(a) \quad L = \log_5 \frac{1}{5} \qquad (d) \quad L = \log_3 243$$

$$(b) \quad L = \log_8 64 \qquad (e) \quad L = \log_{10} 0.001$$

$$(c) \quad L = \log_2 \frac{1}{\sqrt{8}} \qquad (f) \quad L = \log_3 \sqrt[5]{3}$$

Ans. (a) -1; (b) 2; (c) $-3/2$; (d) 5; (e) -3; (f) 1/5

1.26 Find the number N.

$$(a) \quad \log_2 N = -2 \qquad (d) \quad \log_{10} N = 0$$

$$(b) \quad \log_7 N = -1 \qquad (e) \quad \log_4 N = 2$$

$$(c) \quad \log_5 N = -\tfrac{1}{3} \qquad (f) \quad \log_9 N = \tfrac{3}{2}$$

Ans. (a) 1/4; (b) 1/7; (c) $1/\sqrt[3]{5}$; (d) 1; (e) 16; (f) 27

1.27 Given $\log 2 = 0.301030$ and $\log 5 = 0.69897$, evaluate the following without the use of a pocket calculator

$$(a) \quad \log 20 \qquad (d) \quad \log \sqrt{5}$$

$$(b) \quad \log 25 \qquad (e) \quad \log 0.005$$

$$(c) \quad \log 10 \qquad (f) \quad \log \sqrt[3]{4}$$

Ans. (a) 1.301030; (b) 1.39794; (c) 1; (d) 0.349485; (e) $0.69897 - 3$; (f) 0.20068667

CHARACTERISTIC AND MANTISSA; ANTILOGARITHMS

1.28 Find the characteristic and mantissa of:

$$(a) \quad 12.5 \qquad (d) \quad 0.234$$

$$(b) \quad 80\ 808 \qquad (e) \quad 0.0001785$$

$$(c) \quad 0.001 \qquad (f) \quad 1$$

Ans. (a) 1, 0.096910013; (b) 4, 0.907454358; (c) -3, 0; (d) -1, 0.369215857; (e) -4, 0.25163822; (f) 0, 0

1.29 Given $\log 25 = 1.39794$, find the antilog of

$$(a) \quad 3.39794 \qquad (b) \quad 0.39794 \qquad (c) \quad 0.39794 - 2$$

Ans. (a) 2500; (b) 2.5; (c) 0.025

1.30 Find the antilog of:

$$(a) \quad 3.3 \qquad (c) \quad -3.801$$

$$(b) \quad 1.8013245 \qquad (d) \quad -0.008356$$

Ans. (a) 1995.262315; (b) 63.2884559; (c) 0.000158125; (d) 0.980943514

COMPUTING WITH LOGARITHMS

1.31 Solve the equations of Problem 1.24 by logarithms.

1.32 Solve the following exponential equations:

(a) $50(1.035)^n = 200$ (f) $3^x = 5(2^x)$

(b) $500 = 20(2.06)^x - 150$ (g) $126(0.75)^x = 30$

(c) $808(1.092)^{-n} = 90$ (h) $1 + 2^x = 81$

(d) $(1.0463)^{-n} = 0.3826$ (i) $\dfrac{3^x + 1}{2} = 21$

(e) $(1.02)^n - 1 = 0.5314$ (j) $\dfrac{5^x - 1}{5} = 5$

Ans. (a) 40.29758337; (b) 4.816952083; (c) 24.93728565; (d) 21.22762776; (e) 21.52150537;
(f) 3.969362296; (g) 4.988439193; (h) 6.321928095; (i) 3.380238966; (j) 2.024369199

1.33 Solve the following equations for n and check your answers by substitution:

(a) $\dfrac{(1.083)^n - 1}{0.083} = 21$ (d) $\dfrac{1 - (1.087)^{-n}}{0.087} = 4.5$

(b) $\dfrac{(1.11)^n - 1}{0.11} = 11$ (e) $\dfrac{1 - (1.0975)^{-n}}{0.0975} = 6.03812$

(c) $\dfrac{(1.005)^n - 1}{0.005} = 10$ (f) $\dfrac{1 - (1.025)^{-n}}{0.025} = 3$

Ans. (a) 12.65507765; (b) 7.598623985; (c) 9.782407637; (d) 5.954792501; (e) 9.549892917;
(f) 3.157282008

ROUNDING

1.34 Round each of the following numbers to three significant figures:

(a) 528.5   (d) -0.085201 (g) 0.23456
(b) 13.555   (e) 6 835 920   (h) 9875.738
(c) 1.7534   (f) -20.05 (i) -305.97

Ans. (a) 528; (b) 13.6; (c) 1.75; (d) -0.0852; (e) 6 840 000; (f) -20.0; (g) 0.235; (h) 9880;
(i) -306

1.35 Round each of the numbers in Problem 1.34 to four significant figures.

Ans. (a) 528.5; (b) 13.56; (c) 1.753; (d) -0.08520; (e) 6 836 000; (f) -20.05; (g) 0.2346;
(h) 9876; (i) -306.0

1.36 Round each of the numbers in Problem 1.34 to two significant figures.

Ans. (a) 530; (b) 14; (c) 1.8; (d) -0.085; (e) 6 800 000; (f) -20; (g) 0.23; (h) 9900;
(i) -310

Chapter 2

Progressions

2.1 ARITHMETIC PROGRESSIONS

A sequence of numbers, called *terms*, such that any two consecutive numbers in the sequence are separated by a fixed *common difference*, is an *arithmetic progression*.

5, 8, 11, 14, ... is an arithmetic progression with common difference 3

24, 20, 16, 12, ... is an arithmetic progression with common difference -4

Consider the arithmetic progression with first term t_1 and common difference d

$$t_1, \ t_1 + d, \ t_1 + 2d, \ t_1 + 3d, \ ...$$

The nth term of the progression is

$$t_n = t_1 + (n-1)d \tag{2.1}$$

We may write the sum of the first n terms of the progression in two ways:

$$S_n = t_1 + (t_1 + d) + (t_1 + 2d) + \cdots + (t_n - 2d) + (t_n - d) + t_n$$

and $\quad S_n = t_n + (t_n - d) + (t_n - 2d) + \cdots + (t_1 + 2d) + (t_1 + d) + t_1$

Adding the two expressions term by term, we obtain

$$2S_n = (t_1 + t_n) + (t_1 + t_n) + \cdots + (t_1 + t_n) + (t_1 + t_n) = n(t_1 + t_n)$$

or

$$S_n = \frac{n}{2}(t_1 + t_n) \tag{2.2}$$

SOLVED PROBLEMS

2.1 Find the 15th term and the sum of the first 15 terms of the arithmetic progression $-1, 2, 5, 8, ...$

Here, $t_1 = -1$, $d = 3$, and $n = 15$; then

$$t_{15} = t_1 + 14d = -1 + 14(3) = 41 \qquad S_{15} = \frac{15}{2}(t_1 + t_{15}) = \frac{15}{2}(-1 + 41) = 300$$

2.2 (*a*) An arithmetic progression has $t_1 = 7$, $t_n = 77$, $S_n = 420$; find n and d. (*b*) An arithmetic progression has $t_3 = 18$, $t_6 = 42$; find t_1 and S_6.

(*a*) $\qquad t_n = 7 + (n-1)d = 77 \qquad S_n = \frac{n}{2}(7 + 77) = 420$

From the second equation, $n = 10$. Substituting for n in the first equation, we obtain

$$7 + 9d = 77 \qquad \text{or} \qquad d = \frac{70}{9}$$

(b) $t_3 = t_1 = 2d = 18 \qquad t_6 = t_1 + 5d = 42$

Subtracting the first equation from the second, we obtain $3d = 24$ or $d = 8$. Substituting for d in the first equation gives $t_1 = 18 - 2(8) = 2$. Then,

$$S_6 = \frac{6}{2}(2 + 42) = 132$$

2.3 An individual borrows \$5000, agreeing to reduce the principal by \$200 at the end of each month and to pay 15% interest per annum, that is, $1\frac{1}{4}\%$ per month, on all unpaid balances. Find the sum of all interest payments.

In all there are $5000/200 = 25$ payments. Interest payments are calculated as $1\frac{1}{4}\%$ of the balances $5000, 4800, 4600, ..., 200$; they form the arithmetic progression $62.50, 60, 57.50, ..., 2.50$. The sum of all interest payments is the sum of 25 terms of an arithmetic progression with $t_1 = 62.50$ and $t_{25} = 2.50$.

$$S_{25} = \frac{25}{2}(62.50 + 2.50) = \$812.50$$

2.4 In drilling a well, the cost is \$5.50 for the first 10 cm, and for each succeeding 10 cm, the cost is \$1 more than that for the preceding 10 cm. How deep a well may be drilled for \$1000?

Here, $t_1 = 5.50$, $d = 1$, and $S_n = 1000$; thus,

$$1000 = \frac{n}{2}\{5.50 + [5.50 + (n-1)1]\}$$

which reduces to the quadratic equation

$$n^2 + 10n - 2000 = 0 \qquad \text{or} \qquad (n + 50)(n - 40) = 0$$

The positive root is $n = 40$; a well 4 m deep may be drilled.

2.2   GEOMETRIC PROGRESSIONS

A sequence of numbers, called *terms*, such that any two consecutive numbers in the sequence are in a fixed *common ratio*, is a *geometric progression*.

$2, 2x, 2x^2, 2x^3, ...$ is a geometric progression with common ratio x

$8, -4, 2, -1, ...$ is a geometric progression with common ratio $-\frac{1}{2}$

The geometric progression with first term t_1 and common ratio r is

$$t_1, \ t_1 r, \ t_1 r^2, \ t_1 r^3, \ ...$$

The nth term of the progression is

$$t_n = t_1 r^{n-1} \qquad\qquad (2.3)$$

Let S_n denote the sum of the first n terms of the progression:

$$S_n = t_1 + t_1 r + t_1 r^2 + t_1 r^3 + \cdots + t_1 r^{n-2} + t_1 r^{n-1}$$

Then
$$rS_n = t_1 r + t_1 r^2 + t_1 r^3 + t_1 r^4 + \cdots + t_1 r^{n-1} + t_1 r^n$$

Forming the difference $S_n - rS_n$ term by term, we obtain

$$
\begin{aligned}
S_n - rS_n &= t_1 - t_1 r^n \\
S_n(1-r) &= t_1(1-r^n) \\
S_n &= t_1\frac{1-r^n}{1-r} = t_1\frac{r^n-1}{r-1}
\end{aligned}
\tag{2.4}
$$

It is convenient to use the first form of (2.4) when $|r| < 1$, and the second form when $|r| > 1$.

SOLVED PROBLEMS

2.5 Find the 10th term and the sum of the first 10 terms of the geometric progressions
(a) 1, 3, 9, 27, ...; (b) $(1.05)^{-1}, (1.05)^{-2}, (1.05)^{-3}, (1.05)^{-4}, \ldots$

(a) Here $t_1 = 1$, $r = 3$, and $n = 10$; thus

$$t_{10} = t_1 r^9 = 1(3)^9 = 19\,683 \qquad S_{10} = t_1\frac{r^{10}-1}{r-1} = 1\frac{3^{10}-1}{3-1} = 29\,524$$

(b) Here, $t_1 = (1.05)^{-1}$, $r = (1.05)^{-1}$, and $n = 10$; thus,

$$t_{10} = t_1 r^9 = (1.05)^{-1}[(1.05)^{-1}]^9 = (1.05)^{-10} = 0.61391325$$

$$S_{10} = t_1\frac{1-r^{10}}{1-r} = (1.05)^{-1}\frac{1-(1.05)^{-10}}{1-(1.05)^{-1}}$$

$$= (1.05)^{-1}\frac{1-(1.05)^{-10}}{1-(1.05)^{-1}} \times \frac{1.05}{1.05} = \frac{1-(1.05)^{-10}}{1.05-1}$$

$$= \frac{1-(1.05)^{-10}}{0.05} = 7.721734929$$

2.6 (a) A geometric progression has $t_1 = 12$, $r = \frac{1}{2}$, $t_n = \frac{3}{8}$; find n and S_n. (b) A geometric progression has $t_2 = \frac{7}{4}$ and $t_5 = 14$; find t_{10} and S_{10}.

(a) We solve for n.

$$
\begin{aligned}
\tfrac{3}{8} &= 12(\tfrac{1}{2})^{n-1} \\
(\tfrac{1}{2})^5 &= (\tfrac{1}{2})^{n-1} \\
n-1 &= 5 \\
n &= 6
\end{aligned}
$$

Then $\qquad S_6 = 12\dfrac{1-(\tfrac{1}{2})^6}{\tfrac{1}{2}} = 23.625$

(b) $\qquad t_2 = t_1 r = \tfrac{7}{4} \qquad t_5 = t_1 r^4 = 14$
From the first equation we obtain $t_1 = \tfrac{7}{4}r$, and substituting in the second equation for t_1 and solving for r we obtain $r = 2$. Then,

$$t_1 = \tfrac{7}{8} \qquad t_{10} = \tfrac{7}{8}(2)^9 = 448 \qquad S_{10} = \frac{7}{8}\frac{2^{10}-1}{2-1} = 894.125$$

2.7 The value of a certain machine at the end of each year is 80% as much as its value at the beginning of the year. If the machine originally costs $10\,000, find its value at the end of 10 years.

Given $t_1 = 10\ 000$, $r = 0.80$, $n = 11$, we calculate

$$t_{11} = 10\ 000(0.80)^{10} = \$1073.74$$

2.8 Each stroke of a vacuum pump extracts 5% of the air remaining in a container. What decimal fraction of the original air remains after 40 strokes?

Given $t_1 = 0.95$, $r = 0.95$, $n = 40$, we calculate

$$t_{40} = (0.95)^{40} = 0.128512157 \approx 12.85\%$$

2.3 INFINITE GEOMETRIC PROGRESSIONS

Consider the geometric progression

$$1, \ \frac{1}{2}, \ \frac{1}{4}, \ \frac{1}{8}, \ \frac{1}{16}, \cdots$$

whose first term is $t_1 = 1$ and whose common ratio is $r = \frac{1}{2}$. The sum of the first n terms is

$$S_n = \frac{1 - (\frac{1}{2})^n}{1 - \frac{1}{2}} = \frac{1}{1 - \frac{1}{2}} - \frac{(\frac{1}{2})^n}{1 - \frac{1}{2}} = 2 - \left(\frac{1}{2}\right)^{n-1}$$

We note that, for any n, the difference $2 - S_n = (\frac{1}{2})^{n-1}$ is positive, and it becomes smaller and smaller as n increases. As n increases without bound (becomes infinite), we say that the sum S_n approaches 2 as a limit, and write

$$\lim_{n \to \infty} S_n = 2$$

For the general geometric progression,

$$t_1, \ t_1 r, \ t_1 r^2, \ t_1 r^3, \ \ldots$$

the sum of the first n terms may be written as

$$S_n = t_1 \frac{1 - r^n}{1 - r} = \frac{t_1}{1 - r} - \frac{t_1 r^n}{1 - r}$$

When $-1 < r < 1$, as n increases without bound, the term in r^n approaches 0 and S_n approaches $t_1/(1 - r)$. In this case, we call

$$S = \frac{t_1}{1 - r} \qquad (-1 < r < 1) \tag{2.5}$$

the *sum of the infinite geometric progression.*

SOLVED PROBLEMS

2.9 Find the sum of infinite geometric progression

$$1, \ -\frac{1}{3}, \ \frac{1}{9}, \ -\frac{1}{27}, \ \cdots$$

Here, $t_1 = 1$ and $r = -\frac{1}{3}$ $(-1 < r < 1)$.

$$S = \frac{t_1}{1 - r} = \frac{1}{1 - (-\frac{1}{3})} = \frac{1}{\frac{4}{3}} = \frac{3}{4}$$

2.10 Find the sum of infinite geometric progression

$$(1+i)^{-1}, \ (1+i)^{-2}, \ (1+i)^{-3}, \ (1+i)^{-4}, \ \ldots$$

where $i > 0$.

Here, $t_1 = (1+i)^{-1}$ and $r = (1+i)^{-1} < 1$.

$$S = \frac{(1+i)^{-1}}{1-(1+i)^{-1}} = \frac{(1+i)^{-1}}{1-(1+i)^{-1}} \times \frac{1+i}{1+i} = \frac{1}{(1+i)-1} = \frac{1}{i}$$

2.11 Change $0.121212\cdots$ to a proper fraction.

We write $0.121212\cdots = 0.12 + 0.0012 + 0.000012 + \cdots$. This is the sum of an infinite geometric progression with $t_1 = 0.12$ and $r = 0.01$. Then,

$$S = \frac{t_1}{1-r} = \frac{0.12}{1-0.01} = \frac{0.12}{0.99} = \frac{12}{99} = \frac{4}{33}$$

2.12 Change $1.13333\cdots$ to a mixed number.

$$\begin{aligned}
1.13333\cdots &= 1 + \frac{1}{10} + (0.03 + 0.003 + 0.0003 + 0.00003 + \cdots \\
&= 1 + \frac{1}{10} + \frac{0.03}{1-0.1} = 1 + \frac{1}{10} + \frac{0.03}{0.9} = 1 + \frac{1}{10} + \frac{3}{90} \\
&= 1 + \frac{9+3}{90} = 1 + \frac{12}{90} \\
&= 1\frac{2}{15}
\end{aligned}$$

Supplementary Problems

ARITHMETIC PROGRESSIONS

2.13 For the following progressions, find the next term, and also the term and the sum as indicated.

(a) $19, 31, 43, \ldots$. Find the 9th term and the sum to 10 terms.

(b) $\frac{1}{3}, \frac{1}{12}, -\frac{1}{6}, \ldots$. Find the 8th term and the sum to 10 terms.

(c) $9.2, 8, 6.8, \ldots$. Find the 10th term and the sum to 15 terms.

Ans. (a) 55, 115, 730; (b) $-\frac{5}{12}, -\frac{17}{12}, -\frac{95}{12}$; (c) 5.6, -1.6, 12

2.14 Find the sum of (a) the first n positive integers; (b) the first 100 positive even integers; (c) all odd integers from 15 to 219 inclusive; (d) all even integers from 18 to 280, inclusive.
Ans. (a) $n(n+1)/2$; (b) 10 100; (c) 12 051; (d) 19 668

2.15 In an arithmetic progression, (a) given $t_1 = 3$, $t_n = 9$, $n = 7$, find d and t_{10}; (b) given $t_4 = 12$, $t_8 = -4$, find t_1 and d; (c) given $t_n = -11$, $d = -4$, $n = 7$, find t_1 and S_n; (d) given $t_1 = 13$, $d = -3$, $S_n = 20$, find n and t_n.   *Ans.* (a) 1, 12; (b) 24, -4; (c) 13, 7; (d) 8, -8

2.16 A woman borrows \$1500, agreeing to pay \$100 at the end of each month to reduce the outstanding principal, and to pay the interest due on the unpaid balance at the rate 12% per annum (that is, 1% per month). Find the sum of all interest payments.   *Ans.* \$120

2.17 In buying a house a couple agrees to pay \$3000 at the end of the first year, \$3500 at the end of the second year, \$4000 at the end of the third year, and so on. How much do they pay for the house if they make 20 payments in all?   *Ans.* \$155 000

GEOMETRIC PROGRESSIONS

2.18 For each of the following progressions find the next term, the 10th term, and the sum of the first 10 terms.

$$(a) \quad 1, -\tfrac{1}{2}, \tfrac{1}{4}, \ldots \qquad (b) \quad 625, 125, 25, \ldots \qquad (c) \quad 1, 1.08, (1.08)^2, \ldots$$

Ans. (a) $-\tfrac{1}{8}$, $-\tfrac{1}{512}$, 0.666015625; (b) 5, 0.00032, 781.24992; (c) $(1.08)^3$, $(1.08)^9$, 14.48656247

2.19 In a geometric progression, (a) given $r = 10$, $t_8 = 2000$, find t_1 and t_5; (b) given $t_1 = 5$, $r = 2$, $n = 12$, find t_n and S_n; (c) given $t_5 = \tfrac{1}{20}$, $r = \tfrac{1}{4}$, find t_1 and S_5; (d) given $t_1 = 1.03$, $r = 1.03$, find t_{15} and S_{15}.
Ans. (a) 0.0002, 2; (b) 10 240, 20 475; (c) 12.8, 17.05; (d) 1.557967417, 19.1568813

2.20 A person starts a chain letter by giving it to four friends who pass it to four friends each, and so on. If there are no duplications, how many people will have been contacted by 10 sets of letters? *Ans.* 1 398 100

2.21 A rubber ball is dropped from a height of 50 meters to the ground. If it always bounces back one-half of the height from which it falls, find (a) how far it will rise on the 8th rebound; (b) the total distance it has traveled as it hits the ground the 10th time.   *Ans.* (a) 0.1953125 m; (b) 149.8046875 m

2.22 If a person were offered a job paying 1 cent the first day, 2 cents the second, 4 cents the third day, and so on, each day's salary being double that for the preceding day, how much would he receive (a) the 30th day? (b) For a 30-day working period? (c) For each of the three consecutive 10-day working periods?
Ans. (a) \$5 368 709.20; (b) \$10 737 418; (c) \$10.23, \$10 475.52, \$10 726 932.25

2.23 Find the total number of ancestors a person has in 8 generations preceding him, assuming there are no duplicates.   *Ans.* 510

2.24 A certain asset costs \$12 000, and the depreciation at the end of any year is estimated to be 25% of the value at the beginning of the year. What is the estimated value of the asset after 5 years of use?   *Ans.* \$2847.66

INFINITE GEOMETRIC PROGRESSIONS

2.25 Find the sums of the infinite geometric progressions (a) 3, 0.3, 0.03, 0.003, ...; (b) 1, 0.8, $(0.8)^2$, $(0.8)^3$, ...; (c) 100, $100(1.01)^{-1}$, $100(1.01)^{-2}$, $100(1.01)^{-3}$, ...
Ans. (a) 3.3; (b) 5; (c) 10 100

2.26 Change to a proper fraction or mixed number:

$$(a) \quad 0.0363636\cdots \qquad (b) \quad 1.2272727\cdots \qquad (c) \quad 2.0888888\cdots$$

Ans. (a) $\tfrac{2}{55}$; (b) $1\tfrac{5}{22}$; (c) $2\tfrac{4}{45}$

2.27 In problem 2.21, find the total distance the rubber ball travels before coming to rest.
Ans. 150 m

Chapter 3

Simple Interest and Simple Discount

3.1 SIMPLE INTEREST

When an investor lends money to a borrower, the borrower must pay back the money originally borrowed and also the fee charged for the use of the money, called *interest*. From the investor's point of view, interest is income from invested capital. The capital originally invested is called the *principal*. The sum of the principal and the interest due is called the *amount* or *accumulated value*. Any interest transaction can be described by the *rate of interest*, which is the ratio of the interest earned in one time unit to the principal.

At *simple interest*, the interest is computed on the original principal during the whole time, or *term*, of the loan, at the stated annual rate of interest.

We shall use the following notation:

$P \equiv$ principal, or the *present value* of S, or the *discounted value* of S, or the *proceeds*

$I \equiv$ simple interest

$S \equiv$ amount, or the *accumulated value* of P, or the *maturity value* of P

$r \equiv$ rate of interest per year

$t \equiv$ time in years

The simple interest I on principal P for t years at annual rate r is given by

$$I = Prt \tag{3.1}$$

and the amount S is given by

$$S = P + I = P + Prt = P(1 + rt) \tag{3.2}$$

The factor $1 + rt$ in (3.2) is called an *accumulation factor* at simple interest, and the process of calculating S from P via (3.2) is called *accumulation at simple interest*. From (3.2), we have

$$P = \frac{S}{1 + rt} = S(1 + rt)^{-1} \tag{3.3}$$

as the present or discounted value at rate r of S due in t years. The factor $(1 + rt)^{-1}$ in (3.3) is called a *discount factor at simple interest*, and the process of calculating P from S via (3.3) is called *discounting at simple interest*.

The time, t, must be in years. When the time is given in months, then

$$t = \frac{\text{number of months}}{12}$$

When the time is given in days, we may calculate either *exact simple interest*, on the basis of a 365-day year (leap year or not), that is,

$$t = \frac{\text{number of days}}{365}$$

19

or *ordinary simple interest*, on the basis of a 360-day year, that is,

$$t = \frac{\text{number of days}}{360}$$

Ordinary simple interest brings increased revenue to the lender. The general practice in the United States and in international business transactions is to use ordinary simple interest; it is employed throughout this Outline, unless specified otherwise.

SOLVED PROBLEMS

3.1 Find (a) the ordinary and (b) the exact simple interest, on a 60-day loan of $1500 at $14\frac{1}{2}\%$.

We have $P = 1500$ and $r = .0145$.

(a) Using a 360-day year,

$$t = \frac{60}{360} \quad \text{and} \quad I = Prt = 1500(0.145)\left(\frac{60}{360}\right) = \$36.25$$

(b) Using a 365-day year,

$$t = \frac{60}{365} \quad \text{and} \quad I = 1500(0.145)\left(\frac{60}{365}\right) = \$35.75$$

3.2 At what rate of simple interest will $1200 accumulate interest of $72 in 6 months?

We have $P = 1200$, $I = 72$, and $t = 6/12 = 1/2$. From (3.1),

$$r = \frac{I}{Pt} = \frac{72}{1200(1/2)} = 0.12 = 12\%$$

3.3 How long will it take for $500 to accumulate to at least $560 at $13\frac{1}{4}\%$ ordinary simple interest?

We have $P = 500$, $I = 60$, and $r = 0.1325$. From (3.1) we calculate

$$t = \frac{I}{Pr} = \frac{60}{500(0.1325)} = 0.90566038 \text{ years} \approx 326.03774 \text{ days}$$

It will take 327 days to accumulate to at least $560.

3.4 A man borrows $1000 for 220 days at 12.17%. What amount must he repay?

We have $P = 1000$, $r = 0.1217$, and $t = 220/360$. From (3.2),

$$S = P(1 + rt) = 1000\left[1 + (0.1217)\left(\frac{220}{360}\right)\right] = \$1074.37$$

3.5 Eighty days after borrowing money, a person pays back exactly $850. How much was borrowed if the $850 payment includes principal and simple interest at $9\frac{3}{4}\%$?

We have $S = 850$, $r = 0.0975$, and $t = 80/360$. From (3.3),

$$P = S(1 + rt)^{-1} = 850\left[1 + (0.0975)\left(\frac{80}{360}\right)\right]^{-1} = \$831.97$$

3.6 Find the discounted value of $1000 due in 3 months if the rate is 11%.

We have $S = 1000$, $r = 0.11$, and $t = 3/12$. From (3.3),

$$P = S(1 + rt)^{-1} = 1000 \left[1 + (0.11)\left(\frac{3}{12}\right)\right]^{-1} = \$973.24$$

3.7 To encourage prompt payment of invoices, manufacturers and wholesalers offer cash discounts for payments in advance of the due date. The following typical terms may be printed on the sales invoice: $3/10$, $n/30$. Goods billed on this basis are subject to a cash discount of 3% if paid for within 10 days. Otherwise, the full amount must be paid not later than 30 days from the date of invoice. Suppose that a merchant receives an invoice for $2800 with the above terms. (a) What is the highest simple interest rate at which he can borrow in order to take advantage of the cash discount? (b) What profit can the merchant realize by borrowing money at 18% and paying the invoice on the 10th day from the date of invoice?

(a) The cash discount is 3% of 2800, or $84. To take advantage of this cash discount the merchant may borrow $2800 - 84 = \$2716$ for 20 days. The highest interest he should be willing to pay on borrowed money should equal the cash discount, $84. From $P = 2716$, $I = 84$, $t = 20/360$, we calculate

$$r = \frac{I}{Pt} = \frac{84}{2716(20/360)} \approx 0.5567 = 55.67\%$$

(b) Interest on a loan of $2716 for 20 days at 18% is

$$I = 2716(0.18)\left(\frac{20}{360}\right) = \$27.16$$

The profit on the transaction is the difference between the cash discount of $84 and the $27.16 of interest he must pay on a loan, that is, $84 - 27.16 = \$56.84$.

3.8 Judy has a savings account that pays interest at 12% per annum. Interest is calculated on the minimum monthly balance and is paid to the account on December 31. Given the following transactions for her account, opened on January 10, find the interest earned in the first year.

Date	Deposits	Withdrawals	Balance
Jan. 10	$825		$825
Mar. 5	$300		$1125
June 18		$200	$925
Sept. 12		$100	$825
Oct. 3	$250		$1075

We calculate the interest for each month below:

Month	Minimum Balance	Interest for the Month
January	0	0
February	825	$825(0.12)(\frac{1}{12}) = 8.25$
March	825	$825(0.12)(\frac{1}{12}) = 8.25$
April	1125	$1125(0.12)(\frac{1}{12}) = 11.25$
May	1125	$1125(0.12)(\frac{1}{12}) = 11.25$
June	925	$925(0.12)(\frac{1}{12}) = 9.25$
July	925	$925(0.12)(\frac{1}{12}) = 9.25$
August	925	$925(0.12)(\frac{1}{12}) = 9.25$
September	825	$825(0.12)(\frac{1}{12}) = 8.25$
October	825	$825(0.12)(\frac{1}{12}) = 8.25$
November	1075	$1075(0.12)(\frac{1}{12}) = 10.75$
December	1075	$1075(0.12)(\frac{1}{12}) = 10.75$
		TOTAL $104.75

3.2 THE TIME BETWEEN DATES

There are two ways to calculate the number of days between calendar dates.

Exact time is found as the exact number of days including all days except the first. Appendix A gives the serial numbers of the days in the year, and exact time may be obtained as the difference between serial numbers of the given dates. In leap years, the serial number of the day is increased by 1 for all dates after February 28.

Approximate time is found by assuming that each month has 30 days.

When time is given indirectly as the time between dates, we can use either exact or approximate time and compute either exact or ordinary simple interest. Thus, there are **four methods** for computing simple interest between dates: (1) exact time and ordinary interest (the *Banker's Rule*); (2) exact time and exact interest; (3) approximate time and ordinary interest; (4) approximate time and exact interest.

The Banker's Rule is the common method in the United States and international business transactions; the general practice in Canada is to use Method 2. Methods 3 and 4 are used very rarely. In this Outline the Banker's Rule will be used except when otherwise specified. Of the four methods, it usually yields the maximum interest. (This would not be true, for example, for the time interval February 4 to March 2 of any year, where approximate time yields more than exact time.)

SOLVED PROBLEMS

3.9 Find (*a*) the exact and (*b*) the approximate time, from April 18 to November 3 of the same year.

    (*a*) From Appendix A, April 18 is the 108th day of the year and November 3 is the 307th day of the year. The exact time is $307 - 108 = 199$ days.

    (*b*)

Date	Month	Day
November 3	10	33
April 18	4	18
Difference	6	15

where we have "borrowed" 30 days from the 11th month. The approximate time is 6 months and 15 days, or $(6 \times 30) + 15 = 195$ days.

3.10 Find (a) the exact and (b) the approximate time, from May 18, 1996, to April 8, 1997.

(a) From Appendix A, May 18 is the 139th day of the year 1996 (leap year); April 8 is the 98th day of the year 1997. The exact time is $(366 - 139) + 98 = 325$ days.

(b) Starting from January 1, 1996, we write

Date	Month	Day
April 8, 1997	15	38
May 18, 1996	5	18
Difference	10	20

The approximate time is $(10 \times 30) + 20 = 320$ days.

3.11 A sum of $2000 is invested from May 18, 1996, to April 8, 1997, at 16% simple interest. Find the interest earned using the four methods.

From Problem 3.10 the exact time is 325 days and the approximate time is 320 days.

$$\text{Method 1:} \quad I = 2000(0.16)\left(\frac{325}{360}\right) = \$288.89$$

$$\text{Method 2:} \quad I = 2000(0.16)\left(\frac{325}{365}\right) = \$284.93$$

$$\text{Method 3:} \quad I = 2000(0.16)\left(\frac{320}{360}\right) = \$284.44$$

$$\text{Method 4:} \quad I = 2000(0.16)\left(\frac{320}{365}\right) = \$280.55$$

3.12 Using the Banker's Rule, find the simple interest on $3260 at $12\frac{1}{4}\%$ from April 21 to December 24 of the same year.

The exact time is $358 - 111 = 247$ days, and the simple interest is

$$I = 3260(0.1225)\left(\frac{247}{360}\right) = \$274$$

3.13 On January 10, Mr. A borrows $1000 on a demand loan from his bank. Interest is paid at the end of each quarter (March 31, June 30, September 30, December 31) and at the time of the last payment. Interest is calculated at the rate of 12% on the balance of the loan outstanding. Mr. A repaid the loan with the following payments:

March 1	$100
April 17	$300
July 12	$200
August 20	$100
October 18	$300
	$1000

Calculate the interest payments required and the total interest paid.

Using the Banker's Rule, we arrange our calculations in a table.

Dates	Number of Days	Balance	Interest		
Jan. 10-Mar. 1	50	1000	1000(0.12)(50/360)	=	16.67
Mar. 1-March 31	30	900	900(0.12)(30/360)	=	9.00
			March 31 payment		$25.67
Mar. 31-Apr. 17	17	900	900(0.12)(17/360)	=	5.10
Apr. 17-June 30	74	600	600(0.12)(74/360)	=	14.80
			June 30 payment		$19.90
June 30-July 12	12	600	600(0.12)(12/360)	=	2.40
July 12-Aug. 20	39	400	400(0.12)(39/360)	=	5.20
Aug. 20-Sept. 30	41	300	300(0.12)(41/360)	=	4.10
			September 30 payment		$11.70
Sept. 30-Oct. 18	18	300	300(0.12)(18/360)	=	1.80
			October 18 payment		$1.80

The total interest paid is $25.67 + 19.90 + 11.70 + 1.80 = \$59.07$.

3.3 EQUATIONS OF VALUE

All financial decisions must take into account the basic idea that *money has time value*. In a financial transaction, each payment should have an attached date, the date on which it falls due. In other words, mathematics of finance deals with *dated values*.

At a simple interest rate 12%, $100 due in 1 year is considered to be equivalent to $112 due in 2 years, since $100 would accumulate to $112 in the second year. In the same way, $100(1 + 0.12)^{-1} = \$89.29$ would be considered an equivalent sum at present. In general, we compare dated values by the following *definition of equivalence*: $X due on a given date is equivalent at a given simple interest rate r to $Y due t years later if

$$Y = X(1 + rt) \quad \text{or} \quad X = Y(1 + rt)^{-1}$$

To put it another way: When we move money forward, we accumulate, i.e., multiply the sum by an accumulation factor $1 + rt$; when we move money backward, we discount, i.e., multiply the sum by a discount factor $(1 + rt)^{-1}$. (See Problem 3.14.)

Figure 3-1 illustrates dated values equivalent to a given dated value X.

The sum of a set of dated values, due on different dates, has no meaning. We have to replace all the dated values by equivalent dated values *due on the same date*. The sum of the equivalent values is called *the dated value of the set*. (See Problem 3.15.)

One of the most important problems in the mathematics of finance is the replacing of a given set of payments by an equivalent set. We say that *two sets of payments are equivalent* at a given simple interest rate if the dated values of the sets, on any common date, are equal. An equation stating that the dated values, on a common date, of two sets of payments are equal is called an *equation of value* or an *equation of equivalence*. The date used is called the *focal date* or the *comparison date*. In problem solving, the following procedure is suggested:

STEP 1 Make a good time diagram, showing the dated values of one set of payments on one side of the time line and the dated values of the second set of payments on the other side.

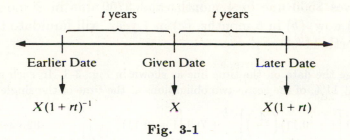

Fig. 3-1

STEP 2 Select a focal date and bring all the dated values to this focal date, using the specified interest rate.

STEP 3 Set up an equation of value at the focal date.

STEP 4 Solve the equation of value, using the appropriate methods of algebra.

In simple-interest problems the answer will vary slightly with the location of the focal date. It is therefore important that the parties involved in the financial transaction agree on the location of the focal date. (See Problems 3.16 and 3.17.) In compound interest (Section 4.7), the answer is independent of the location of the focal date.

SOLVED PROBLEMS

3.14 An obligation of \$1500 is due in 6 months with interest at 11%. At 15% simple interest, find the value of the obligation (*a*) at the end of 3 months, (*b*) at the end of 12 months.

The value of the obligation in 6 months is

$$1500\left[1 + (0.11)\left(\frac{6}{12}\right)\right] = \$1582.50$$

Let \$X be the value of the obligation at the end of 3 months, and \$Y be the value at the end of 12 months, as shown on the time line in Fig. 3-2.

(*a*) $$X = 1582.50\left[1 + (0.15)\left(\frac{3}{12}\right)\right]^{-1} = \$1525.30$$

(*b*) $$Y = 1582.50\left[1 + (0.15)\left(\frac{6}{12}\right)\right] = \$1701.19$$

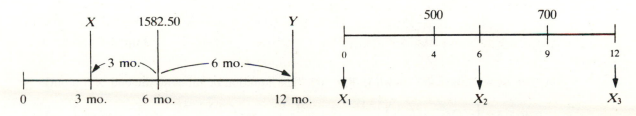

Fig. 3-2   Fig. 3-3

3.15 Ms. Hill owes \$500 due in 4 months and \$700 due in 9 months. What single payment (*a*) now, (*b*) in 6 months, (*c*) in 1 year, will liquidate these obligations if money is worth 11%?

We arrange the data on the time line as shown in Fig. 3-3. In each case, we calculate the dated value, at 11%, of the set of two obligations at the time of the single payment.

(*a*) $X_1 = 500\left[1 + (0.11)\left(\dfrac{4}{12}\right)\right]^{-1} + 700\left[1 + (0.11)\left(\dfrac{9}{12}\right)\right]^{-1} = 482.32 + 646.65 = \$1128.97$

(*b*) $X_2 = 500\left[1 + (0.11)\left(\dfrac{2}{12}\right)\right] + 700\left[1 + (0.11)\left(\dfrac{3}{12}\right)\right]^{-1} = 509.17 + 681.27 = \$1190.44$

(*c*) $X_3 = 500\left[1 + (0.11)\left(\dfrac{8}{12}\right)\right] + 700\left[1 + (0.11)\left(\dfrac{3}{12}\right)\right] = 536.67 + 719.25 = \$1255.92$

3.16 Mrs. Adams has two options available in repaying a loan: she can pay \$200 at the end of 5 months and \$300 at the end of 10 months, or she can pay \$$X$ at the end of 3 months and \$$2X$ at the end of 6 months. If the options are equivalent and money is worth 12%, find X, using as the focal date (*a*) the end of 6 months; (*b*) the end of 3 months.

(*a*) The dated values are shown in Fig. 3-4. The equation of value at the end of 6 months is:

dated value of option 2 = dated value of option 1

$$X\left[1 + (0.12)\left(\frac{3}{12}\right)\right] + 2X = 200\left[1 + (0.12)\left(\frac{1}{12}\right)\right] + 300\left[1 + (0.12)\left(\frac{4}{12}\right)\right]^{-1}$$

$$1.03X + 2X = 202 + 288.46$$
$$3.03X = 490.46$$
$$X = \$161.87$$

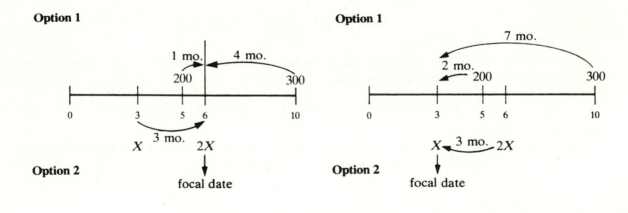

Fig. 3-4   Fig. 3-5

(*b*) The dated values are shown in Fig. 3-5. The equation of value at the end of 3 months is:

$$X + 2X\left[1 + (0.12)\left(\frac{3}{12}\right)\right]^{-1} = 200\left[1 + (0.12)\left(\frac{2}{12}\right)\right]^{-1} + 300\left[1 + (0.12)\left(\frac{7}{12}\right)\right]^{-1}$$

$$X + 1.9417476X = 196.08 + 280.37$$
$$2.9417476X = 476.45$$
$$X = \$161.96$$

Note the slight difference from the result of (*a*).

3.17 Blake borrowed $5000 on January 1, 1995. He paid $2000 on April 30, 1995, and $2000 on August 31, 1995. The final payment was made on December 15, 1995. Find the size of the final payment if the rate of interest was 7% and the focal date was (*a*) December 15, 1995; (*b*) January 1, 1995.

The dated values are shown in Fig. 3-6.

(*a*) Equation of value on December 15, 1995:

$$\underset{\text{dated value of the payments}}{} = \underset{\text{dated value of the debt}}{}$$

$$2000\left[1+(0.07)\left(\frac{229}{360}\right)\right]+2000\left[1+(0.07)\left(\frac{106}{360}\right)\right]+X = 5000\left[1+(0.07)\left(\frac{348}{360}\right)\right]$$

$$2089.06+2041.22+X = 5338.33$$

$$X = \$1208.05$$

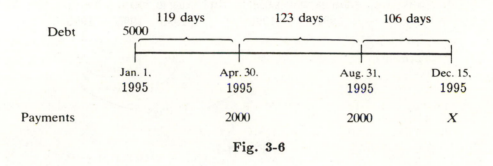

Fig. 3-6

(*b*) Equation of value on January 1, 1995:

$$2000\left[1+(0.07)\left(\frac{119}{360}\right)\right]^{-1}+2000\left[1+(0.07)\left(\frac{242}{360}\right)\right]^{-1}+X\left[1+(0.07)\left(\frac{348}{360}\right)\right]^{-1} = 5000$$

$$1954.77+1910.12+0.9366219\,X = 1135.11$$

$$X = \$1211.92$$

3.4   PARTIAL PAYMENTS

Financial obligations are sometimes liquidated by a series of partial payments during the term of obligation. Then it is necessary to determine the balance due on the final due date. There are two common ways to allow interest credit on short-term transactions.

Merchant's Rule. The entire debt and each partial payment earn interest to the final settlement date. The balance due on the final due date is simply the difference between the accumulated value of the debt and the accumulated value of the partial payments.

United States Rule. The interest on the unpaid balance of the debt is computed each time a partial payment is made. If the payment is greater than the interest due, the difference is used to reduce the debt. If the payment is less than the interest due, it is held without interest until other partial payments are made whose sum exceeds the interest due at the time of the last of these partial payments. (See Problem 3.19.) The balance due on the final date is the outstanding balance after the last partial payment carried to the final due date.

As the two methods result in two different concluding payments, it is important that the parties to a business transaction agree on the method to be used.

SOLVED PROBLEMS

3.18 Gordon borrowed $1000 on January 15, 1995, at 16%. He paid $350 on April 12, 1995; $20 on August 10, 1995; and $400 on October 3, 1995. What is the balance due on December 1, 1995, under the Merchant's Rule?

We arrange all dated values on a time line shown in Fig. 3-7. Simple interest is computed at 16% on the original debt of $1000 for 320 days, on the first partial payment of $350 for 233 days, on the second partial payment of $20 for 113 days, and on the third partial payment of $400 for 59 days.

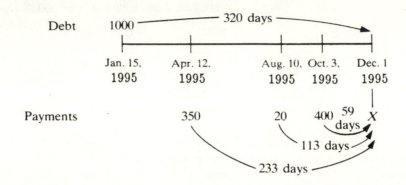

Fig. 3-7

Original debt	1000.00		1st partial payment	350.00
Interest for 320 days	142.22		Interest for 233 days	36.24
Accumulated value	$1142.22		2nd partial payment	20.00
of the debt			Interest for 113 days	1.00
			3rd partial payment	400.00
			Interest for 59 days	10.49
			Accumulated value	$817.73
			of the partial payments	

The balance due on December 1, 1995, is $1142.22 - 817.73 = \$324.49$.

Alternate Solution

We can write an equation of value using December 1, 1995, as the focal date:

$$\text{value of the payments} \ = \ \text{value of the debt}$$

$$350\left[1+(0.16)\left(\frac{233}{360}\right)\right] + 20\left[1+(0.16)\left(\frac{113}{360}\right)\right] + 400\left[1+(0.16)\left(\frac{59}{360}\right)\right] + X \ = \ 1000\left[1+(0.16)\left(\frac{320}{360}\right)\right]$$

$$386.24 + 21.00 + 410.49 + X = 1142.22$$

$$X = \$324.49$$

3.19 What is the balance due on December 1, 1995, in Problem 3.18, under the United States Rule?

All dated values are shown on a time line in Fig. 3-8.

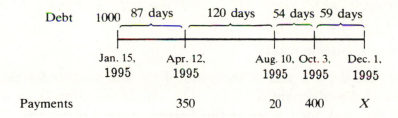

Fig. 3-8

There are four comparison dates. Each time a payment is made, the preceding balance is accumulated at 16% to this point and a new balance is obtained.

Original debt	1000.00
Interest for 87 days	38.67
Amount due on April 12, 1995	1038.67
First partial payment	350.00
Balance due on April 12, 1995	688.67
Interest for 120 days is $38.73; the payment of $20, being less than interest, is held without interest	
Interest for $120 + 54 = 174$ days	53.26
Amount due on October 3, 1995	741.93
Payments of $20 and $400	420.00
Balance due on October 3, 1995	321.93
Interest for 59 days	8.44
Amount due on December 1, 1995	$330.37

The above calculations may be carried out in a shorter form, as shown below.

$$\text{balance due on April 12, 1995} = 1000\left[1 + (0.16)\left(\frac{87}{360}\right)\right] - 350 = \$688.67$$

$$\text{balance due on October 3, 1995} = 688.67\left[1 + (0.16)\left(\frac{174}{360}\right)\right] - 420 = \$321.93$$

$$\text{balance due on December 1, 1995} = 321.93\left[1 + (0.16)\left(\frac{59}{360}\right)\right] = \$330.37$$

3.5 SIMPLE DISCOUNT

Simple Discount at an Interest Rate

In discounting at simple interest by means of (3.3), the difference $D = S - P$ is called the *simple discount* (on S) *at an interest rate* (r). We may interpret D either as the interest I on P which when added to P gives S, or as the *true discount* on S which when subtracted from S gives P.

Simple Discount at a Discount Rate

The *discount rate* d for a year is the ratio of the discount D for the year to the amount S on which the discount is given. The simple discount D on an amount S, also called *bank discount*, for t years at the discount rate d is calculated by means of the formula

$$D = Sdt \tag{3.4}$$

and the discounted value, or proceeds, P of S is given by

$$P = S - D = S - Sdt = S(1 - dt) \tag{3.5}$$

(See Problem 3.21.)

The charge for some short-term loans may be based on the final amount rather than on the present value. The lender calculates the bank discount D on the final amount S that must be paid on the due date and deducts it from S; the borrower receives the proceeds P. For this reason, bank discount is sometimes called *interest in advance*. (See Problem 3.22.) From (3.5),

$$S = \frac{P}{1 - dt} = P(1 - dt)^{-1} \tag{3.6}$$

and (3.6) is used to calculate the maturity value of a loan for specified proceeds. (See Problem 3.23.)

A discount rate d and an interest rate r are *equivalent* (over time t) if they result in the same present value P for an amount S due in the future. In Problem 3.24 we derive a formula for the simple interest rate r equivalent to a simple discount rate d:

$$r = \frac{d}{1 - dt} \tag{3.7}$$

Solving (3.7) for d, we obtain the rate of simple discount equivalent (over time t) to a rate of simple interest r:

$$d = \frac{r}{1 + rt} \tag{3.8}$$

SOLVED PROBLEMS

3.20 Find the present value at 12% simple interest of $1000 due in 5 months. What is the true discount?

We have $S = 1000$, $r = 0.12$, and $t = 5/12$. From (3.3),

$$P = S(1 + rt)^{-1} = 1000[1 + (0.12)(5/12)]^{-1} = \$952.38$$

The true discount is $D = S - P = 1000 - 952.38 = \$47.62$.

3.21 Find the present value at 12% simple discount of $1000 due in 5 months. What is the simple discount?

We have $S = 1000$, $d = 0.12$, and $t = 5/12$. From (3.5),

$$P = S(1 - dt) = 1000\left[1 - (0.12)\left(\frac{5}{12}\right)\right] = \$950$$

The simple discount is $D = S - P = 1000 - 950 = \$50$.

3.22 A bank charges 11% simple interest in advance (that is, 11% bank discount) on short-term loans. Find the sum received by the borrower who requests (a) $900 for 90 days, (b) $1500 from May 3 to October 15.

(a) $S = 900$, $d = 0.11$, and $t = \dfrac{90}{360}$.

$$P = 900\left[1 - (0.11)\left(\frac{90}{360}\right)\right] = \$875.25$$

(b) $S = 1500$, $d = 0.11$, and $t = \dfrac{165}{360}$.

$$P = 1500\left[1 - (0.11)\left(\frac{165}{360}\right)\right] = \$1424.38$$

3.23 A bank charges 12% bank discount on short-term loans. A borrower needs $2000 cash, to be repaid with interest in 9 months. What size loan should he ask for, and how much interest will he pay?

We have $P = 2000$, $d = 0.12$, and $t = 9/12$. From (3.6),

$$S = \frac{2000}{1 - (0.12)(9/12)} = \$2197.80$$

The borrower should ask for $2197.80; the interest on the loan is $197.80.

3.24 Derive (3.7).

The rates d and r are equivalent if they result in the same present value P for an amount S due in t years. Thus, for equivalence, we equate the right-hand sides of (3.3) and (3.5) and solve for r:

$$S(1 - dt) = S(1 + rt)^{-1}$$
$$1 - dt = (1 + rt)^{-1}$$
$$1 + rt = \frac{1}{1 - dt}$$
$$rt = \frac{1 - 1 + dt}{1 - dt}$$
$$r = \frac{d}{1 - dt}$$

3.25 What simple interest rate does the bank in Problem 3.23 realize?

From (3.7), with $d = 0.12$, and $t = 9/12$,

$$r = \frac{d}{1-dt} = \frac{0.12}{1-(0.12)(9/12)} \approx 13.19\%$$

Alternate Solution

From (3.1), with $I = 197.80$, $P = 2000$, and $t = 9/12$,

$$r = \frac{I}{Pt} = \frac{197.80}{2000(9/12)} \approx 13.19\%$$

3.26 A bank bids 96.823 for a 91-day, $1 million Treasury bill. (This means that the bank is willing to pay $968 230 for the bill, which will be worth $1 million 91 days after its issuance.) If the bid is accepted, what yield will the bank get, on (*a*) a bank discount basis? (*b*) A simple interest basis?

(*a*) We have $D = 1\,000\,000 - 968\,230 = \$31\,770$, $S = 1\,000\,000$, and $t = 91/360$. From (3.4),

$$d = \frac{D}{St} = \frac{31\,770}{1\,000\,000(91/360)} = 0.125683516 \approx 12.57\%$$

(*b*) We have $I = 31\,770$, $P = 968\,230$, and $t = 91/360$. From (3.1),

$$r = \frac{I}{Pt} = \frac{31\,770}{968\,230(91/360)} = 0.129807501 \approx 12.98\%$$

Alternatively, from (3.7),

$$r = \frac{d}{1-dt} = \frac{0.125683516}{1-(0.125683516)(91/360)} = 0.129807501 \approx 12.98\%$$

3.27 The bank in Problem 3.26 held the bill for 60 days and sold it for $990 000. Find (*a*) the interest rate earned by the bank; (*b*) the yield the new buyer will get, on a discount basis, if he holds the bill to maturity.

(*a*) The bank earned interest $I = 990\,000 - 968\,230 = \$21\,770$ over 60 days on the investment of $968 230. Therefore,

$$r = \frac{I}{Pt} = \frac{21\,770}{968\,230(60/360)} \approx 13.49\%$$

(*b*) We have $D = 10\,000$, $S = 1\,000\,000$, and $t = 31/360$. Thus,

$$d = \frac{D}{St} = \frac{10\,000}{1\,000\,000(31/360)} \approx 11.61\%$$

3.28 A company issues a $500 000 piece of commercial paper to mature on June 1. This is purchased by an investment firm on March 3. What is the purchase price if the discount rate is $11\frac{1}{4}\%$?

From (3.5), with $S = 500\,000$, $d = 0.1125$, and $t = 90/360$,

$$P = 500\,000\left[1 - (0.1125)\left(\frac{90}{360}\right)\right] = \$485\,937.50$$

3.6 PROMISSORY NOTES

A *promissory note* is a written promise by a debtor, called the *maker* of the note, to pay to, or to the order of, the creditor, called the *payee* of the note, a sum of money, with or without interest, on a specified date. An example of an interest-bearing note is given in Fig. 3-9.

$1500.00   New York City, May 11, 1995

Ninety Days after date, I promise to pay to the order of J. D. Green Fifteen hundred and 00/100 dollars for value received with interest at 8% per annum.

(Signed)   *J. B. Smith*

Fig. 3-9

The *face value* is the sum stated in the note ($1500 in Fig. 3-9). The *term* is the period stated in the note (90 days in Fig. 3-9). The *maturity date* is the date on which the debt is to be paid. If the term is given in months, approximate time is used to obtain the maturity date; if the term is given in days, exact time is used. In Fig. 3-9, the maturity date is 90 days after May 11, 1995, which is August 9, 1995 (3 months after May 11, 1995, would be August 11, 1995). The *maturity value* of a note is the sum to be paid on the maturity date. For an interest-bearing note, the maturity value is the accumulated value of the face value for the term of the note. Ordinary simple interest is used to obtain the maturity value. For a noninterest-bearing note, the maturity value equals the face value. A promissory note may be sold one or more times before its maturity date. The buyer may either (i) discount the maturity value of the note for the time from the date of sale to the maturity date, at his discount rate, with the seller receiving the proceeds of the sale, as determined by (3.5); or (ii) specify the interest rate he wants to realize on the investment, the proceeds being determined by (3.3). (See Problems 3.30 and 3.31.)

SOLVED PROBLEMS

3.29 Find the maturity value of the note in Fig. 3-9.

The maturity value S of the note is the accumulated value of $1500 for 90 days at 8% simple interest. From (3.2),

$$S = P(1 + rt) = 1500 \left[1 + (0.08) \left(\frac{90}{360} \right) \right] = \$1530.00$$

3.30 On July 2, 1995, Mr. Green sold the note in Fig. 3-9 to a bank at 9% bank discount rate. (*a*) How much money did Mr. Green receive? (*b*) What rate of interest did Mr. Green realize on his investment? (*c*) What rate of interest did the bank realize on its investment, if it holds the note till maturity?

(a) We arrange the data on a time line, as shown in Fig. 3-10. From (3.5) we calculate the proceeds on July 2, 1995:

$$P = S(1 - dt) = 1530.00 \left[1 - (0.09)\left(\frac{38}{360}\right)\right] = \$1515.46$$

The bank pays Mr. Green $1515.46 and obtains possession of the note.

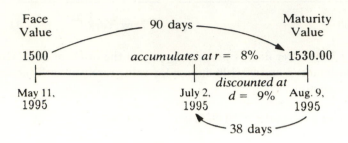

Fig. 3-10

(b) Mr. Green realized a profit of $15.46 on his investment of $1500 for the 52 days he held the note. The rate of interest Mr. Green realized was

$$r = \frac{I}{Pt} = \frac{15.46}{1500(52/360)} = 7.14\%$$

(c) The bank received $1530.00 from Mr. Smith on August 9, 1995, thereby realizing a profit of $1530.00 - 1515.46 = \$14.54$ on its investment of $1515.46 for 38 days. The rate of interest the bank realized was

$$r = \frac{I}{Pt} = \frac{14.54}{1515.46(38/360)} = 9.09\%$$

3.31 On July 10, a debtor signs a note for $1000 due in 5 months with interest at 14%. On October 18, the holder of the note sells it to a bank which discounts notes at a simple interest rate of 15%. Find the proceeds of the sale.

We arrange the data on a time line, Fig. 3-11. The maturity value of the note is

$$S = 1000 \left[1 + (0.14)\left(\frac{5}{12}\right)\right] = \$1058.33$$

The proceeds on October 18 are

$$P = 1058.33 \left[1 + (0.15)\left(\frac{53}{360}\right)\right]^{-1} = \$1035.46$$

3.32 On August 16, a retailer buys goods worth $2000. If he pays cash he will get a 5% cash discount, so he signs a 60-day noninterest-bearing note at his bank, which discounts notes at 6%. What should the face value of this note be to give the retailer the exact amount needed to pay cash for the goods.

Because the cash discount is $5\% \times 2000 = \$100$, the retailer needs $1900 in cash. He will sign a noninterest-bearing note whose maturity value S (the face value) is calculated by (3.6),

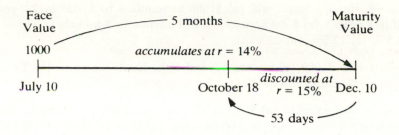

Fig. 3-11

given $P = 1900$, $d = 0.06$, and $t = 60/360$.

$$S = \frac{P}{1 - dt} = \frac{1900}{1 - (0.06)(60/360)} = \$1919.19$$

3.33 A company borrowed \$50 000 on May 1, 1995, and signed a promissory note bearing interest at 11% for 3 months. On the maturity date, the company paid the interest in full and gave a second note for 3 months without interest and for such an amount that when it was discounted at a 12% discount rate on the day it was signed, the proceeds were just sufficient to pay the debt. Find (a) the amount of interest paid on the first note, (b) the face value of the second note.

(a) Interest on the first note on August 1, 1995:

$$I = Prt = 50\ 000(0.11)\left(\frac{3}{12}\right) = \$1375$$

(b) We have $P = 50\ 000$, $d = 0.12$, and $t = 3/12$. From (3.6), the maturity value of the second note (equals the face value) is

$$S = \frac{P}{1 - dt} = \frac{50\ 000}{1 - (0.12)(3/12)} = \$51\ 546.39$$

Supplementary Problems

SIMPLE INTEREST

3.34 Find the ordinary and the exact simple interest on (a) a 90-day loan of \$500 at $8\frac{1}{2}$%, (b) a 118-day loan of \$600 at 16%. *Ans.* (a) \$10.63, \$10.48; (b) \$31.47, \$31.04

3.35 Find the maturity value of (a) a \$2500 loan for 18 months at 12% simple interest, (b) a \$1200 loan for 120 days at 8.5% ordinary simple interest, and (c) a \$10 000 loan for 64 days at 15% exact simple interest. *Ans.* (a) \$2950; (b) \$1234; (c) \$10 263.01

3.36 A loan shark made a loan of $100 to be repaid with $120 at the end of one month. What was the annual interest rate? *Ans.* 240%

3.37 At what rate of simple interest will (*a*) $1000 accumulate to $1420 in $2\frac{1}{2}$ years? (*b*) Money double itself in 8 years? (*c*) $500 accumulate $10 interest in 2 months?
Ans. (*a*) 16.8%; (*b*) 12.5%; (*c*) 12%

3.38 How long will it take $1000 (*a*) to earn $100 at 15% simple interest? (*b*) To accumulate to at least $1200 at $13\frac{1}{2}$% simple interest? *Ans.* (*a*) 8 months; (*b*) 534 days

3.39 Find the accumulated value of (*a*) $500 at 11% over 60 days, (*b*) $1000 at 15% over 55 days, using both ordinary and exact simple interest. *Ans.* (*a*) $509.17, $509.04; (*b*) $1022.92, $1022.60

3.40 Find the discounted value of (*a*) $500 due in 82 days at 9%, (*b*) $1000 due in 55 days at 15%, using both ordinary and exact simple interest. *Ans.* (*a*) $489.96, $490.09; (*b*) $977.60, $977.90

3.41 What principal will accumulate to (*a*) $5100 in 6 months at 9% simple interest? (*b*) $580 in 120 days at 18% exact simple interest? *Ans.* (*a*) $4880.38; (*b*) $547.59

3.42 A couple borrows $10 000. The annual interest rate is $10\frac{1}{2}$%, payable monthly, and the monthly payment is $200. How much of the first payment goes to interest and how much to principal? *Ans.* $87.50, $112.50

3.43 A bank pays 10% per annum on savings accounts. Interest is credited quarterly on March 31, June 30, September 30, and December 31, based on the minimum quarterly balance. If a person opens an account with a deposit of $200 on January 1 and withdraws $100 on August 8, how much interest is earned in the first year? *Ans.* $15.70

3.44 A retailer receives an invoice for $8000 for a shipment of furniture, with terms 3/10, *n*/40. (*a*) What is the highest simple interest rate at which he can afford to borrow money in order to take advantage of the cash discount? (*b*) If the retailer can borrow money at the simple interest rate 21%, find his profit resulting from the cash discount when he pays the invoice on the 10th day. *Ans.* (*a*) 37.11%; (*b*) $104.20

3.45 A cash discount of 4% is given if a bill is paid 30 days in advance of its due date. What is the highest simple interest rate at which you can afford to borrow money in order to take advantage of the cash discount? *Ans.* 50%

THE TIME BETWEEN DATES

3.46 Find the exact and approximate times (*a*) between March 15 and September 3 of the same year; (*b*) from October 2, 1996, to June 15, 1997. *Ans.* (*a*) 172 d, 168 d; (*b*) 256 d, 253 d

3.47 On April 7, 1996, a woman borrows $1000 at 8%. She repays the debt on November 22, 1996. Find the amount of simple interest, using the four methods. *Ans.* $50.89, $50.19, $50.00, $49.32

3.48 A sum of $5000 is invested from November 3, 1994, to February 8, 1995, at 15% simple interest. Find the amount of interest earned, using the four methods.
Ans. $202.08, $199.32, $197.92, $195.21

3.49 Using the Banker's Rule, find the amount of simple interest (*a*) on $1800 from October 2, 1996, to June 15, 1997, at $9\frac{3}{4}$%; (*b*) on $80 000 from March 21, 2004 to July 24, 2004, at $14\frac{1}{2}$%; (*c*) on $2500 from August 15, 1995, to May 1, 1996, at 23.09%; (*d*) on $200 from September 30, 1995, to July 7, 1997, at 18%. *Ans.* (*a*) $124.80; (*b*) $4027.78; (*c*) $416.90; (*d*) $64.60

3.50 Find the difference in simple interest on a $20 000 loan at a simple interest rate of 12% from April 15 to the next February 4, using the exact time/ordinary interest and exact time/exact interest. *Ans.* $26.94

3.51 Find the total interest paid in Problem 3.13, using exact time and exact interest. *Ans.* $58.27

3.52 Jessica borrowed $1500 from her Credit Union on a demand loan on August 16. Interest on the loan, calculated on the unpaid balance, is charged to her account on the 1st of each month. Jessica made a payment of $300 on September 17, a payment of $500 on October 7, a payment of $400 on November 12 and repaid the balance on December 15. The rate of interest on the loan on August 16 was 12% per annum. The rate was changed to 11.5% on September 25 and 12.5% on November 20. Calculate the interest payments required and the total interest paid, using exact time/exact interest.
Ans. $7.89, $13.32, $7.78, $4.32, $1.44, total = $34.75

EQUATIONS OF VALUE

3.53 Find the equivalent payments for each of the debts.

	Debts	Equivalent Payments	Focal Date	Rate
(a)	$1200 due in 80 days	In full	Today	12%
(b)	$200 due in 3 months, $800 due in 9 months	In full	6 months hence	8%
(c)	$600 due 2 months ago, $400 due in 3 months	$500 today and the balance in 6 months	Today	13%
(d)	$800 due today	Two equal payments due in 4 and 7 months	7 months hence	12%
(e)	$2000 due 300 days ago	Three equal payments due today, in 60 days and in 120 days	Today	11%

Ans. (a) $1168.83; (b) $988.31; (c) $532.94; (d) $421.67; (e) $740.96

3.54 If money is worth 13% simple interest, find the values of a debt of $1500 due in 8 months with interest at $14\frac{1}{2}\%$ (a) today, (b) 4 months from now, (c) 1 year from now.
Ans. (a) $1513.80; (b) $1576.68; (c) $1716.28

3.55 Debts of $500 due 20 days ago and $400 due in 50 days are to be settled by a payment of $600 now and a final payment 90 days from now. Find the value of the final payment at a simple interest rate of 11% with a focal date at the present. *Ans.* $305.21

3.56 Paula owes $100 due in 6 months and $150 due in 1 year. She and the lender agree that she can pay off both debts today using a simple interest rate of 16% and putting the focal date now. How much will be paid in cash today? *Ans.* $221.90

3.57 Carl owes $300 due in 3 months and $500 due in 8 months. What single payment (a) now, (b) in 6 months, (c) in 1 year will liquidate these obligations, if money is worth 8% and the focal date is the time of the single payment? *Ans.* (a) $768.80; (b) $799.42; (c) $831.33

3.58 Jackie borrows $1000 at 11%. She is to repay the loan in equal payments at 3 months, 6 months, and 9 months. Find the size of the payments, putting the focal date (a) at the present, (b) at the end of 9 months. *Ans.* (a) $351.51; (b) $351.18

3.59 At 12% simple interest, find the value today of the following set of obligations: $800 due in 4 months with interest at 15%, $1200 due in 6 months with interest at $10\frac{1}{4}\%$, and $900 due in 1 year with interest at 12%. *Ans.* $2897.78

3.60 Solve Problem 3.59 using 1 year from today as the focal date. *Ans.* $2903.92

3.61 Paul borrows $4000 at 18% simple interest. He is to repay the loan by paying $1000 at the end of 3 months and making two equal payments at the end of 6 months and 9 months. Find the size of the equal payments, using as the focal date (*a*) the end of 6 months, (*b*) the present. *Ans.* (*a*) $1693.97; (*b*) $1692.01

3.62 Last night Marion won $5000 in a lottery. She was given two options. She can take $5000 today or $X every six months (beginning six months from now) for 2 years. If the options are equivalent and the simple interest rate is 10%, find X using a focal date of 2 years. *Ans.* $1395.35

3.63 A loan of $2500 taken out on April 2 requires equal payments on May 25, July 20, and September 10 and a final payment of $500 on October 15. If the focal date is October 15, what is the size of the equal payments at 9% simple interest using exact time/exact interest? *Ans.* $691.87

3.64 Julio borrows $2000 at 14%. He is to repay the debt in 4 equal installments, one at the end of each 3-month period for 1 year. Find the size of the payments, given the focal date (*a*) at the present, (*b*) at the end of 1 year. *Ans.* (*a*) $543.05; (*b*) $541.57

3.65 Amy borrows $800 at 16%. She agrees to pay off the debt with payments of size $X, $2X, and $4X in 3 months, 6 months and 9 months, respectively. Find X using focal date (*a*) at the present, (*b*) in 3 months, (*c*) in 6 months, (*d*) in 9 months. *Ans.* (*a*) $125.30; (*b*) $125.55; (*c*) $125.47; (*d*) $125.14

PARTIAL PAYMENTS

3.66 A loan of $1000 is due in one year with interest at $14\frac{1}{4}$%. The debtor pays $200 in 3 months and $400 in 7 months. Find the balance due in one year by (*a*) the Merchant's Rule, (*b*) the United States Rule. *Ans.* (*a*) $497.37; (*b*) $503.54

3.67 A debt of $5000 is due in six months with interest at 10%. Partial payments of $3000 and $1000 are made in 2 and 4 months, respectively. What is the balance due on the final settlement date (*a*) by the Merchant's Rule; (*b*) by the United States Rule? *Ans.* (*a*) $1133.33; (*b*) $1136.68

3.68 A loan of $1400 is due in one year with simple interest at 12%. Partial payments of $400 in 2 months, $30 in 6 months, and $600 in 8 months are made. Find the balance due in one year using the United States Rule. *Ans.* $478.07

3.69 On February 4, 1995, Peter borrowed $3000 at 11%. He paid $1000 on April 21, 1995; $600 on May 12, 1995; and $700 on June 11, 1995. What is the balance due on August 15, 1995, (*a*) by the Merchant's Rule? (*b*) By the United States Rule? *Ans.* (*a*) $809.24; (*b*) $812.36

3.70 A debt of $1200 is to be paid off by payments of $500 in 45 days, $300 in 100 days, and a final payment of $436.92. Interest is at $r = 11$% and the Merchant's Rule was used to calculate the final payment. In how many days should the final payment be made using exact time/exact interest? *Ans.* 175 days

3.71 Melisa borrows $1000 on May 8 at $18\frac{1}{2}$% simple interest. She pays $500 on July 17 and $400 on September 29. What is her balance on October 31, (*a*) by the Merchant's Rule? (*b*) By the United States Rule? *Ans.* (*a*) $156.62; (*b*) $158.92

3.72 Roger borrows $4500 at 9% simple interest on July 3, 1993. He pays $1250 on October 27, 1993, and $2500 on January 7, 1994. Find the balance due on May 1, 1994, using exact time/exact interest (*a*) by the Merchant's Rule, (*b*) by the United States Rule. *Ans.* (*a*) $957.50; (*b*) $965.08

SIMPLE DISCOUNT

3.73 Find the bank discount on (*a*) $2000 for 120 days at 15%, (*b*) $10 000 for 91 days at 9.83%, (*c*) $5000 from April 21 to May 17 at $12\frac{1}{2}\%$. *Ans.* (*a*) $100; (*b*) $248.48; (*c*) $45.14

3.74 At $12\frac{3}{4}\%$ bank discount, find the value today of (*a*) $1500 due in 9 months, (*b*) $10 000 due in 182 days. *Ans.* (*a*) $1356.56; (*b*) $9355.42

3.75 Find the value today of the obligations in Problem 3.74 at $12\frac{3}{4}\%$ simple interest. *Ans.* (*a*) $1369.08; (*b*) $9394.45

3.76 A bank charges $13\frac{1}{2}\%$ interest in advance on short-term loans. Find the sums received by borrowers who request (*a*) $1500 for 6 months, (*b*) $2800 from June 1 to September 18, (*c*) $5000 from March 15 to November 7. *Ans.* (*a*) $1398.75; (*b*) $2685.55; (*c*) $4555.62

3.77 What rates of simple interest did the borrowers pay in Problem 3.76? *Ans.* (*a*) 14.48%; (*b*) 14.08%; (*c*) 14.82%

3.78 What is the interest rate equivalent to a discount rate of (*a*) 10% for 1 year? (*b*) 15% for 9 months. *Ans.* (*a*) 11.11%; (*b*) 16.90%

3.79 Robert borrows $1000 for 8 months from a lender who charges a 16% discount rate. (*a*) How much money does Robert receive? (*b*) What size of loan should Robert ask for in order to receive $1000 cash? (*c*) What is the equivalent simple interest rate he pays on the loan? *Ans.* (*a*) $893.33; (*b*) $1119.40; (*c*) 17.91%

3.80 A company issued a $2 million piece of commercial paper to mature on August 1. The paper is purchased by an investor on June 3. Find the purchase price if based on a discount rate of 11%. *Ans.* $1 963 944.44

3.81 The investor in Problem 3.80 sells the paper on July 12 at a price that will yield the new buyer 10%, on a discount basis. Find the purchase price. *Ans.* $1 988 888.89

3.82 An investor purchased a $100 000, 182-day Treasury bill at a bid price of 94.100. Find the yield (*a*) on a bank discount basis, (*b*) on a simple interest basis, if the investor holds the bill to maturity. *Ans.* (*a*) 11.67%; (*b*) 12.40%

3.83 The investor in Problem 3.82 held the bill for 80 days and then sold it in the secondary market at $10\frac{3}{4}\%$ bank discount. Find (*a*) the proceeds of the sale, (*b*) the interest rate earned by the investor during the period he held the bill, (*c*) the yield the new buyer will get, on an interest basis, if he holds the bill to maturity. *Ans.* (*a*) $96 954.17; (*b*) 13.65%; (*c*) 11.09%

PROMISSORY NOTES

3.84 Find the maturity date and maturity value of each of the following promissory notes:

	Face	Date	Term	Interest Rate
(*a*)	1800	Feb. 4	5 months	$11\frac{1}{2}\%$
(*b*)	3000	Apr. 21	3 months	$18\frac{1}{4}\%$
(*c*)	2000	May 12	120 days	9%
(*d*)	1500	June 11	60 days	$12\frac{3}{4}\%$

Ans. (*a*) July 4, $1886.25; (*b*) July 21, $3136.88; (*c*) Sept. 9, $2060; (*d*) Aug. 10, $1531.88

3.85 A bank charges 10% bank discount on short-term loans. Find the face value of the noninterest-bearing note given the bank, if the borrower receives (*a*) $1200 for 3 months, (*b*) $2000 for 60 days, (*c*) $3000 from July 5 to September 19, (*d*) $1500 from January 3 to September 19. *Ans.* (*a*) $1230.77; (*b*) $2033.90; (*c*) $3064.70; (*d*) $1616.28

3.86 Find the proceeds when each of the following notes is discounted:

	Face	Date	Term	Interest Rate	Date of Discount	Discount Rate
(a)	$2000	Sept. 1	60 days	11%	Oct. 1	$9\frac{1}{2}$%
(b)	$5000	Apr. 21	3 months	16%	June 11	16%
(c)	$800	Mar. 20	6 months	-	July 7	$13\frac{1}{4}$%
(d)	$3000	Feb. 4	240 days	-	May 12	12%

Ans. (a) $2020.55; (b) $5107.56; (c) $777.92; (d) $2857

3.87 In Problem 3.86, find the rate of simple interest earned by the purchaser if he holds the note until maturity. *Ans.* (a) 9.57%; (b) 16.29%; (c) 13.62%; (d) 12.60%

3.88 In Problem 3.86 (a) and (b), find the rate of simple interest earned by the seller of the note, assuming he was the original payee of the note. *Ans.* (a) 12.33%; (b) 15.18%

3.89 On April 21, a retailer buys goods amounting to $5000. If he pays cash he will get a 4% cash discount. To take advantage of this cash discount, he signs a 90-day noninterest-bearing note at his bank, which discounts notes at a discount rate of 9%. What should the face value of this note be to give him the exact amount needed to pay cash for the goods. *Ans.* $4910.49

3.90 Mr. Rose owes Mr. Chen $1000. Mr. Chen agrees to accept as payment a noninterest-bearing note for 90 days; the note can be immediately discounted at a local bank which charges a discount rate of 10%. What should the face of the note be so that Mr. Chen will receive $1000 as proceeds? *Ans.* $1025.64

3.91 A 90-day note for $800 bears interest at 10%. It is sold 60 days before maturity to a bank that uses a 12% simple interest rate for discounting notes. What are the proceeds? *Ans.* $803.92

3.92 A bank wishes to earn 9% simple interest in discounting notes. What discount rate should it use, if the term of the discount is (a) 3 months? (b) 120 days? *Ans.* (a) 8.80%; (b) 8.74%

3.93 The First National Bank discounts at 13% bank discount a noninterest-bearing note for $10 000 due in 90 days. On the same day, the note is discounted again, at 12% bank discount, by a Federal Reserve Bank. Find the profit made by the First National Bank. *Ans.* $25

3.94 A 90-day note promises to pay Ms. Chiu $2000 plus simple interest at 13%. After 51 days it is sold to a bank that discounts notes at a 12% simple interest rate. (a) How much money does Ms. Chiu receive? (b) What rate of interest does Ms. Chiu realize on her investment? *Ans.* (a) $2038.50; (b) 13.59%

3.95 An investor lends $5000 and receives a promissory note promising repayment of the loan in 90 days with 12% simple interest. This note is immediately sold to a bank which charges 10% simple interest. (a) How much does the bank pay for the note? (b) What is the investor's profit? (c) What is the bank's profit on this investment when the note matures? *Ans.* (a) $5024.39; (b) $24.39; (c) $125.61

3.96 A merchant buys goods worth $2000 and signs a 90-day noninterest-bearing promissory note. Find the proceeds if the supplier sells the note to a bank that uses a 13% simple interest rate. How much profit did the supplier make if the goods cost $1500? *Ans.* $437.05

Chapter 4

Compound Interest
and Compound Discount

4.1 ACCUMULATED VALUE

If the interest due is added to the principal at the end of each interest period and thereafter earns interest, the interest is said to be *compounded*. The sum of the original principal and total interest is called the *compound amount* or *accumulated value*. The difference between the accumulated value and the original principal is called the *compound interest*. The interest period, the time between two successive interest computations, is also called the *conversion period*.

Interest may be converted into principal annually, semiannually, quarterly, monthly, weekly, daily, or continuously. The number of times interest is converted in one year, or compounded per year, is called the *frequency of conversion*. The rate of interest is usually stated as an annual interest rate, referred to as the *nominal* rate of interest. The phrase "interest at 12%" or "money worth 12%" means 12% compounded annually; otherwise, the frequency of conversion is indicated, e.g., 16% compounded semiannually, 10% compounded daily. When compounding daily, most U.S. banks use a 365-day year. (See Problem 4.4.)

The following notation will be used:

P ≡ original principal, or the present value of S, or the discounted value of S

S ≡ compound amount of P, or the accumulated value of P

n ≡ total number of interest (or conversion) periods involved

m ≡ number of interest periods per year, or the frequency of compounding

j_m ≡ nominal (yearly) interest rate which is compounded (payable, convertible) m times per year (in some textbooks the nominal interest rate j_m is denoted by $i^{(m)}$ or r)

i ≡ interest rate per interest period

The interest rate per period, i, equals j_m/m. For example, $j_{12} = 12\%$ means that a nominal (yearly) rate of 12% is converted (compounded, payable) 12 times per year, $i = 1\% = 0.01$ being the interest rate per month.

Let P represent the principal at the beginning of the first interest period and i the interest rate per conversion period. We shall calculate the accumulated values at the ends of successive interest periods for n periods.

At the end of the 1st period:

interest due	Pi
accumulated value	$P + Pi = P(1 + i)$

At the end of the 2nd period:

interest due	$[P(1 + i)]i$
accumulated value	$P(1 + i) + [P(1 + i)]i = P(1 + i)(1 + i) = P(1 + i)^2$

At the end of the 3rd period:

interest due	$[P(1 + i)^2]i$
accumulated value	$P(1 + i)^2 + [P(1 + i)^2]i = P(1 + i)^2(1 + i) = P(1 + i)^3$

Continuing in this manner, we see that the successive accumulated values,

$$P(1 + i), \ P(1 + i)^2, \ P(1 + i)^3, \ \dots$$

form a geometric progression whose nth term is

$$S = P(1 + i)^n \tag{4.1}$$

where S is the accumulated value of P at the end of n interest periods. Formula (4.1) is the *fundamental compound interest formula*. The process of calculating S from P is called *accumulation*, and the factor $(1 + i)^n$ is called the *accumulation factor* or the *accumulated value of \$1*.

The accumulated value S of principal P at rate j_m for t years is, by (4.1),

$$S = P(1 + i)^n = P\left(1 + \frac{j_m}{m}\right)^{mt} = P\left[\left(1 + \frac{j_m}{m}\right)^m\right]^t \tag{4.2}$$

The accumulated value under continuous compounding is obtained by letting $m \to \infty$ in (4.2):

$$S = \lim_{m \to \infty} P\left[\left(1 + \frac{j_m}{m}\right)^m\right]^t = P\left[\lim_{m \to \infty} \left(1 + \frac{j_m}{m}\right)^m\right]^t \tag{4.3}$$

Now, in any text on calculus one will find the equation

$$\lim_{m \to \infty} \left(1 + \frac{x}{m}\right)^m = e^x$$

where $e = 2.718 \cdots$ is the base of natural logarithms. Thus, (4.3) becomes

$$S = P[e^{j_\infty}]^t = Pe^{j_\infty t} \tag{4.4}$$

in which j_m (which is independent of m) has been redesignated j_∞. (See Problem 4.6.) Actuaries use δ for $j_\infty \equiv i^{(\infty)}$ and call it the *force of interest*.

The fundamental compound interest formula, (4.1), also applies to the geometric growth of population, trees, sales, and so on. (See Problem 4.8.)

SOLVED PROBLEMS

4.1 Find (a) the simple interest on $1000 for 2 years 12%, ($b$) the compound interest on $1000 for 2 years at 12% compounded semiannually (that is, $j_2 = 12\%$).

(a) $$I = Prt = 1000(0.12)(2) = \$240$$

(b) Since the conversion period is 6 months, interest is earned at the rate of 6% per period, and there are 4 interest periods in 2 years.

At the End of Period	Interest	Accumulated Value
1	1000(0.06) = $60	$1060.00
2	1060(0.06) = $63.60	$1123.60
3	1123.60(0.06) = $67.42	$1191.02
4	1191.02(0.06) = $71.46	$1262.48

The compound interest is $1262.48 - 1000 = \$262.48$.

Alternate Solution

From (4.1), with $P = 1000$, $i = 0.06$, and $n = 4$,

$$S = P(1 + i)^n = 1000(1.06)^4 = \$1262.48$$

and the compound interest is $S - P = \$262.48$.

4.2 Find the compound interest on $1000 at ($a$) $j_{12} = 6\%$ for 5 years, (b) $j_{12} = 15\%$ for 30 years.

(a) We have $P = 1000$, $i = 0.06/12 = 0.005$, and $n = 5 \times 12 = 60$. From (4.1),

$$S = P(1 + i)^n = 1000(1.005)^{60} = \$1348.85$$

The compound interest is $S - P = \$348.85$.

(b) We have $P = 1000$, $i = 0.15/12 = 0.0125$, and $n = 30 \times 12 = 360$. From (4.1),

$$S = 1000(1.0125)^{360} = \$87\ 541.00$$

The compound interest is $S - P = \$86\ 541.00$, which is more than 86 times the original investment of $1000. If the investment had been at 15% simple interest, the interest earned would have been only

$$I = 1000(0.15)(30) = \$4500$$

This illustrates the power of compound interest at a high rate of interest for a long period of time.

4.3 Tabulate and graph the growth of $100 at compound interest rates $j_{12} = 6\%$, 8%, 10%, 12% and times 5, 10, 15, 20, 25, 30, 35, 40, 45, 50 years.

See Table 4-1 and Fig. 4-1.

Table 4-1

Years	n	$j_{12} = 6\%$, $i = 0.005$	$j_{12} = 8\%$, $i = 0.08/12$	$j_{12} = 10\%$ $i = 0.10/12$	$j_{12} = 12\%$, $i = 0.01$
5	60	134.89	148.98	164.53	181.67
10	120	181.94	221.96	270.70	330.04
15	180	245.41	330.69	445.39	599.58
20	240	331.02	492.68	732.81	1089.26
25	300	446.50	734.02	1205.69	1978.85
30	360	602.26	1093.57	1983.74	3594.96
35	420	812.36	1629.26	3263.87	6530.96
40	480	1095.75	2427.34	5370.07	11 864.77
45	540	1478.00	3616.36	8835.42	21 554.69
50	600	1993.60	5387.82	14 536.99	39 158.34

4.4 A person deposited $1000 into a retirement savings plan on February 4, 1978. How much money will be in the plan on February 4, 1998, at 11.4% compounded daily, assuming (*a*) approximate time (1 year = 360 days)? (*b*) exact time (1 year = 365 days)?

(*a*) We have $P = 1000$, $i = 0.114/360$, $n = 20 \times 360 = 7200$.

$$S = P(1+i)^n = 1000\left(1 + \frac{0.114}{360}\right)^{7200} = \$9773.15$$

(*b*) We have $P = 1000$, $i = 0.114/365$, $n = 20 \times 365 = 7300$.

$$S = P(1+i)^n = 1000\left(1 + \frac{0.114}{365}\right)^{7300} = \$9773.20$$

The result differs from that of (*a*) by only 5 cents. Most U.S. banks use exact time when compounding daily, as will this Outline.

4.5 John has a savings account which pays interest at $13\frac{3}{4}\%$ per annum. The interest is calculated daily on the minimum daily balance and is paid into the account at the end of each month. Given the following transactions for the account, opened on March 15, find the interest earned by the end of July.

Date	Deposit	Withdrawal
March 15	800	
April 30	300	
July 7		200

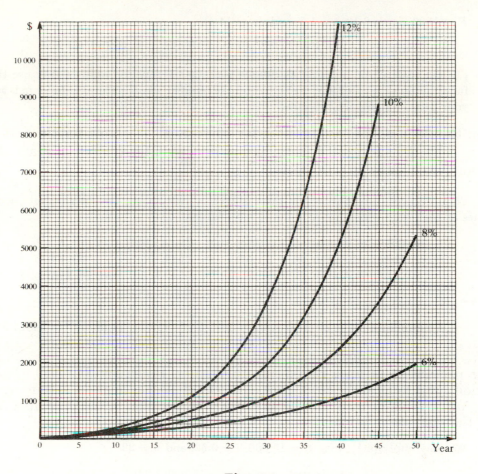

Fig. 4-1

The computation may be presented as follows:

$$\text{interest from March 15 to March 31} = 800(0.1375)\left(\frac{16}{365}\right) = \$4.82$$

$$\text{interest from April 1 to April 30} = 804.82(0.1375)\left(\frac{30}{365}\right) = \$9.10$$

$$\text{interest from May 1 to May 31} = 1113.92(0.1375)\left(\frac{31}{365}\right) = \$13.01$$

$$\text{interest from June 1 to June 30} = 1126.93(0.1375)\left(\frac{30}{365}\right) = \$12.74$$

$$\text{interest from July 1 to July 6} = 1139.67(0.1375)\left(\frac{6}{365}\right) = \$2.58$$

$$\text{interest from July 7 to July 31} = 939.67(0.1375)\left(\frac{25}{365}\right) = \underline{\$8.85}$$

$$\text{total interest earned} = \$51.10$$

4.6 Fifteen hundred dollars is invested for 18 months at a nominal rate of 13%. Find the accumulated value if interest is compounded (*a*) monthly, (*b*) continuously.

(*a*) $P = 1500$, $i = 0.13/12$, $n = 18$, and, from (4.1),

$$S = P(1+i)^n = 1500\left(1 + \frac{0.13}{12}\right)^{18} = \$1821.06$$

(b) $P = 1500$, $j_\infty = 0.13$, $t = 1.5$, and, from (4.4),

$$S = Pe^{j_\infty t} = 1500e^{(0.13)(1.5)} = \$1822.97$$

(To calculate e^x on a pocket calculator without the function e^x, you may use the inverse function, $\ln x$ [or $2.3025851 \log x$], to find the number whose natural logarithm is x. Consult your owner's manual for details.)

4.7 Two thousand dollars are invested for 10 years at $j_2 = 10\%$ for the first 3 years, at $j_4 = 8\%$ for the next 4 years, and at $j_{12} = 9\%$ for the last 3 years. Find the accumulated value after 10 years.

This problem can be solved in three stages by calculating the intermediate accumulated values at each date which has an interest rate change.

Accumulated value after 3 years   = $2000(1.05)^6 = \$2680.19$
Accumulated value after 7 years   = $2680.19(1.02)^{16} = \$3679.33$
Accumulated value after 10 years   = $3679.33(1.0075)^{36} = \$4814.94$

This solution may also be expressed as:

Accumulated value after 10 years $= 2000(1.05)^6(1.02)^{16}(1.0075)^{36} = \$4814.94$

4.8 The population of East Euclid was 15 000 on December 31, 1980. During the period 1980 to 1990 the town grew at a rate of 2% per annum. Assuming the rate of growth remains constant, estimate (a) the population on December 31, 2000; (b) the increase in population in the year 1998.

(a) We may apply (4.1), with $P = 15\,000$, $i = 0.02$, and $n = 20$.

$$\text{estimated population} = 15\,000(1.02)^{20} \approx 22\,289$$

(b) $\text{estimated population on December 31, 1997} = 15\,000(1.02)^{17} \approx 21\,004$

Then the estimated increase in population in the year 1998 is 2% of 21 004, or 420.

4.2   EQUIVALENT RATES

For a given nominal rate of interest, the accumulated value S increases with increasing frequency of conversion. (See Problem 4.9.) Two nominal rates with different frequencies of conversion are said to be *equivalent* if they yield the same accumulated value at the end of one year (and hence, at the end of any number of years). (See Problem 4.10.)

For a given nominal rate of interest j_m (compounded m times per year), we define the *annual effective interest rate* to be that rate j which, if compounded annually, will yield the same amount of interest per year. (See Problem 4.11.)

SOLVED PROBLEMS

4.9 Find the accumulated values and the amounts of compound interest earned on an investment of $10 000 for 10 years at a nominal rate of 12% compounded with frequencies $m = 1, 2, 4, 12, 52, 365$.

See Table 4-2.

Table 4-2

m	i	n	S	Interest
1	0.12	10	\$31 058.48	\$21 058.48
2	0.06	20	\$32 071.36	\$22 071.36
4	0.03	40	\$32 620.38	\$22 620.38
12	0.01	120	\$33 003.87	\$23 003.87
52	0.12/52	520	\$33 155.30	\$23 155.30
365	0.12/365	3650	\$33 194.62	\$23 194.62

4.10 Find the rate (*a*) j_{12} equivalent to $j_1 = 10.08\%$, (*b*) j_2 equivalent to $j_4 = 12\%$, (*c*) j_4 equivalent to $j_\infty = 9\%$, (*d*) per month equivalent to 5% per half-year.

In each case we equate the accumulated values of \$1 at the end of one year.

(*a*) \$1 at $j_{12} = 12i$ will accumulate to $(1+i)^{12}$; \$1 at $j_1 = 10.08\%$ will accumulate to 1.1008.

$$
\begin{aligned}
(1+i)^{12} &= 1.1008 \\
1+i &= (1.1008)^{1/12} \\
i &= (1.1008)^{1/12} - 1
\end{aligned}
$$

and
$$j_{12} = 12[(1.1008)^{1/12} - 1] = 0.096422513 \approx 9.64\%$$

(*b*) \$1 at $j_2 = 2i$ will accumulate to $(1+i)^2$; \$1 at $j_4 = 12\%$ will accumulate to $(1.03)^4$.

$$
\begin{aligned}
(1+i)^2 &= (1.03)^4 \\
1+i &= (1.03)^2 \\
i &= (1.03)^2 - 1
\end{aligned}
$$

and
$$j_2 = 2[(1.03)^2 - 1] = 0.1218 = 12.18\%$$

(*c*) \$1 at $j_4 = 4i$ will accumulate to $(1+i)^4$; \$1 at $j_\infty = 9\%$ will accumulate to $e^{0.09}$.

$$
\begin{aligned}
(1+i)^4 &= e^{0.09} \\
1+i &= e^{0.09/4} \\
i &= e^{0.09/4} - 1
\end{aligned}
$$

and
$$j_4 = 4(e^{0.09/4} - 1) = 0.091020137 \approx 9.10\%$$

(*d*) \$1 at i per month will accumulate to $(1+i)^{12}$; \$1 at 5% per half-year will accumulate to $(1.05)^2$.

$$
\begin{aligned}
(1+i)^{12} &= (1.05)^2 \\
1+i &= (1.05)^{1/6} \\
i &= (1.05)^{1/6} - 1 = 0.008164846 \approx 0.82\%
\end{aligned}
$$

4.11 Find the annual effective interest rate j corresponding to (*a*) 15% compounded continuously; (*b*) a nominal rate j_m.

(*a*) \$1 at rate j for 1 year will accumulate to $1+j$; \$1 at rate $j_\infty = 15\%$ for 1 year will accumulate to $e^{0.15}$.

$$
\begin{aligned}
1+j &= e^{0.15} \\
j &= e^{0.15} - 1 = 0.161834243 \approx 16.18\%
\end{aligned}
$$

(b) \$1 at rate j for 1 year will accumulate to $1+j$; \$1 at rate j_m for 1 year will accumulate to $[1+(j_m/m)]^m$.

$$1+j = \left(1+\frac{j_m}{m}\right)^m$$

$$j = \left(1+\frac{j_m}{m}\right)^m - 1$$

4.12 What simple interest rate r is equivalent to $j_{365} = 12\%$ if money is invested for 3 years?

\$1 at rate r for 3 years will accumulate to $1+3r$; \$1 at rate $j_{365} = 12\%$ for 3 years will accumulate to $[1+(0.12/365)]^{1095}$.

$$1+3r = \left(1+\frac{0.12}{365}\right)^{1095}$$

$$3r = \left(1+\frac{0.12}{365}\right)^{1095} - 1$$

$$r = \frac{1}{3}\left[\left(1+\frac{0.12}{365}\right)^{1095} - 1\right] = 0.14441487 \approx 14.44\%$$

4.13 A savings and loan association offers guaranteed investment certificates paying interest at $j_{12} = 11\frac{1}{4}\%$, $j_2 = 11\frac{1}{2}\%$, and $j_1 = 11\frac{3}{4}\%$. Which option is the best?

We calculate the effective annual rate of interest j for each j_m. From Problem 4.10(b),

$$\text{for } j_{12} = 11\tfrac{1}{4}\%: \quad j = \left(1+\frac{0.1125}{12}\right)^{12} - 1 = 0.118485937$$

$$\text{for } j_2 = 11\tfrac{1}{2}\%: \quad j = \left(1+\frac{0.115}{2}\right)^{2} - 1 = 0.11830625$$

$$\text{for } j_1 = 11\tfrac{3}{4}\%: \quad j = j_1 = 0.1175$$

Guaranteed investment certificates at $j_{12} = 11\frac{1}{4}\%$ give the best rate of return.

4.14 A sum of money is left invested for 3 years. In the first year, it earns interest at $j_{12} = 15\%$; in the second year, the rate of interest is $j_4 = 10\%$; and in the third year, the rate of interest changes to $j_{365} = 12\%$. Find the effective rate of interest j that would accumulate the same amount of interest at the end of 3 years. This level equivalent rate is also called the *geometric mean rate of return* for an investment.

\$1 at rate j for 3 years will accumulate to $(1+j)^3$; \$1 at the given rates for 3 years will accumulate to

$$\left(1+\frac{0.15}{12}\right)^{12}\left(1+\frac{0.10}{4}\right)^{4}\left(1+\frac{0.12}{365}\right)^{365}$$

Thus

$$(1+j)^3 = \left(1+\frac{0.15}{12}\right)^{12}\left(1+\frac{0.10}{4}\right)^{4}\left(1+\frac{0.12}{365}\right)^{365}$$

$$(1+j)^3 = 1.444583388$$

$$j = (1.444583388)^{1/3} - 1 = 0.130440058 \approx 13.04\%$$

4.3 DISCOUNTED VALUE

In business transactions it is frequently necessary to determine what principal P now will accumulate at a given compound interest rate i to a specified amount S at a specified date (n interest periods from now). From fundamental formula (4.1),

$$P = S(1+i)^{-n} \tag{4.5}$$

In (4.5), P is called the *discounted value* of S, or *present value* of S, or *proceeds*. The process of finding P from S is called *discounting*. The difference $S - P$ is called the *compound discount* on S; it is compound discount at an interest rate. (Compound discount at a discount rate will be considered in Section 4.8.) The factor $(1+i)^{-n}$ is called the *discount factor* or the *discounted value of 1*.

If interest is compounded continuously at rate j_∞, then (4.4) gives

$$P = Se^{-j_\infty t} \tag{4.6}$$

SOLVED PROBLEMS

4.15 How much would have to be deposited today in an investment fund paying $j_{12} = 10.4\%$ to have \$2000 in 3 years' time?

From (4.5), with $S = 2000$, $i = 0.104/12$, and $n = 3 \times 12 = 36$,

$$P = 2000\left(1 + \frac{0.104}{12}\right)^{-36} = \$1465.93$$

4.16 Find the present value of \$8000 due in 5 years at 7% compounded (*a*) quarterly, (*b*) daily, (*c*) continuously.

(*a*) We have $S = 8000$, $i = 0.07/4 = 0.0175$, $n = 5 \times 4 = 20$, and, from (4.5),

$$P = S(1+i)^{-n} = 8000(1.0175)^{-20} = \$5654.60$$

(*b*) We have $S = 8000$, $i = 0.07/365$, $n = 5 \times 365 = 1825$, and, from (4.5),

$$P = S(1+i)^{-n} = 8000\left(1 + \frac{0.07}{365}\right)^{-1825} = \$5637.69$$

(*c*) We have $S = 8000$, $j_\infty = 0.07$, $t = 5$, and, from (4.6),

$$P = Se^{-j_\infty t} = 8000e^{-(0.07)(5)} = \$5637.50$$

4.17 A note for \$2500 dated August 1, 1993, is due with compound interest at $j_{12} = 15\frac{1}{4}\%$ four years after date. On November 1, 1994, the holder of the note has it discounted by a lender who charges $j_4 = 13\frac{1}{2}\%$. Find the proceeds and the compound discount.

We arrange the data on a time diagram, Fig. 4-2.

Maturity value: $S = 2500\left(1 + \dfrac{0.1525}{12}\right)^{48} = \$4583.43$

Proceeds: $P = 4583.43\left(1 + \dfrac{0.135}{4}\right)^{-11} = \$3181.41$

Compound discount: $S - P = 4583.43 - 3181.41 = \$1402.02$

50   COMPOUND INTEREST AND COMPOUND DISCOUNT   [CHAP. 4

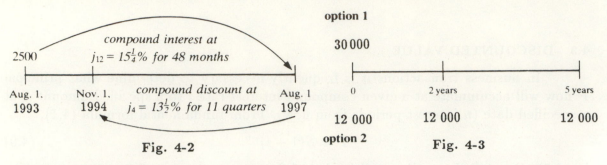

Fig. 4-2   Fig. 4-3

4.18 A person can buy a lot for \$30 000 now; or for \$12 000 now, \$12 000 in 2 years, and \$12 000 in 5 years. Which option is better, if money can be invested at (a) $j_{12} = 12\%$? (b) $j_{12} = 8\%$ for the first 3 years and $j_4 = 6\%$ for the next 2 years?

We arrange the data on a time diagram, Fig. 4-3.

(a) Discounted value for option 2:

$$12\ 000 + 12\ 000 \left(1 + \frac{0.12}{12}\right)^{-24} + 12\ 000 \left(1 + \frac{0.12}{12}\right)^{-60} = 12\ 000 + 9450.79 + 6605.40$$
$$= \$28\ 056.19$$

The payments option is better, since its discounted value is less than \$30 000.

(b) Discounted value for option 2:

$$12\ 000 + 12\ 000 \left(1 + \frac{0.08}{12}\right)^{-24} + 12\ 000 \left(1 + \frac{0.06}{4}\right)^{-8} \left(1 + \frac{0.08}{12}\right)^{-36}$$
$$= 12\ 000 + 10\ 231.16 + 8386.26 = \$30\ 617.42$$

The cash option is better.

4.19 The management of a company must decide between two proposals, on the basis of the following information:

Proposal	Investment Now	Net Cash Inflow at the End of		
		Year 1	Year 2	Year 3
A	80 000	95 400	39 000	12 000
B	100 000	35 000	58 000	80 000

Advise management regarding the proposal that should be selected, assuming that on projects of this type the company can earn $j_1 = 14\%$.

We arrange the data on a time diagram, Fig. 4-4.

Net discounted value of proposal A:

$$-80\ 000 + 95\ 400(1.14)^{-1} + 39\ 000(1.14)^{-2} + 12\ 000(1.14)^{-3}$$
$$= -80\ 000 + 83\ 684.21 + 30\ 009.23 + 8099.66 = \$41\ 793.10$$

Net discounted value of proposal B:

$$-100\ 000 + 35\ 000(1.14)^{-1} + 58\ 000(1.14)^{-2} + 80\ 000(1.14)^{-3}$$
$$= -100\ 000 + 30\ 701.75 + 44\ 629.12 + 53\ 997.72 = \$29\ 328.53$$

Management should select proposal A, which was the higher net discounted value

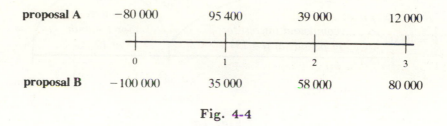

proposal A   −80 000   95 400   39 000   12 000

proposal B   −100 000   35 000   58 000   80 000

Fig. 4-4

4.4 ACCUMULATED AND DISCOUNTED VALUES FOR FRACTIONAL INTEREST PERIODS

Formulas (4.1) and (4.5) were developed on the assumption that n is an integer. Theoretically, they may be used whether n is an integer or a fraction. When we calculate the accumulated value or the discounted value using (4.1) or (4.5) for a fractional part of an interest conversion period, we are employing the *exact* or *theoretical method* of accumulating or discounting. (See Problems 4.20 and 4.21.)

In practice, the exact method is rarely used. Instead, we use compound interest for the number of complete interest periods and simple interest (at the stated nominal yearly rate) for the fractional part of the interest period. This method is called the *approximate* or *practical* method of accumulating or discounting; it yields slightly larger values than does the exact method. (See Problems 4.22, 4.23, and 4.24.) Unless stated otherwise, the approximate method will be used in this Outline.

SOLVED PROBLEMS

4.20 Find the accumulated value of $1000 over 5 years 7 months at $j_2 = 13\frac{1}{2}\%$, using the exact method.

We have $P = 1000$, $i = 0.135/2$, $n = 2 \times 5\frac{7}{12} = 67/6$. From (4.1),

$$S = P(1+i)^n = 1000\left(1 + \frac{0.135}{2}\right)^{67/6} = \$2073.84$$

4.21 Find the discounted value of $2800 due in 3 years 7 months if money is worth 10% effective and the exact method is used.

We have $S = 2800$, $i = 0.10$, $n = 43/12$, and, from (4.5),

$$P = S(1+i)^{-n} = 2800(1.10)^{-43/12} = \$1989.91$$

4.22 Solve Problem 4.20 using the approximate method.

We arrange the data on a time diagram, Fig. 4-5. First, $1000 is accumulated for 11 periods at $i = 0.135/2$, and then the resulting value is accumulated for an additional month at the simple interest rate $13\frac{1}{2}\%$.

$$S = 1000\left(1 + \frac{0.135}{2}\right)^{11}\left[1 + (0.135)\left(\frac{1}{12}\right)\right] = \$2074.46$$

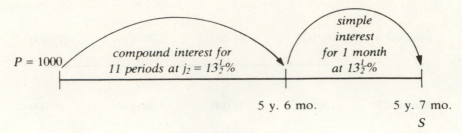

Fig. 4-5

4.23 Solve Problem 4.21 using the approximate method.

We arrange the data on a time diagram, Fig. 4-6. We first discount \$2800 for 4 periods at $i = 0.10$ (in general, we discount for the smallest number of complete periods containing the given time), and then accumulate the discounted value for 5 months at the simple interest rate 10%.

$$P = 2800(1.10)^{-4}\left[1 + (0.10)\left(\frac{5}{12}\right)\right] = \$1992.12$$

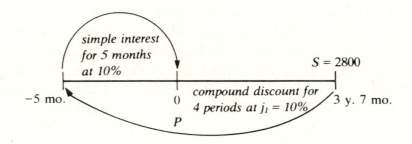

Fig. 4-6

4.24 A promissory note for \$2000 dated April 5, 1994, is due on October 1, 1998, with interest at $j_1 = 12\%$. On June 7, 1995, the holder of the note has it discounted at a bank which charges $j_4 = 14\%$. Find the proceeds and the compound discount.

To find the maturity value S of the note, we first accumulate \$2000 at $i = 0.12$ for 4 periods and then accumulate this value for 179 days (from April 5, 1998, to October 1, 1998) at the simple interest rate 12%.

$$S = 2000(1.12)^4\left[1 + (0.12)\left(\frac{179}{360}\right)\right] = \$3334.81$$

To find the proceeds P, we first discount \$3334.81 for 14 periods at $i = 0.14/4 = 0.035$ and then accumulate this value for 67 days (from April 1, 1995, to June 7, 1995) at the simple interest rate 14%.

$$P = \$3334.81(1.035)^{-14}\left[1 + (0.14)\left(\frac{67}{360}\right)\right] = \$2113.86$$

The compound discount is then $S - P = 3334.81 - 2113.86 = \$1220.95$.

4.5   FINDING THE RATE

When P, S, and n are given, we may solve (4.1) for the unknown interest rate i, either directly or by use of logarithms. [See Problems 4.25(a), 4.26, and 4.27.]

If interest is compounded continuously, we may solve (4.4) for the unknown rate j_∞, using natural logarithms. [See Problems 4.25(b) and 4.28.]

If i is the annual rate of interest being paid in the market place and r is the annual rate of inflation, then 1 invested at the beginning of the year will grow to $(1+i)$ at the end of the year. However, its purchasing power is only equal to $(\dfrac{1+i}{1+r})$. Hence, the real rate of return is

$$i_{real} = \frac{1+i}{1+r} - 1 = \frac{i-r}{1+r}$$

For low inflation rates r, $1+r$ is close to 1 and i_{real} is close to $i-r$. In order to make a meaningful comparison of interest and inflation, both rates should refer to the same one-year period. (See Problem 4.29.)

SOLVED PROBLEMS

4.25 Solve (a) (4.1) for the unknown rate i, (b) (4.4) for the unknown rate j_∞.

(a)
$$S = P(1+i)^n$$
$$(1+i)^n = \frac{S}{P}$$
$$1+i = \left(\frac{S}{P}\right)^{1/n}$$
$$i = \left(\frac{S}{P}\right)^{1/n} - 1$$

(b)
$$S = Pe^{j_\infty t}$$
$$e^{j_\infty t} = \frac{S}{P}$$
$$j_\infty t = \ln\frac{S}{P}$$
$$j_\infty = \frac{1}{t}\ln\frac{S}{P}$$

4.26 An investment fund advertises that it will triple your money in 10 years. What rate of interest compounded monthly is implied?

With $P = x$, $S = 3x$, and $n = 120$, we solve (4.1) for i:

$$3x = x(1+i)^{120}$$
$$(1+i)^{120} = 3$$
$$1+i = 3^{1/120}$$
$$i = 3^{1/120} - 1$$

and
$$j_{12} = 12(3^{1/120} - 1) = 0.110365662 \approx 11.04\%$$

Using Logarithms

$$(1+i)^{120} = 3$$
$$120\log(1+i) = \log 3$$
$$\log(1+i) = \frac{1}{120}\log 3$$
$$\log(1+i) = 0.00397601$$
$$1+i = 1.009197139$$
$$i = 0.009197139$$

and
$$j_{12} = 0.110365662 \approx 11.04\%$$

4.27 From 1988 to 1993, the earnings per share of common stock of a company increased from \$4.71 to \$9.38. What was the compounded annual rate of increase?

Given $P = 4.71$, $S = 9.38$, and $n = 5$, we solve (4.1) for i:

$$9.38 = 4.71(1+i)^5 \qquad \text{or} \qquad i = \left(\frac{9.38}{4.71}\right)^{1/5} - 1 = 0.147721154 \approx 14.77\%$$

4.28 At what nominal rate compounded continuously will an investment increase 50% in value in 3 years?

With $P = x$, $S = 1.5x$, and $t = 3$, we solve (4.4) for j_∞:

$$1.5x = xe^{j_\infty(3)}$$
$$e^{3j_\infty} = 1.5$$
$$3j_\infty = \ln 1.5$$
$$j_\infty = \frac{\ln 1.5}{3} = 0.135155036 \approx 13.52\%$$

4.29 Jackie invested \$1000 for one year at $j_1 = 10\%$. The annual inflation rate for that year was 4%. (*a*) What was the real annual rate of return on her investment? (*b*) What was her real annual after-tax return, if she paid tax at a 40% marginal rate?

(*a*) The real annual rate of return $= \dfrac{0.10 - 0.04}{1 + 0.04} = \dfrac{0.06}{1.04} = 0.057692308 \approx 5.77\%$

(*b*) The real annual after-tax rate of return $= \dfrac{0.10(0.60) - 0.04}{1 + 0.04} = \dfrac{0.02}{1.04} = 0.019230769 \approx 1.92\%$

4.6 FINDING THE TIME

When P, S, and i are known, we may find the unknown n, using one of the following methods:

(i) Logarithms may be used to solve the exponential equation (4.1) for n. If compound interest is allowed for the fractional part of a conversion period, a logarithmic solution gives the correct value of n. (See Problem 4.30.)

(ii) Linear interpolation may be used to approximate the value of n. If simple interest is required for the fractional part of a conversion period, linear interpolation gives the correct value of n. (See Problem 4.31.)

If interest is compounded continuously, we may solve (4.4) for the unknown t, using natural logarithms. [See Problem 4.32(*a*).]

SOLVED PROBLEMS

4.30 How long will it take \$2000 to accumulate \$800 interest at 10% compounded quarterly, if compound interest is allowed for the fractional part of a conversion period?

Let n represent the number of quarter-years. Given $P = 2000$, $S = 2800$, $i = 0.025$, we solve (4.1) for n:

$$2800 = 2000(1.025)^n$$
$$(1.025)^n = 1.4$$
$$n \log 1.025 = \log 1.4$$
$$n = \frac{\log 1.4}{\log 1.025} = 13.62643323 \text{ quarters}$$

Using approximate time (1 month = 30 days), $n \approx 3$ years 4 months 26 days.

4.31 Solve Problem 4.30 assuming that simple interest is allowed for the fractional part of a conversion period.

We arrange the data in an interpolation table, using accumulation factors rounded to four decimal places:

$$0.0345 \left\{ 0.0215 \left\{ \begin{array}{c|c} (1.025)^n & n \\ \hline 1.3785 & 13 \\ 1.4000 & n \\ 1.4130 & 14 \end{array} \right\} x \right\} 1$$

We have
$$\frac{x}{1} = \frac{0.0215}{0.0345} \qquad \text{or} \qquad x = 0.623188406$$

Then, $n = 13 + x = 13.62318841$ quarters. Using approximate time,

$$n \approx 3 \text{ years 4 months 26 days}$$

Alternate Solution

The accumulated value of $2000 for 13 periods at $j_4 = 10\%$ is

$$2000(1.025)^{13} = \$2757.02$$

Now we calculate the time for $2757.02 to accumulate $2800 - 2757.02 = \$42.98$ simple interest at rate 10%:

$$t = \frac{I}{Pr} = \frac{42.98}{2757.02(0.1)} = 0.155892957 \text{ years}$$

Using approximate time (1 year = 360 days), $t \approx 56$ days = 1 month 26 days. Thus the total time is 13 quarters 1 month 26 days, or 3 years 4 months 26 days.

4.32 By what date will $800 deposited on February 4, 1994, be worth at least $1200 (*a*) at 12% compounded continuously? (*b*) at 12% compounded daily?

(*a*) Given $P = 800$, $S = 1200$, $j_\infty = 0.12$, we solve (4.4) for t, using natural logarithms:

$$\begin{aligned}
1200 &= 800e^{0.12t} \\
e^{0.12t} &= 1.5 \\
0.12t &= \ln 1.5 \\
t &= \frac{\ln 1.5}{0.12} = 3.378875901 \text{ years}
\end{aligned}$$

Using exact time (1 year = 365 days), $t \approx 1234$ days. On June 22, 1997, the deposit will be worth at least $1200.

(*b*) Given $P = 800$, $S = 1200$, $i = 0.12/365$, we solve (4.1) for n, using common logarithms:

$$\begin{aligned}
1200 &= 800\left(1 + \frac{0.12}{365}\right)^n \\
\left(1 + \frac{0.12}{365}\right)^n &= 1.5 \\
n \log\left(1 + \frac{0.12}{365}\right) &= \log 1.5 \\
n &= \frac{\log 1.5}{\log\left(1 + \frac{0.12}{365}\right)} = 1233.492437 \approx 1234 \text{ days}
\end{aligned}$$

The date is the same as that found in (*a*).

4.33 If an investment doubles in value in 6 years at a certain rate of interest compounded monthly, how long will it take for the same investment to triple in value?

First we solve (4.1) for the unknown quantity $1+i$, given $P = x$, $S = 2x$, and $n = 6 \times 12 = 72$.

$$\begin{aligned}
2x &= x(1+i)^{72} \\
(1+i)^{72} &= 2 \\
1+i &= 2^{1/72}
\end{aligned}$$

Then we solve (4.1) for the unknown n, given $P = x$, $S = 3x$, and $1 + i = 2^{1/72}$.

$$
\begin{aligned}
3x &= x2^{n/72} \\
2^{n/72} &= 3 \\
\frac{n}{72} \log 2 &= \log 3 \\
n &= \frac{72 \log 3}{\log 2} = 114.1173001 \text{ months}
\end{aligned}
$$

Using approximate time (1 month = 30 days), $n \approx 9$ years 6 months 4 days.

4.7 EQUATIONS OF VALUE

In Section 3.3 we dealt with equations of value at a simple interest rate; most of the principles and procedures carry over to a compound interest rate. Thus, we have the *definition of equivalence*: $X due on a given date is equivalent at a given compound interest rate i to $Y due n periods later, if

$$
Y = X(1 + i)^n \qquad \text{or} \qquad X = Y(1 + i)^{-n}
$$

Figure 4-7 displays dated values equivalent to a given dated value X.

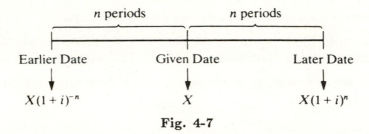

Fig. 4-7

Two important properties of equivalence at compound interest are:

1. At a given compound interest rate, if X is equivalent to Y, and Y is equivalent to Z, then X is equivalent to Z (the *transitivity property*). (See Problems 4.34 and 4.35.)

2. If two sets of payments are equivalent on one focal date (i.e., the sum of the equivalent dated values of the members of one set is equal to the corresponding sum for the other set), then they are equivalent on any focal date. (See Problem 4.37.)

Neither of these properties holds at simple interest. At compound interest, property 2 implies that the focal date for an equation of value (defined as in the case of simple interest) may be chosen arbitrarily.

The date on which a set of obligations due at different future dates can be equitably discharged by making a single payment equal to the sum of the several debts is called the *average due date of the debts*. The time from the present to that date is called the *equated time*.

The following practical rule for finding equated time is frequently used (see Problems 4.40 and 4.41): (i) Multiply each debt by the time (years) to elapse before it falls due. (ii) Add all products obtained in (i) and divide by the sum of the debts.

SOLVED PROBLEMS

4.34 Prove that a given compound interest rate, if X is equivalent to Y, and Y is equivalent to Z, then X is equivalent to Z.

Refer to Fig. 4-8. If X is equivalent to Y, then

$$Y = X(1+i)^{n_2-n_1}$$

If Y is equivalent to Z, then

$$Z = Y(1+i)^{n_3-n_2}$$

Eliminating Y between the two equations, we obtain

$$Z = X(1+i)^{n_2-n_1}(1+i)^{n_3-n_2} = X(1+i)^{n_3-n_1}$$

Thus Z is equivalent to X.

Fig. 4-8

4.35 An obligation of \$2500 falls due at the end of 7 years. If money is worth $j_{12} = 10\%$, find an equivalent debt at the end of (a) 3 years, (b) 10 years.

We arrange the data on a time diagram, Fig. 4-9. Time is shown in interest periods, i.e. months. By definition of equivalence,

$$(a) \quad X = 2500\left(1+\frac{0.10}{12}\right)^{-48} = \$1678.58 \qquad (b) \quad Y = 2500\left(1+\frac{0.10}{12}\right)^{36} = \$3370.45$$

Note that X is equivalent to Y:

$$X = Y(1+i)^{-84} = 3370.45\left(1+\frac{0.10}{12}\right)^{-84} = \$1678.58$$

or

$$Y = X(1+i)^{84} = 1678.58\left(1+\frac{0.10}{12}\right)^{84} = \$3370.45$$

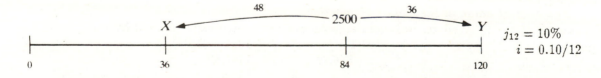

Fig. 4-9

4.36 A person owes \$1000 due at the end of 18 months and \$1500 due at the end of 4 years. If money is worth $j_4 = 6\%$, what single payment (a) now, (b) in 2 years, will liquidate these obligations?

(a) Let X be the single payment now. The time diagram is shown in Fig. 4-10. Using today as the focal date,

$$X = 1000(1.015)^{-6} + 1500(1.015)^{-16} = 914.54 + 1182.05 = \$2096.59$$

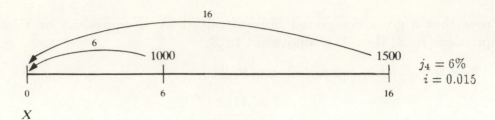

Fig. 4-10

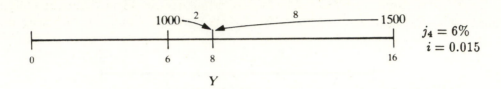

Fig. 4-11

(b) Let Y be the single payment in 2 years. The time diagram is shown in Fig. 4-11. Using the end of 2 years as the focal date,

$$Y = 1000(1.015)^2 + 1500(1.015)^{-8} = 1030.22 + 1331.57 = \$2361.79$$

Note that the dated values X and Y are equivalent:

$$Y = X(1.015)^8 = 2096.59(1.015)^8 = \$2361.79$$

4.37 A consumer buys goods worth \$1500, paying \$500 down and \$500 at the end of 6 months. If the store charges interest at $j_{12} = 18\%$ on the unpaid balance, what final payment will be necessary at the end of one year?

Let X denote the required payment. We arrange the data on a time diagram, Fig. 4-12. The down payment of \$500 may be subtracted from \$1500, resulting in a debt of \$1000. Any focal date may be chosen in writing the equation of value.

Equation of value at the end of 12 months

$$
\begin{aligned}
\text{dated value of the payments} &= \text{dated value of the debt} \\
500(1.015)^6 + X &= 1000(1.015)^{12} \\
546.72 + X &= 1195.62 \\
X &= \$648.90
\end{aligned}
$$

Equation of value at the present

$$
\begin{aligned}
\text{dated value of the payments} &= \text{dated value of the debt} \\
500(1.015)^{-6} + X(1.015)^{-12} &= 1000 \\
457.27 + 0.83638743X &= 1000 \\
0.83638743X &= 542.73 \\
X &= \$648.90
\end{aligned}
$$

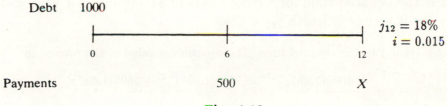

Fig. 4-12

4.38 A man stipulates in his will that \$50 000 from his estate is to be placed in a fund from which his three daughters are each to receive the same amount when aged 21. When the man dies, the girls are aged 19, 15, and 13. How much will each receive, if the fund earns interest at $j_2 = 12\%$?

Let X be the required payment. The 19-year-old will receive \$$X$ in 2 years, the 15-year-old in 6 years, and the 13-year-old in 8 years. We arrange the data in Fig. 4-13, selecting the present as the focal date for an equation of value.

$$\text{value of the payments} = \text{value of the estate}$$
$$X(1.06)^{-4} + X(1.06)^{-12} + X(1.06)^{-16} = 50\ 000$$
$$0.792093663X + 0.496969364X + 0.393646284X = 50\ 000$$
$$1.682709311X = 50\ 000$$
$$X = \$29\ 713.99$$

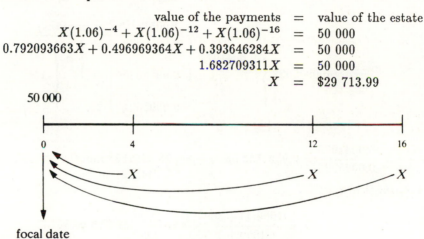

Fig. 4-13

4.39 If money is worth 10\% compounded continuously, what equal payments X at the end of 6 months and 2 years will equitably replace the obligations: \$1000 due now and \$2000 with interest from today at $j_2 = 11\%$ due in 3 years?

The maturity value of the second obligation is $2000(1.055)^6 = \$2757.69$. We arrange the data on a time diagram, Fig. 4-14, where time is shown in years. An equation of value at the present gives

$$Xe^{-(0.10)(1/2)} + Xe^{-(0.10)(2)} = 1000 + 2757.69e^{-(0.10)(3)}$$
$$0.951229424X + 0.818730753X = 1000 + 2042.95$$
$$1.769960177X = 3042.95$$
$$X = \$1719.22$$

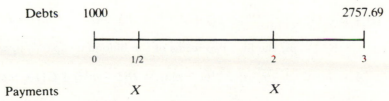

Fig. 4-14

4.40 What is the equated time for paying debts of $1000 due in 1 year and $3000 due in 2 years if money is worth $j_{12} = 6\%$?

Let n (years) be the equated time. The equation of value at the present is

$$4000(1.005)^{-12n} = 1000(1.005)^{-12} + 3000(1.005)^{-24}$$

Then

$$(1.005)^{-12n} = \frac{1000(1.005)^{-12} + 3000(1.005)^{-24}}{4000} = 0.9008656$$

Using logarithms:

$$-12n = \frac{\log .09008656}{\log 1.005}$$

$$n = 1.7443333 \text{ years} \approx 20.93 \text{ months}$$

Using interpolation:

	$(1.005)^{-12n}$	n (months)
	0.9050629	20
	0.9008656	n
	0.9005601	21

$$0.0045028 \left\{ \begin{array}{l} 0.0041973 \left\{ \begin{array}{l} \\ \\ \end{array} \right\} x \\ \\ \end{array} \right\} 1$$

We have $\dfrac{x}{1} = \dfrac{0.0041973}{0.0045028} = 0.9321533$ or $x = 20.9321533$ months

$$x = 1.7443461 \text{ years}$$

Using the practical rule:

$$x = \frac{1000(1) + 3000(2)}{1000 + 3000} = \frac{7000}{4000} = 1.75 \text{ years}$$

4.41 Derive the practical rule for finding the equated time.

Let the debts be $\$A$ due in a years from today, $\$B$ due in b years from today, and $\$C$ due in c years from today as shown in Fig. 4-15. Let $i = j_m/m$ be the rate per interest period and let n (in years) denote the equated time.

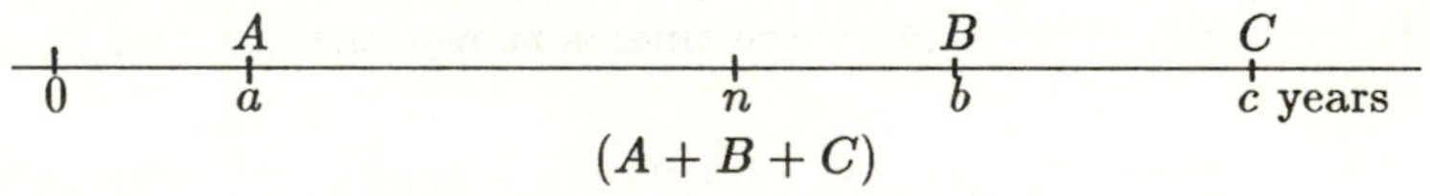

Fig. 4-15

Using today as focal date, the equation of value is

$$(A + B + C)(1 + i)^{-mn} = A(1 + i)^{-ma} + B(1 + i)^{-mb} + C(1 + i)^{-mc}$$

Replacing $(1 + i)^{-p}$ by $1 - pi$, the first two terms of the binomial expansion, we have

$$(A + B + C)(1 - mni) = A(1 - mai) + B(1 - mbi) + C(1 - mci)$$

Then

$$(A + B + C)mni = (Aa + Bb + Cc)mi$$

and

$$n = \frac{(Aa + Bb + Cc)mi}{(A + B + C)mi} = \frac{Aa + Bb + Cc}{A + B + C}$$

$$= \frac{\text{sum of products of each debt and time (years) to elapse before it falls due}}{\text{sum of debts}}$$

4.8 COMPOUND DISCOUNT AT A DISCOUNT RATE

In Section 3.5 we dealt with simple discount at a discount rate. Similarly, we can introduce compound discount at a discount rate. Let $d^{(m)}$ be the nominal rate of discount compounded m times per year. Then the discount rate per conversion period is $d^{(m)}/m$ and the discounted value P of a future amount S due in n periods is

$$P = S\left(1 - \frac{d^{(m)}}{m}\right)^n \tag{4.7}$$

From (4.7) the accumulated value S of a principal P for n periods at $d^{(m)}$ is

$$S = P\left(1 - \frac{d^{(m)}}{m}\right)^{-n} \tag{4.8}$$

For a given nominal rate of discount $d^{(m)}$ (compounded m times per year), we define the *annual effective discount rate* to be that rate d which, if compounded annually, will yield the same discounted value per year (See Problem 4.44.)

SOLVED PROBLEMS

4.42 Find the discounted value of $1000 due in 2 years at (*a*) $d^{(12)} = 12\%$, (*b*) $d^{(365)} = 7\%$.

(*a*) We have $S = 1000$, $n = 2 \times 12 = 24$, discount rate per month $= \frac{0.12}{12} = 0.01$, and from (4.7),

$$P = S\left(1 - \frac{d^{(m)}}{m}\right)^n = 1000(1 - .01)^{24} = \$785.68$$

(*b*) We have $S = 1000$, $n = 2 \times 365 = 730$, discount rate per day $= \frac{0.07}{365}$, and from (4.7),

$$P = S\left(1 - \frac{d^{(m)}}{m}\right)^n = 1000\left(1 - \frac{0.07}{365}\right)^{730} = \$869.35$$

4.43 Find the accumulated value of $500 at the end of 3 years at (*a*) 8% simple discount, (*b*) $d^{(2)} = 8\%$.

(*a*) We have $P = 500$, $t = 3$, $d = 0.08$, and from (3.6),

$$S = \frac{P}{1 - dt} = P(1 - dt)^{-1} = 500[1 - (0.08)(3)]^{-1} = \$657.89$$

(*b*) We have $P = 500$, $n = 3 \times 2 = 6$, discount rate per half-year $= \frac{0.08}{2} = 0.04$, and from (4.8),

$$S = P\left(1 - \frac{d^{(m)}}{m}\right)^{-n} = 500(1 - 0.04)^{-6} = \$638.77$$

4.44 Determine the annual effective discount rate d corresponding to (a) $d^{(m)}$, (b) $d^{(12)} = 9\%$.

In each case we equate discounted values of \$1 due in 1 year.

(a)
$$1 - d = \left(1 - \frac{d^{(m)}}{m}\right)^m$$

and
$$d = 1 - \left(1 - \frac{d^{(m)}}{m}\right)^m$$

(b)
$$1 - d = \left(1 - \frac{0.09}{12}\right)^{12}$$

and
$$d = 1 - \left(1 - \frac{0.09}{12}\right)^{12} = 0.086378765 \approx 8.64\%$$

4.45 Find the rate (a) $d^{(12)}$ equivalent to $d^{(2)} = 6\%$, (b) $d^{(4)}$ equivalent to $j_{12} = i^{(12)} = 12\%$, (c) d equivalent to $j_\infty \equiv i^{(\infty)} = 9\%$.

In each case we equate the accumulated values of \$1 at the end of one year.

(a)
$$\left(1 - \frac{d^{(12)}}{12}\right)^{-12} = (1 - 0.03)^{-2}$$
$$1 - \frac{d^{(12)}}{12} = (0.97)^{1/6}$$

and
$$d^{(12)} = 12[1 - (0.97)^{1/6}] = 0.060764049 \approx 6.08\%$$

(b)
$$\left(1 - \frac{d^{(4)}}{4}\right)^{-4} = (1.01)^{12}$$
$$1 - \frac{d^{(4)}}{4} = (1.01)^{-3}$$

and
$$d^{(4)} = 4[1 - (1.01)^{-3}] = 0.117639408 \approx 11.76\%$$

(c)
$$(1 - d)^{-1} = e^{0.09}$$
$$1 - d = e^{-0.09}$$

and
$$d = 1 - e^{-0.09} = 0.086068815 \approx 8.61\%$$

Supplementary Problems

ACCUMULATED VALUE

4.46 Find the accumulated value of \$500 for one year at (a) $j_{12} = 8\%$, (b) $j_{12} = 12\%$, (c) $j_{12} = 16\%$. In each case, compare with the accumulated value at a simple interest rate.
   Ans. (a) \$541.50, \$540; (b) \$563.41, \$560; (c) \$586.14, \$580

4.47 Find the accumulated value of $100 for 5 years at 16% compounded (a) annually, (b) semiannually, (c) quarterly, (d) monthly, (e) weekly, (f) daily, (g) continuously.
Ans. (a) $210.03; (b) $215.89; (c) $219.11; (d) $221.38; (e) $222.28; (f) $222.51; (g) $222.55

4.48 Find the compound interest earned on (a) $500 for 2 years 2 months at $11\frac{1}{4}$% compounded monthly, (b) $1000 for 6 years at 9% compounded semiannually, (c) $850 for 3 years at 8.2% compounded continuously, (d) $10 000 for 1 year at 10.7% compounded daily, (e) $220 for 18 months at $18\frac{1}{2}$% compounded quarterly.
Ans. (a) $137.29; (b) $695.88; (c) $237.06; (d) $1129.16; (e) $68.56

4.49 A person deposits $1000 into a savings account that earns interest at 12.25% compounded daily. How much interest will be earned (a) during the first year? (b) During the second year?
Ans. (a) $130.30; (b) $147.27

4.50 Parents put $1000 into a savings account at birth of their son. If the account earns interest at 6% compounded monthly, how much money will be in the account when their son is 18 years old? *Ans.* $2936.77

4.51 In 1492, Queen Isabella sponsored Christopher Columbus' journey by giving him $10 000. If she had placed this money in a bank account at $j_1 = 3$%, how much money would have been in the account in 1992? *Ans.* $26.219 billion

4.52 In the 1970 census the U.S. population was estimated at 203 million. If this population grew at the rate of 3% per annum, what would be its size in the year 2000? *Ans.* 493 million

4.53 The XYZ Company has had an increase in sales of 4% per annum. If sales in 1993 are $680 000, what would be the estimated sales for 1998? *Ans.* $830 000

4.54 Carol deposited $500 into her savings account on January 1, 1990. What will be in the account on January 1, 1995, if $j_2 = 8$% for 1990, $j_2 = 9$% for 1991 and 1992, and $j_2 = 6$% for 1993 and 1994? *Ans.* $725.86

4.55 Melinda has a savings account that earns interest at 12% per annum. She opened her account with $1000 on January 1. How much interest will she earn during the first year, if the interest is (a) compounded daily? (b) Calculated daily using exact simple interest and paid into the account on June 30 and December 31? (c) Calculated daily using exact simple interest and paid into the account at the end of each month? (d) Compounded continuously?
Ans. (a) $127.47; (b) $123.60; (c) $126.81; (d) $127.50

EQUIVALENT RATES

4.56 Find the annual effective rate equivalent to (a) 16% compounded quarterly, (b) 18% compounded monthly, (c) $9\frac{1}{4}$% compounded daily, (d) 12% compounded continuously.
Ans. (a) 16.99%; (b) 19.56%; (c) 9.69%; (d) 12.75%

4.57 Find the nominal rate j_m equivalent to the annual effective rate j, if (a) $j = 6$%, $m = 2$; (b) $j = 9$%, $m = 4$; (c) $j = 10$%, $m = 12$; (d) $j = 17$%, $m = 365$; (e) $j = 8$%, $m = 52$; (f) $j = 11.82$%, $m = \infty$.
Ans. (a) 5.91%; (b) 8.71%; (c) 9.57%; (d) 15.70%; (e) 7.70%; (f) 11.17%

4.58 Find (a) j_4 equivalent to $j_2 = 8$%, (b) j_2 equivalent to $j_4 = 8$%, (c) j_4 equivalent to $j_{12} = $ ~~8.05%~~ 18%, (d) j_2 equivalent to $j_{52} = 11$%, (e) j_{12} equivalent to $j_2 = 18\frac{1}{4}$%, (f) j_{365} equivalent to $j_4 = 12.79$%, (g) j_{12} equivalent to $j_\infty = 15$%, (h) j_∞ equivalent to $j_1 = 13$%.
Ans. (a) 7.92%; (b) 8.08%; (c) 18.27%; (d) 11.30%; (e) 17.59%; (f) 12.59%; (g) 15.09%; (h) 12.22%

4.59 What simple interest rate is equivalent to $j_{12} = 13\frac{1}{2}$% if money is invested for 2 years?
Ans. 15.40%

4.60 For a given nominal rate $j_{12} = 12i$, find the equivalent rate (a) j_1, (b) j_2, (c) j_4, (d) j_{52}, (e) j_{365}, (f) j_∞. *Ans.* (a) $(1 + i)^{12} - 1$; (b) $2[(1 + i)^6 - 1]$; (c) $4[(1 + i)^3 - 1]$; (d) $52[(1 + i)^{12/52} - 1]$; (e) $365[(1 + i)^{12/365} - 1]$; (f) $\ln(1 + i)^{12}$

4.61 Olivia must borrow \$1000 for 2 years. She is offered the money at (a) 13% compounded monthly, (b) $13\frac{1}{4}$% compounded semiannually, (c) $14\frac{1}{2}$% simple interest. Which offer should she accept? *Ans.* (c)

4.62 Which rate gives the best, and which the worst, return on your investment? (a) $j_{12} = 15\%$, $j_2 = 15\frac{1}{2}\%$, $j_{365} = 14.9\%$; (b) $j_{12} = 16\%$, $j_2 = 16\frac{1}{2}\%$, $j_{365} = 15.9\%$. *Ans.* (a) j_2 best, j_{365} worst; (b) j_{365} best, j_2 worst

4.63 A bank pays 12% per annum on its savings accounts. At the end of every three years, a 2% bonus is paid on the balance at that time. Find the annual effective rate of interest earned by an investor, if the deposit is withdrawn (a) in 2 years, (b) in 3 years, (c) in 4 years. *Ans.* (a) 12%; (b) 12.74%; (c) 12.56%

4.64 If j_2 and $j_1 = j_2 + 0.0025$ are equivalent rates of interest, find j_2. *Ans.* 10%

4.65 A trust company offers guaranteed investment certificates paying $j_2 = 8.9\%$ and $j_1 = 9\%$. Which option yields the higher annual effective rate of interest? *Ans.* $j_2 = 8.9\%$

4.66 A sum of money is left invested for 3 years. In the first year it earns interest at $j_{12} = 15\%$. In the second year, the rate of interest earned is $j_4 = 10\%$, and in the third year the rate of interest changes to $j_{365} = 12\%$. Find the level rate of interest, j_1, that would give the same accumulated value at the end of three years. *Ans.* 13.04%

4.67 What compound interest rate j_4 is equivalent over a 3-year period to a simple interest rate of 6% the first year, followed by a simple discount rate of 8% for the next 2 years? *Ans.* 7.83%

4.68 Find a simple discount rate equivalent to 8% compounded continuously if money is invested for 4 years. *Ans.* 6.85%

4.69 What simple interest rate is equivalent to the force of interest $\delta = 7\%$ if money is invested for 5 years? *Ans.* 8.38%

4.70 Bank A has an annual effective interest rate of 10%. Bank B has a nominal interest rate of $9\frac{3}{4}\%$. What is the minimum frequency of compounding for bank B in order that the rate at bank B be at least as attractive as that at bank A? *Ans.* 4

4.71 An insurance company says you can pay for your life insurance by paying \$100 at the beginning of each year or \$51.50 at the beginning of each half-year. They say the rate of interest underlying this calculation is $j_2 = 3\%$. What is the true value of j_2? *Ans.* 12.37%

4.72 In general, the annual effective rate of interest is the ratio of the amount of interest earned during the year to the amount of principal invested at the beginning of the year.

(a) Show that, at a simple interest rate r, the annual effective rate of interest for the nth year is $\dfrac{r}{1 + r(n - 1)}$, which is a decreasing function of n. Thus a constant rate of simple interest implies a decreasing annual effective rate of interest.

(b) Show that, at a compound interest rate i per year, the annual effective rate of interest for the nth year is i, which is independent of n. Thus a constant rate of compound interest implies a constant annual effective rate of interest.

4.73 A fund earns interest at the nominal rate of 8.04% compounded quarterly. At the end of each quarter, just after interest is credited, an expense charge equal to 0.50% of the fund is withdrawn. Find the annual effective yield realized by the fund. *Ans.* 6.14%

DISCOUNTED VALUE

4.74 Find the present value of (a) $100 000 due in 25 years, if money is worth $j_{12} = 12\%$; (b) $2500 due in 10 years, if money is worth $j_2 = 9.6\%$; (c) $800 due in 3 years, if money is worth 12% compounded daily; (d) $3000 with interest at $j_2 = 16\frac{1}{2}\%$ due in 5 years, if money is worth 15% effective; (e) $5000 due in 15 months, if money is worth 11% compounded continuously. *Ans.* (a) $5053.45; (b) $978.85; (c) $558.17; (d) $3295.42; (e) $4357.67

4.75 On her 20th birthday a young woman receives $1000 as a result of a deposit her parents made on the day she was born. How large was that deposit, if it earned interest at (a) $j_2 = 6\%$, (b) $j_4 = 12\%$? *Ans.* (a) $306.56; (b) $93.98

4.76 An obligation of $2000 is due December 31, 2000. What is the value of this obligation on June 30, 1996, at $j_4 = 13\frac{1}{4}\%$? *Ans.* $1112.44

4.77 A note dated October 1, 1993, calls for the payment of $800 in 7 years. On October 1, 1995, it is sold at a price that will yield the purchaser $j_4 = 16\%$. How much is paid for the note? *Ans.* $365.11

4.78 A note for $1000 dated January 1, 1993, is due with compound interest at 13% compounded semiannually 5 years after date. On July 1, 1994, the holder of the note has it discounted by a lender who charges $14\frac{1}{2}\%$ compounded quarterly. Find the proceeds and the compound discount. *Ans.* $1140.23, $736.91

4.79 Find the compound discount if $1000 due in 5 years with interest at $j_1 = 14\frac{1}{2}\%$ is discounted at a nominal rate of 16% compounded (a) annually, (b) semiannually, (c) quarterly, (d) monthly, (e) weekly, (f) daily, (g) continuously. *Ans.* (a) $1031.01; (b) $1056.44; (c) $1069.84; (d) $1079.04; (e) $1082.64; (f) $1083.57; (g) $1083.73

4.80 You can buy a building lot for $180 000 cash, or for payments of $100 000 now, $50 000 in 1 year, and $50 000 in 2 years. Which option is better for you, if money is worth (a) 16% compounded monthly? (b) 12% compounded monthly? *Ans.* (a) Payments; (b) cash

4.81 A note for $2500 dated January 1, 1994, is due with interest at $j_{12} = 12\%$ forty months later. On May 1, 1994, the holder of the note has it discounted by Financial Consultants Inc. at $j_4 = 13\frac{1}{4}\%$. The same day, the note is sold by Financial Consultants to a bank that discounts notes at $j_1 = 13\%$. What is the profit made by Financial Consultants? *Ans.* $62.19

4.82 A young couple own a block of land worth $29 000. They are offered a 20% deposit and 2 equal payments of $15 000 each at the end of years 2 and 4. If money is worth $j_2 = 8\%$ for the first 2 years and $j_4 = 12\%$ for the next 2 years, should they accept the offer? *Ans.* Do not accept the offer, PV=$28 743.92

4.83 A person can buy a lot for $13 000 cash outright or $6000 down, $6000 in 2 years and $6000 in 5 years. Which option is better if money can be invested at (a) $j_{12} = 18\%$? (b) $j_4 = 12\%$ for the first 3 years and $j_4 = 14\%$ for the next 2 years? *Ans.* (a) Payments; (b) cash

FRACTIONAL INTEREST PERIODS

4.84 Find the accumulated value of $1500 for 16 months at $j_4 = 18\%$, using (a) compound interest; (b) simple interest for the fractional part of an interest period. *Ans.* (a) $1896.90; (b) $1897.31

4.85 Find the discounted value of $5000 due in 8 years 10 months if interest is at rate $j_2 = 12.73\%$ and (a) compound interest, (b) simple interest, is used for the fractional part of an interest period. *Ans.* (a) $1680.84; (b) $1681.55

4.86 On July 7, 1993, Madeleine borrowed $1200 at 12% compounded monthly. How much would she have to repay on September 18, 1996? *Ans.* $1757.85

4.87 A noninterest-bearing note for $850 is due on December 5, 1996. On August 7, 1993, the holder of the note has it discounted by a lender who charges $j_2 = 15\frac{1}{4}\%$. What are the proceeds? *Ans.* $521.75

4.88 A note for $1200 dated August 21, 1983, is due with interest at $j_{12} = 14\frac{3}{4}\%$ in 2 years. On June 18, 1984, the holder of the note has it discounted by a lender who charges $16\frac{1}{4}\%$ compounded quarterly. Find the proceeds and the compound discount. *Ans.* $1335.06, $273.80

4.89 (*a*) If $0 < i < 1$, prove that

$$(1+i)^t < 1 + it \quad (0 < t < 1)$$
$$(1+i)^t > 1 + it \quad (t > 1)$$

(*Hint*: Use the binomial theorem and the error property of an alternating series whose terms decrease in magnitude.) (*b*) Give a geometric illustration of the relationships in (*a*) by graphing $(1+i)^t$ and $1 + it$. (*c*) Infer from (*a*) that the approximate accumulated and discounted values are greater than the exact accumulated and discounted values when fractional parts of interest periods are involved.

FINDING THE RATE

4.90 Find the nominal rate compounded quarterly at which $2000 will accumulate to $3000 in 3 years 9 months. *Ans.* 10.96%

4.91 Find j_{12} at which $100 will accumulate $50 interest in 4 years 7 months. *Ans.* 8.88%

4.92 John deposited $1000 three and a half years ago. He has $1581.72 in his account now. What rate of interest compounded semiannually did he earn on his deposit? *Ans.* 13.54%

4.93 (*a*) At what annual effective rate of interest will money triple in 15 years? (*b*) At what rate compounded quarterly will an investment grow 50% in 4 years? (*c*) At what rate compounded daily will an investment double in value in 5 years? (*d*) At what rate compounded continuously will a deposit of $1000 accumulate interest of $250 in 30 months? *Ans.* (*a*) 7.60%; (*b*) 10.27%; (*c*) 13.87%; (*d*) 8.93%

4.94 The population of a county was 200 000 in 1970, and 250 000 in 1980. Estimate the increase in population of the county between 1990 and 1995. *Ans.* 36 886

4.95 On January 1 $500 000 is deposited in Fund X and $50 000 is deposited in Fund Y. No previous deposits exist. Fund X earns compound interest at rate of i per year. Fund Y earns simple interest at $(i + .01)$ per year. On April 1 one additional deposit of $50 000 is made to Fund Y. No additional deposits are made to Fund X. On December 31 the sum of the values of Funds X and Y is $642 000. Find i. *Ans.* 7%

4.96 Calculate the real annual rate of return for the following pairs of annual interest rates i and annual inflation rates r: (*a*) $i = 6\%$, $r = 2\%$; (*b*) $i = 8\%$, $r = 4\%$; (*c*) $i = 10\%$, $r = 6\%$. *Ans.* (*a*) 3.92%; (*b*) 3.85%; (*c*) 3.77%

FINDING THE TIME

4.97 Assuming that compound interest is allowed for the fractional part of a conversion period, how long will it take for: (*a*) $1800 to accumulate to $2200 at $j_4 = 8\%$? (*b*) $100 to accumulate to $130 at $j_2 = 9\%$? (*c*) $500 to accumulate to $800 at $j_{12} = 12\%$? *Ans.* (*a*) 2 years 6 months 12 days; (*b*) 2 years 11 months 23 days; (*c*) 3 years 11 months 7 days

4.98 Solve Problem 4.97 if simple interest is required for the fractional part of a conversion period. *Ans.* Same as in Problem 4.97.

4.99 Using logarithms, find the time it will take for (a) a deposit to double in value at $j_1 = 19.56\%$, (b) an investment to double in value at 15% compounded daily, (c) $800 to grow to $1500 at $j_2 = 9.8\%$, (d) an investment to triple in value at 15% compounded continuously. *Ans.* (a) 3 years 10 months 17 days; (b) 1687 days; (c) 6 years 6 months 25 days; (d) 7 years 118 days

4.100 $500 was deposited on January 1, 1980, in an account paying 12% compounded semiannually. On January 1, 1983, $400 was deposited in another account paying $15\frac{1}{2}\%$ compounded annually. Find the time when the two accounts will be of equal value, if the exact method is used for fractions of an interest period. *Ans.* 20.780279 years from January 1, 1983

4.101 Determine how long $100 must be left to accumulate at $j_{12} = 18\%$ for it to amount to twice the accumulated value of another $100 deposited at the same time at $j_2 = 10\%$. *Ans.* 8.5486117 years

4.102 If the cost of living rises 8% a year, how long will it take for the purchasing power of $1 to fall to 60¢? *Ans.* 6.6374573 years

4.103 If an investment doubles at a certain rate of interest compounded continuously in 5 years, how long will it take for this investment to triple in value? *Ans.* 7.9248125 years

4.104 By what date will $1000 deposited on November 20, 1995, at $12\frac{1}{2}\%$ compounded daily be worth at least $1250? *Ans.* September 2, 1997

4.105 Account A starts now with $100 and pays $j_1 = 4\%$. After 2 years an additional $25 is deposited in account A. Account B is opened 1 year from now with a deposit of $95 and pays $j_1 = 8\%$. When (in years and days) will account B have $1\frac{1}{2}$ times the accumulated value in Account A if simple interest is allowed for part of a year and exact time/exact interest is used? *Ans.* 19 years 233 days

4.106 If the population of the world doubles every 25 years, how long does it take to increase by 50%? Assume population growth takes place continuously at a uniform rate. *Ans.* 14.624063 years

4.107 The force of interest is $\delta = 10\%$. At what time should a single payment of $2500 be made so as to be equivalent to payments of $1000 in 1.25 years and $1500 in 6.5 years? *Ans.* 4.0612599 years

4.108 How long will it take $250 to accumulate to $400, if the force of interest is $\delta = 7\%$ for the first 2 years and $\delta = 8\%$ thereafter. *Ans.* 6.1250454 years

EQUATIONS OF VALUE

4.109 An obligation of $500 falls due at the end of 3 years. Find an equivalent debt at the end of (a) 3 months, (b) 3 years 9 months, given $j_4 = 12\%$. *Ans.* (a) $361.21; (b) $546.37

4.110 What sum of money, due at the end of 5 years, is equivalent to $1800 due at the end of 12 years, if money is worth $j_2 = 11\frac{3}{4}\%$? *Ans.* $809.40

4.111 Anita owes $200 due in 6 months and $300 due in 15 months. What single payment (a) now, (b) in 12 months will liquidate these obligations, if money is worth $j_{12} = 15\%$? *Ans.* (a) $434.63; (b) $504.51

4.112 $800 is due at the end of 4 years, and $700 at the end of 8 years. If money is worth $j_2 = 12\%$, find an equivalent single amount at the end of (a) 2 years, (b) 6 years, (c) 10 years. Show that your answers are equivalent. *Ans.* (a) $981.55; (b) $1564.45; (c) $2493.49

4.113 A debt of \$2000 is due at the end of 8 years. If \$1000 is paid at the end of 3 years, what single payment at the end of 7 years would liquidate the debt, if money is worth $j_2 = 12\%$?
Ans. \$186.14

4.114 David borrows \$4000 at $j_4 = 12\%$. He promises to pay \$1000 at the end of one year, \$2000 at the end of 2 years, and the balance at the end of 3 years. What will the final payment be?
Ans. \$2185.25

4.115 A bank account showed the following deposits and withdrawals:

	Deposits	Withdrawals
January 1, 1993	\$200	
July 1, 1993	\$150	
January 1, 1994		\$250
July 1, 1994	\$100	

If the account earns $j_2 = 6\%$, find the balance in the account on January 1, 1995. *Ans.* \$226.78

4.116 A debt of \$1000 with interest at $j_4 = 10\%$ will be repaid by a payment of \$200 at the end of 3 months and three equal payments at the ends of 6, 9, and 12 months. What will these payments be? *Ans.* \$288.86

4.117 On January 1, 1993, Mr. Smith borrowed \$5000 to be repaid in a lump sum with interest at $j_4 = 9\%$ on January 1, 1999. It is now January 1, 1995. Mr. Smith would like to pay \$500 today and complete the liquidation with equal payments on January 1, 1997, and January 1, 1999. If money is now worth $j_4 = 8\%$, what will these payments be? *Ans.* \$3611.27

4.118 To pay off a loan of \$5000 at $j_{12} = 15\%$, Mrs. Jones agrees to make three payments in two, five, and ten months, respectively. The second payment is to be double the first, and the third payment is to be triple the first. What are the sizes of the payments?
Ans. \$908.34, \$1816.68, \$2725.02

4.119 You are given two loans, with each loan to be repaid by a single payment in the future. Each payment will include both principal and interest. The first loan is repaid by a \$3000 payment at the end of four years. The interest is accrued at 10% per annum compounded semiannually. The second loan is repaid by a \$4000 payment at the end of five years. The interest is accrued at 8% per annum compounded semiannually. These two loans are to be consolidated. The consolidated loan is to be repaid by two equal installments of X, with interest at 12% per annum compounded semiannually. The first payment is due immediately and the second payment is due one year from now. Calculate X. *Ans.* \$2504.12

4.120 You are given the following data on three series of payments:

	Payment at end of year			Accumulated value
	6	12	18	at end of year 18
Series A	240	200	300	X
Series B	0	360	700	$X + 100$
Series C	Y	600	0	X

Assume interest is compounded annually. Calculate Y. *Ans.* \$106.67

4.121 A debt of \$5000 is due at the end of 5 years. It is proposed that \$$X$ be paid now with another \$$X$ paid in 10 years time to liquidate the debt. Calculate the value of X if the annual effective interest rate is 12% for the first 6 years and 8% for the next 4 years. *Ans.* \$2067.29

4.122 A company wishes to replace the following three debts:

$20 000 due on July 1, 1991
$30 000 due on January 1, 1994 and
$35 000 due on July 1, 1997,

with a single debt of $Y payable on January 1, 1994. Calculate the value of Y if $j_2 = 12\%$ prior to January 1, 1994, and $j_2 = 10\%$ after January 1, 1994. *Ans.* $81 638.36

4.123 On September 1, 1992, Paul borrowed $3000, agreeing to pay interest at 12% compounded quarterly. He paid $900 on March 1, 1993, and $1200 on December 1, 1993.

(a) What equal payments on June 1, 1994, and December 1, 1994, will be needed to settle the debt?

(b) If Paul paid $900 on March 1, 1993, $1200 on December 1, 1993 and $900 on March 1, 1994, what would be his outstanding balance on September 1, 1994?
Ans. (a) $706.89 (b) $459.58

4.124 Find the equated time for paying two debts of $1200 each, one due in 6 months and the other in 1 year, if money is worth 6% compounded monthly. *Ans.* 0.75 year

COMPOUND DISCOUNT AT A DISCOUNT RATE

4.125 Find the discounted value of $2000 due in 3 years at (a) $d^{(2)} = 8\%$, (b) $d^{(12)} = 10.3\%$.
Ans. (a) $1565.52; (b) $1466.40

4.126 Find the accumulated value of $3000 for 2 years at (a) $d^{(52)} = 10\%$, (b) $d^{(365)} = 10\%$.
Ans. (a) $3664.91; (b) $3664.31

4.127 Find the annual effective discount rate corresponding to (a) $d^{(4)} = 8.2\%$, (b) $d^{(365)} = 6\%$.
Ans. (a) 7.95%; (b) 5.82%

4.128 Find $d^{(4)}$ equivalent to (a) $d^{(12)} = 9.8\%$, (b) $d^{(52)} = 8.5\%$, (c) $j_4 = 8\%$, (d) $j_\infty = 7.3\%$.
Ans. (a) 9.72%; (b) 8.42%; (c) 7.84%; (d) 7.23%

4.129 Calculate the value at the end of 4 years of an $18 000 car, using an annual effective rate of depreciation of 30%. (*Hint*: This is the same as an annual effective rate of compound discount of 30%.) *Ans.* $4321.80

4.130 A noninterest-bearing note of amount X is due in 3 months. A finance company calculates the value of the note today to be $3825. Find X under each of the following interest calculation methods: (a) compound interest at an annual effective interest rate of 9%, (b) compound discount at an annual effective discount rate of 9%, (c) a simple interest rate of 9%, (d) a simple discount rate of 9%.
Ans. (a) $3908.30; (b) $3916.26; (c) $3911.06; (d) $3913.04

4.131 Let $i = \dfrac{j_m}{m} = \dfrac{i^{(m)}}{m}$ be an effective interest rate per period; $d = \dfrac{d^{(m)}}{m}$ be an effective discount rate per period; and $v = (1 + i)^{-1}$. Show that: (a) $v = 1 - d$, (b) $d = iv$, (c) $(1 + \dfrac{i^{(m)}}{m})^m = (1 - \dfrac{d^{(m)}}{m})^{-m}$.

4.132 Find $i^{(4)}$ equivalent to $d^{(12)} = 6\%$ for 2 years followed by $d^{(365)} = 5\%$ for 2 years. *Ans.* 5.55%

4.133 Show that equivalent nominal interest and discount rates $i^{(m)}$ and $d^{(m)}$ satisfy the relationships

(a) $d^{(m)} = \dfrac{i^{(m)}}{1 + \frac{i^{(m)}}{m}}$ and (b) $i^{(m)} = \dfrac{d^{(m)}}{1 - \frac{d^{(m)}}{m}}$

Chapter 5

Simple Annuities

5.1 DEFINITIONS AND NOTATION

An *annuity* is a sequence of payments, usually equal (see Section 6.5 for annuities where payments vary), made at equal intervals of time. Premiums on insurance, mortgage payments, interest payments on bonds, payments of rent, payments on installment purchases, and dividends are all annuities.

The time between successive payments of an annuity is called the *payment interval*. The time from the beginning of the first payment interval to the end of the last payment interval is called the *term* of an annuity. When the term of an annuity is fixed, i.e., the dates of the first and last payments are fixed, the annuity is called an *annuity certain*. When the term of the annuity depends on some uncertain event, the annuity is called a *contingent annuity*. Bond interest payments form an annuity certain; life-insurance premiums form a contingent annuity (they cease with the death of the insured). Unless otherwise specified, the word "annuity" will refer to an annuity certain.

When the payments are made at the ends of the payment intervals, the annuity is called an *ordinary annuity*. When the payments are made at the beginning of the payment intervals, the annuity is called an *annuity due*. A *deferred annuity* is an annuity whose first payment is due at some later date.

When the payment interval and interest conversion period coincide, the annuity is called a *simple annuity*; otherwise, it is a *general annuity*.

We define the *accumulated value* of an annuity as the equivalent dated value of the set of payments due, at the end of the term. Similarly, the *discounted value* of an annuity is defined as the equivalent dated value of the set of payments due, at the beginning of the term.

We shall use the following notation:

R $\equiv$ periodic payment of the annuity

n $\equiv$ number of interest conversion periods during the term of an annuity (or, in the case of a simple annuity, the total number of payments)

i $\equiv$ interval rate per conversion period

S $\equiv$ accumulated value, or the amount, of an annuity

A $\equiv$ discounted value, or the present value, of an annuity

In this chapter, we shall deal only with simple annuities; that is, annuities whose payments are made at the end of the interest conversion periods.

5.2 ACCUMULATED VALUE OF AN ORDINARY SIMPLE ANNUITY

The accumulated value S of an ordinary simple annuity of n payments of $\$R$ each is the equivalent dated value of the set of these payments due, at the end of the term of the annuity (which is the date of the last payment). In Fig. 5-1 we display an ordinary simple annuity, with the interest period as the time unit.

In Problem 5.1, we derive the formula for the accumulated value S of an ordinary simple annuity:

$$S = R\, s_{\overline{n}|i} \equiv R\frac{(1+i)^n - 1}{i} \tag{5.1}$$

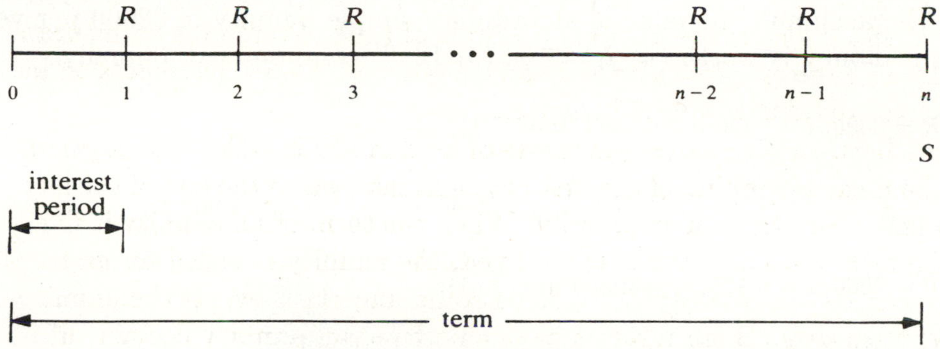

Fig. 5-1

Here $s_{\overline{n}|i}$, read "s angle n at i," is the accumulated value of an ordinary simple annuity of n payments of \$1 each; it is called the *accumulated value of \$1 per period*, or an *accumulation factor for n payments*. Traditionally, textbooks in the Mathematics of Finance provided Compound Interest Tables, listing the values of $s_{\overline{n}|i}$ for certain values of i and n. We calculate the factors $s_{\overline{n}|i}$ directly on a calculator, and use all the digits of the display to achieve the highest possible accuracy.

When (5.1) is solved for R, we obtain

$$R = \frac{S}{s_{\overline{n}|i}} = S\frac{i}{(1+i)^n - 1}$$

as the periodic payment of an ordinary simple annuity whose accumulated value S is given. (See Problem 5.8.)

SOLVED PROBLEMS

5.1 Derive (5.1).

Consider an ordinary simple annuity of n payments of \$1 each, as shown in Fig. 5-2. Let $s_{\overline{n}|i}$ denote the accumulated value of this annuity. The equation of value at the end of the term is

$$s_{\overline{n}|i} = 1 + (1+i)^1 + (1+i)^2 + \cdots + (1+i)^{n-1}$$

This is a geometric progression of n terms with first term $t_1 = 1$ and ratio $r = 1 + i$. Thus,

$$s_{\overline{n}|i} = t_1\frac{r^n - 1}{r - 1} = \frac{(1+i)^n - 1}{(1+i) - 1} = \frac{(1+i)^n - 1}{i}$$

The accumulated value S of an ordinary simple annuity of n payments of \R each is then $R\, s_{\overline{n}|i}$.

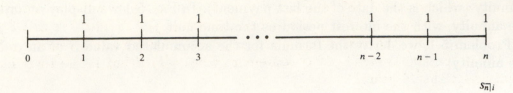

Fig. 5-2

5.2 Find the accumulated value of an ordinary simple annuity of $2000 per year for 5 years if money is worth (a) $j_1 = 9\%$, (b) $12\frac{1}{2}\%$ compounded annually.

(a) $R = 2000$, $i = 0.09$, $n = 5$; and from (5.1),

$$S = 2000 s_{\overline{5}|.09} = 2000 \frac{(1.09)^5 - 1}{0.09} = \$11\ 969.42$$

(b) $R = 2000$, $i = 0.125$, $n = 5$; and from (5.1),

$$S = 2000 s_{\overline{5}|.125} = 2000 \frac{(1.125)^5 - 1}{0.125} = \$12\ 832.52$$

5.3 John is repaying a debt with payments of $250 a month. If he misses his payments for July, August, September, and October, what payment will be required in November to put him back on schedule, if interest is at $j_{12} = 14.4\%$?

John will have to pay four overdue payments with interest, together with the November payment. These payments have accumulated value

$$250 s_{\overline{5}|.012} = 250 \frac{(1.012)^5 - 1}{0.012} = \$1280.36$$

5.4 Diane deposits $300 every 3 months into a savings account that pays interest at $j_4 = 8\%$. How much money is in her account just after the deposit on March 1, 1997, if the first deposit was made on March 1, 1993?

We arrange the data in Fig. 5-3, noting that the ordinary annuity starts one interest period before the first deposit, i.e., on December 1, 1992. Thus, the accumulated value S of her deposits on March 1, 1997, is

$$S = 300 s_{\overline{17}|.02} = 300 \frac{(1.02)^{17} - 1}{0.02} = \$6003.62$$

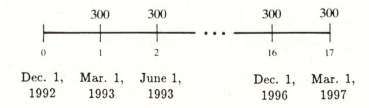

Fig. 5-3

5.5 Frank has deposited $1000 at the end of each year into a retirement savings plan for the last 10 years. His deposits earned interest at $j_1 = 8\%$ for the first 3 years; at $j_1 = 10\frac{1}{4}\%$ for the next 4 years; and at $j_1 = 9\%$ for the last 3 years. (a) What is the accumulated value of his retirement plan? (b) What is the total interest earned for the 10 years?

(a) We arrange the data as in Fig. 5-4. To obtain S we write an equation of value at 10:

$$S = 1000s_{\overline{3}|.08}(1.1025)^4(1.09)^3 + 1000s_{\overline{4}|.1025}(1.09)^3 + 1000s_{\overline{3}|.09}$$
$$= 6211.49 + 6032.38 + 3278.10 = \$15\ 521.97$$

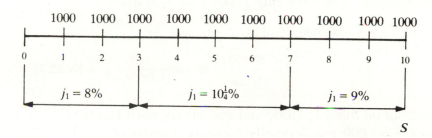

Fig. 5-4

(b) The total interest earned for the 10 years $= 15\ 521.97 - 10\ 000 = \$5521.97$.

5.6 Prove
$$s_{\overline{m+n}|i} = (1+i)^n s_{\overline{m}|i} + s_{\overline{n}|i}$$

$$
\begin{aligned}
s_{\overline{m+n}|i} &= \frac{(1+i)^{m+n} - 1}{i} = \frac{(1+i)^{m+n} - (1+i)^n + (1+i)^n - 1}{i} \\
&= \frac{(1+i)^{m+n} - (1+i)^n}{i} + \frac{(1+i)^n - 1}{i} \\
&= (1+i)^n \frac{(1+i)^m - 1}{i} + \frac{(1+i)^n - 1}{i} = (1+i)^n s_{\overline{m}|i} + s_{\overline{n}|i}
\end{aligned}
$$

5.7 Jane opens a savings account with a deposit of $2000 on February 1, 1993, and makes monthly deposits of $200 for 5 years, starting March 1, 1993. Starting March 1, 1998, she makes monthly withdrawals of $400 for 3 years. Find the balance in her account just after the last withdrawal (on February 1, 2001) if $j_{12} = 6\%$.

We arrange the data as in Fig. 5-5. To obtain the balance, X, on February 1, 1993, we write an equation of value on that date:

$$X = 200(1.005)^{96} + 200\ s_{\overline{60}|.005}(1.005)^{36} - 400\ s_{\overline{36}|.005}$$
$$= 3228.29 + 16\ 698.49 - 15\ 734.44 = \$4192.34$$

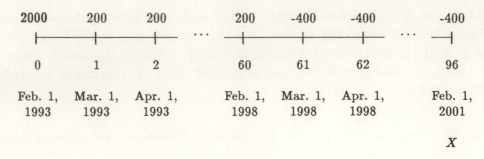

Fig. 5-5

5.8 It is estimated that a machine will need to be replaced 10 years from now at a cost of \$80 000. How much must be set aside each year to provide that money if the company's savings earn interest at $j_1 = 8\%$?

We have $S = 80\ 000$, $i = 0.08$, $n = 10$, and

$$R = \frac{80\ 000}{s_{\overline{10}|.08}} = 80\ 000 \frac{0.08}{(1.08)^{10} - 1} = \$5522.36$$

5.9 Starting on June 1, 1995, and continuing until December 1, 2000, a company will need \$250 000 semiannually to retire a series of bonds. What equal semiannual deposits in a fund paying $j_2 = 10\%$ beginning on June 1, 1990, and continuing until December 1, 2000, are necessary to retire the bonds as they fall due?

Letting X denote the required semiannual deposit, we arrange the data as in Fig. 5-6. To calculate X we may write an equation of value on Dec. 1, 1995:

$$X\, s_{\overline{22}|.05} = 250\ 000 s_{\overline{12}|.05}$$

$$X = 250\ 000 \frac{s_{\overline{12}|.05}}{s_{\overline{22}|.05}} = \$103\ 343.97$$

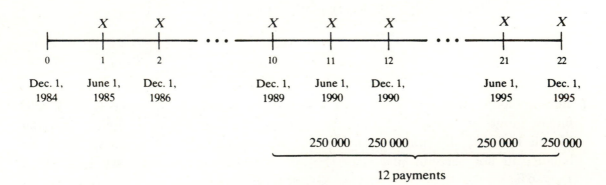

Fig. 5-6

5.3   DISCOUNTED VALUE OF AN ORDINARY SIMPLE ANNUITY

The discounted value, A, of an ordinary simple annuity of n payments is the equivalent dated value of the set of these payments due, at the beginning of the term, that is, 1 period before the first payment. Since A and S are both dated values of the same set of payments, they must be equivalent to each other:

$$A = S(1 + i)^{-n}$$

Substituting for S from (5.1), we have

$$A = R\, s_{\overline{n}|i}(1 + i)^{-n} = R\frac{(1 + i)^n - 1}{i}(1 + i)^{-n} = R\frac{1 - (1 + i)^{-n}}{i} \equiv R\, a_{\overline{n}|i} \qquad (5.2)$$

(See also Problem 5.63.)

The factor $a_{\overline{n}|i}$ (read "a angle n at i") is called a *discount factor for n payments*, or the *discounted value of \$1 per period*. We calculate $a_{\overline{n}|i}$ directly on a calculator, using all the digits of the display to achieve the highest possible accuracy.

When (5.2) is solved for R, we obtain

$$R = \frac{A}{a_{\overline{n}|i}} = A\frac{i}{1 - (1 + i)^{-n}}$$

as the periodic payment of an ordinary simple annuity whose discounted value A is given. (See Problem 5.17.)

SOLVED PROBLEMS

5.10 Find the present value of an annuity of \$380 at the end of each month for 3 years, at (a) $j_{12} = 12\%$, (b) $j_{12} = 10.38\%$.

(a) $R = 380$, $i = 0.01$, $n = 36$; and from (5.2),

$$A = 380a_{\overline{36}|.01} = 380\frac{1 - (1.01)^{-36}}{0.01} = \$11\,440.85$$

(b) $R = 380$, $i = 0.1038/12 = 0.00865$, $n = 36$; and from (5.2),

$$A = 380a_{\overline{36}|.00865} = 380\frac{1 - (1.00865)^{-36}}{0.00865} = \$11\,711.81$$

5.11 Jeff buys a used car by paying \$1500 down and \$182.50 a month for 3 years. (a) What was the cash price of the car, if the interest rate on the loan is $j_{12} = 18\%$? (b) What was the total interest on the loan?

(a) Letting C denote the cash price, we arrange the data as in Fig. 5-7. The cash price of the car is the down payment plus the discounted value of the ordinary simple annuity of 36 monthly payments of \$182.50 each.

$$
\begin{aligned}
C &= 1500 + 182.50a_{\overline{36}|.015} = 1500 + 182.50\frac{1 - (1.015)^{-36}}{0.015} \\
&= 1500 + 5048.07 = \$6548.07
\end{aligned}
$$

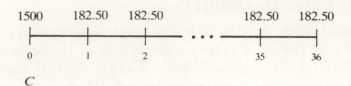

Fig. 5-7

(b)   The total interest   =   total paid on the loan − amount of the loan
                           = $36 \times 182.50 - 5048.07$
                           = $6570 - 5048.07 = \$1521.93$

5.12 Prove and interpret: $(1+i)a_{\overline{n}|i} = a_{\overline{n-1}|i} + 1$.

$$(1+i)a_{\overline{n}|i} = (1+i)\frac{1-(1+i)^{-n}}{i} = \frac{1+i-(1+i)^{-(n-1)}}{i} = \frac{1-(1+i)^{-(n-1)}}{i} + 1 = a_{\overline{n-1}|i} + 1$$

Each side of this equation represents the value of an annuity of n payments of $1 each at the time of the first payment.

5.13 Mrs. Cheung signed a contract that calls for a down payment of $2000 and for payments of $250 a month for 5 years. Money is worth $j_{12} = 12\%$. (a) What is the cash value of the contract? (b) If Mrs. Cheung missed the first 6 payments, what must she pay at the time of the 7th payment to discharge her entire indebtedness? (c) If, at the beginning of the 3rd year (after the 24th payment has been made), the contract is sold to a buyer at a price that will yield him $j_{12} = 15\%$, what does the buyer pay?

(a)   Let C denote the cash value of the contract. Then, as in Problem 5.11,

$$C = 2000 + 250a_{\overline{60}|.01} = 2000 + 250\frac{1-(1.01)^{-60}}{0.01} = 2000 + 11\,238.76 = \$13\,238.76$$

(b)   Let X denote the required payment. Mrs. Cheung must pay the accumulated value of the first 7 payments plus the discounted value of the remaining $60 - 7 = 53$ payments in order to discharge her indebtedness completely at the time of the 7th payment (see Fig. 5-8).

$$X = 250s_{\overline{7}|.01} + 250a_{\overline{53}|.01} = 250\frac{(1.01)^{7}-1}{0.01} + 250\frac{1-(1.01)^{-53}}{0.01}$$
$$= 1803.38 + 10\,246.09 = \$12\,049.47$$

Alternatively, we can calculate X by finding the discounted value of all 60 payments and then accumulating it to the end of the 7th period:

$$X = 250a_{\overline{60}|.01}(1.01)^{7} = \$12\,049.47$$

(c)   Let Y be the price the buyer must pay. Then Y is the discounted value of the remaining $60 - 24 = 36$ payments at $j_{12} = 15\%$.

$$Y = 250a_{\overline{36}|.0125} = \$7211.82$$

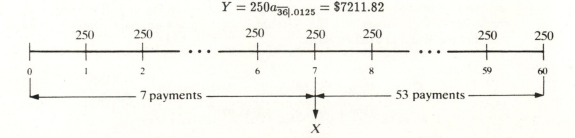

Fig. 5-8

5.14 To settle a debt with interest at 12% compounded semiannually, Leonard agrees to make 15 payments of $400 at the end of each half-year and a final payment of $292.39 six months later. What is the debt?

We arrange the data on a time diagram, Fig. 5-9. The debt X may be found as the discounted value of an ordinary annuity of 15 payments of $400 per half-year plus the discounted value of $292.39 due in 8 years (16 half-years).

$$X = 400a_{\overline{15}|.06} + 292.39(1.06)^{-16} = 3884.90 + 115.10 = \$4000$$

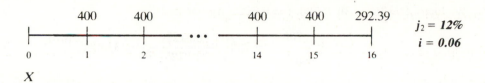

Fig. 5-9

5.15 An annuity with payments at the end of each month pays $200 for 2 years, then $300 for the next year, and then $400 for the following 2 years. Find the discounted value of these payments at $j_{12} = 10\%$.

We arrange the data as in Fig. 5-10. To calculate A we write an equation of value at 0, using $i = 0.10/12$.

$$\begin{aligned} A &= 200a_{\overline{24}|i} + 300a_{\overline{12}|i}(1+i)^{-24} + 400a_{\overline{24}|i}(1+i)^{-36} \\ &= 4334.17 + 2796.11 + 6429.65 = \$13\ 559.93 \end{aligned}$$

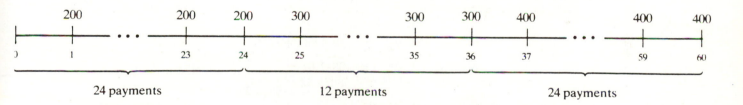

Fig. 5-10

5.16 A company is considering the possibility of acquiring new equipment for $40 000. The salvage value is estimated to be $5000 at the end of 6 years. Maintenance costs will be $400 per month, payable at the end of each month. The company could lease the equipment for $1200 per month, payable at the end of each month. Under the 6-year lease agreement, the lessor would pay the maintenance costs. If the company can earn $j_{12} = 18\%$ on its capital, advise the company whether to buy or to lease.

For each alternative we calculate the *net present value* (NPV), the difference between the present value of cash inflows and the present value of cash outflows.

$$\text{NPV of BUY} = 5000(1.015)^{-72} - (40\,000 + 400a_{\overline{72}|.015})$$
$$= 1711.65 - 57\,537.87 = -\$55\,826.22$$

$$\text{NPV of LEASE} = -1200a_{\overline{72}|.015} = -\$52\,613.60$$

The negative values of NPV represent costs to the company; since NPV of LEASE is smaller in magnitude than NPV of BUY, the company should lease the equipment.

5.17 A television set worth \$780 may be purchased by paying \$80 down and the balance in monthly installments for 2 years. Find the monthly installment if the dealer charges 15% compounded monthly, and the first installment is due in one month.

We have $A = 780 - 80 = 700$, $i = 0.0125$, $n = 24$ and

$$R = \frac{700}{a_{\overline{24}|.0125}} = 700\frac{0.0125}{1 - (1.0125)^{-24}} = \$33.94$$

5.18 Jackie has made semiannual deposits of \$500 for 5 years into a savings fund paying interest at $j_2 = 6\frac{1}{4}\%$. What semiannual deposits for the next 2 years will bring the fund up to \$10\,000?

Letting X denote the required deposit, we arrange the data as in Fig. 5-11. To calculate X we write an equation of value at 14, using $i = 0.0625/2 = 0.03125$

$$500s_{\overline{10}|i}(1+i)^4 + X\,s_{\overline{4}|i} = 10\,000$$
$$6520.17 + X\,s_{\overline{4}|i} = 10\,000$$
$$X = \frac{3479.83}{s_{\overline{4}|i}} = \$830.22$$

Use of another focal date, for example, 10, will give the same answer.

$$500s_{\overline{10}|i} + X\,a_{\overline{4}|i} = 10\,000(1+i)^{-4}$$
$$5765.05 + X\,a_{\overline{4}|i} = 8841.87$$
$$X = \frac{3076.82}{a_{\overline{4}|i}} = \$830.22$$

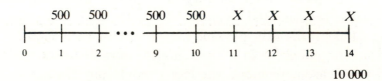

Fig. 5-11

5.4   OTHER SIMPLE ANNUITIES

Annuities Due

As defined in Section 5.1, an *annuity due* is an annuity whose periodic payment is due at the beginning of each payment interval. The term of an annuity due starts at the time of the first payment and ends one payment period after the date of the last payment. Figure 5-12 shows the simple case (payment interval and interest period coincide) of an annuity due of n payments.

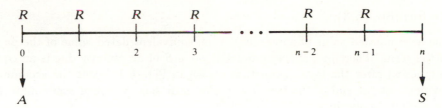

Fig. 5-12

It is easy to recognize an annuity due as a "slipped" ordinary annuity. Thus we can simply write formulas for the accumulated value S and discounted value A of an annuity due:

$$S = R\, s_{\overline{n}|i}(1 + i) \tag{5.3}$$

(see Problem 5.19) and

$$A = R\, a_{\overline{n}|i}(1 + i) \tag{5.4}$$

(see Problem 5.20). In some textbooks the right sides of (5.3) and (5.4) appear as $R\, \ddot{s}_{\overline{n}|i}$ and $\ddot{a}_{\overline{n}|i}$, respectively.

Deferred Annuities

A *deferred annuity* is an annuity whose first payment is due sometime after the end of the first interest period. It is customary to analyze any deferred annuity as an *ordinary* deferred annuity; that is, as an ordinary annuity whose term is deferred for, say, k periods (Fig. 5-13). Note that the first payment of the ordinary annuity is at time $k + 1$, because the term of an ordinary annuity starts one period before its first payment. Thus, when the time of the first payment is given, one determines the period of deferment by moving back one interest period.

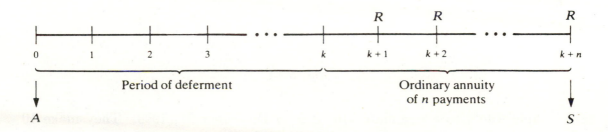

Fig. 5-13

To find the discounted value A of an ordinary deferred annuity, we compute the discounted value of n payments one period before the first payment and discount this sum for k periods:

$$A = R\, a_{\overline{n}|i}(1 + i)^{-k} \tag{5.5}$$

In some textbooks the right side of (5.5) appears as $R_{k|}a_{\overline{n}|i}$.

The most efficient way to solve an annuity problem is to make a time diagram, determine the type of annuity, and then apply the proper formula.

In more complicated situations students remember that any problem can be broken down into single payments and annuities, which can be moved to any focal date to set up an equation of value.

SOLVED PROBLEMS

5.19 Derive formula (5.3).

The accumulated value of an annuity is the equivalent dated value of the payments at the end of the term. Therefore, the accumulated value S of an annuity due is an equivalent value due one period after the last payment, as shown in Fig. 5-12. But the accumulated value of the payments at the end of the $(n-1)$st period is $R\,s_{\overline{n}|i}$. We then accumulate $R\,s_{\overline{n}|i}$ for one interest period, to obtain

$$S = R\,s_{\overline{n}|i}(1+i)$$

5.20 Derive formula (5.4).

To find A, we recall that the discounted value of an annuity was defined as the equivalent dated value of the payments at the beginning of the term. The discounted value of the n payments in Fig. 5-12 one period before the 1st payment is $R\,a_{\overline{n}|i}$. We then accumulate $R\,a_{\overline{n}|i}$ for one interest period to obtain

$$A = R\,a_{\overline{n}|i}(1+i)$$

5.21 Jackie deposits \$200 at the beginning of each month for 5 years in an account paying interest at $10\frac{1}{2}\%$ compounded monthly. How much is in her account at the end of 5 years?

We arrange the data as in Fig. 5-14. We have $R = 200$, $i = 0.105/12 = 0.00875$, $n = 60$; from (5.3),

$$S = R\,s_{\overline{n}|i}(1+i) = 200s_{\overline{60}|.00875}(1.00875) = \$15\ 831.10$$

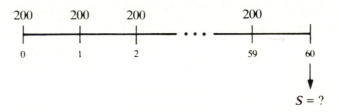

Fig. 5-14

5.22 A couple wants to accumulate \$10 000 by December 31, 1999. They make 10 equal annual deposits starting January 1, 1990. If interest is at $j_1 = 12\%$, what annual deposit is needed?

We have $S = 10\ 000$, $i = 0.12$, $n = 10$; from (5.3),

$$R = \frac{S}{s_{\overline{n}|i}(1+i)} = \frac{10\ 000}{s_{\overline{10}|.12}(1.12)} = \$508.79$$

5.23 A life insurance policy allows the option of paying your premium yearly in advance or monthly in advance. If the monthly premium is $15, what annual premium would be equivalent at $j_{12} = 12\%$?

We arrange the data as in Fig. 5-15. We have $R = 15$, $i = 0.01$, $n = 12$; from (5.4)

$$A = R\, a_{\overline{n}|i}(1+i) = 15a_{\overline{12}|.01}(1.01) = \$170.51$$

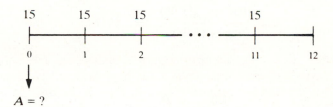

Fig. 5-15

5.24 A used car sells for $9550. Brent wishes to pay for it in 18 monthly installments, the first due on the day of purchase. If 18% compounded monthly is charged, find the size of the monthly payment.

We have $A = 9550$, $i = 0.18/12 = 0.015$, $n = 18$; from (5.4),

$$R = \frac{A}{a_{\overline{n}|i}(1+i)} = \frac{9550}{a_{\overline{18}|.015}(1.015)} = \$600.34$$

5.25 According to Mr. Novak's will, his $100 000 life insurance benefit is to be invested at $j_1 = 13\%$, and from this fund his widow will receive $15 000 each year, the first payment immediately, so long as she lives. On the payment date following the death of his wife, the balance of the fund is to be donated to a local charity. If his wife died 4 years 3 months later, how much did the charity receive?

We arrange the data on a time diagram, Fig. 5-16, for convenience writing all amounts in units of $1000. We write an equation of value using the end of 5 years as the focal date and solve it for the required sum X.

$$
\begin{aligned}
15s_{\overline{5}|.13}(1.13) + X &= 100(1.13)^5 \\
109.8405868 + X &= 184.2435179 \\
X &= 74.40292793
\end{aligned}
$$

The charity received $74 402.93

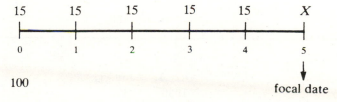

Fig. 5-16

5.26 A man aged 40 deposits $1000 at the beginning of each year for 25 years into a registered retirement savings account paying interest at $j_1 = 9\%$. Starting on his 65th birthday, he makes 15 equal annual withdrawals from the fund at the beginning of each year. During this period the fund pays interest at $j_1 = 7\%$. Find the amount of each withdrawal.

Arranging the data as in Fig. 5-17, we set up an equation of value at his 65th birthday.

$$1000s_{\overline{25}|.09}(1.09) = R\,a_{\overline{15}|.07}(1.07)$$

$$R = \frac{1000s_{\overline{25}|.09}(1.09)}{a_{\overline{15}|.07}(1.07)} = \$9473.53$$

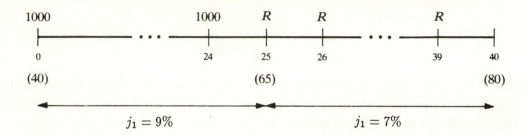

Fig. 5-17

5.27 Find the discounted value of an ordinary annuity deferred 5 years, if it pays $1000 a year for 10 years and interest is at $j_1 = 8\%$.

We have $R = 1000$, $i = 0.08$, $n = 10$, $k = 5$; from (5.5),

$$A = R\,a_{\overline{n}|i}(1 + i)^{-k} = 1000a_{\overline{10}|.08}(1.08)^{-5} = \$4566.77$$

5.28 Find the value on January 1, 1995, of quarterly payments of $100 for 10 years, if the first payment is on January 1, 1997, and interest is 7% per annum compounded quarterly.

The time diagram is shown in Fig. 5-18. We have $R = 100$, $i = 0.0175$, $n = 8$, $k = 37$; from (5.5),

$$A = R\,a_{\overline{n}|i}(1 + i)^{-k} = 100a_{\overline{40}|.0175}(1.0175)^{-7} = \$2532.43$$

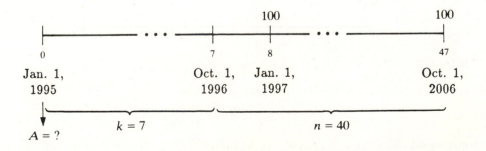

Fig. 5-18

5.29 What sum of money should be set aside at a child's birth to provide eight semi-annual payments of $1500 for a university education, if the first payment is to be made on the child's 19th birthday? The fund will earn interest at $j_2 = 9\%$.

The time diagram is shown in Fig. 5-19. We have $R = 1500$, $i = 0.045$, $n = 8$, $k = 37$; from (5.5),

$$A = R\, a_{\overline{n}|i}(1+i)^{-k} = 1500 a_{\overline{8}|.045}(1.045)^{-37} = \$1941.16$$

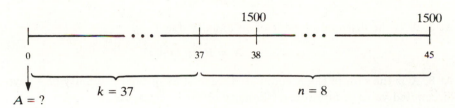

Fig. 5-19

5.30 Mrs. Wong changes employers at age 46. She is given $8500 as her vested benefits in the company's pension plan. She invests this money in a registered retirement savings plan paying $j_1 = 8\%$ and leaves it there until her ultimate retirement at age 60. She plans on 25 annual withdrawals from this fund, the first on her 61st birthday. Find the size of these withdrawals.

As shown in Fig. 5-20, $A = 8500$, $i = 0.08$, $n = 25$, $k = 14$. Substituting into (5.5) and solving for R,

$$R\, a_{\overline{25}|.08}(1.08)^{-14} = 8500$$
$$R = \frac{8500(1.08)^{14}}{a_{\overline{25}|.08}} = \$2338.80$$

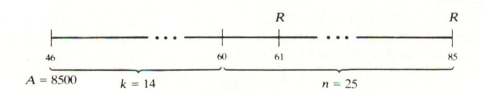

Fig. 5-20

5.31 What sum of money should be set aside to provide an income of $500 a month for 3 years, if the money earns interest at $j_{12} = 15\%$ and the first payment is to be received (a) one month from now? (b) Immediately? (c) Two years from now?

We have $R = 500$, $n = 36$, $i = 0.15/12 = 0.0125$.

(a) The discounted value A of an ordinary annuity is

$$A = R\, a_{\overline{n}|i} = 500 a_{\overline{36}|.0125} = \$14\,423.63$$

(b) The discounted value A of an annuity due is

$$A = R\, a_{\overline{n}|i}(1+i) = 500 a_{\overline{36}|.0125}(1.0125) = \$14\,603.93$$

(*c*) The discounted value A of a deferred annuity with term of deferment $k = 23$ periods is

$$A = R\, a_{\overline{n}|i}(1+i)^{-k} = 500a_{\overline{36}|.0125}(1.0125)^{-23} = \$10\ 838.99$$

5.32 Find expressions for the single sums X equivalent to the set of seven payments of Fig. 5-21 at times (*a*) 1, (*b*) 5, (*c*) 8, (*d*) 12, and (*e*) 15, assuming a rate i per period.

(*a*) At 1, X is the discounted value of a deferred annuity with deferment $k = 3$ periods.

$$X = R\, a_{\overline{7}|i}(1+i)^{-3}$$

(*b*) At 5, X is the discounted value of an annuity due.

$$X = R\, a_{\overline{7}|i}(1+i)$$

(*c*) At 8, X is the sum of the accumulated value of an ordinary annuity of 4 payments and the discounted value of an ordinary annuity of three payments.

$$X = R\, s_{\overline{4}|i} + R\, a_{\overline{3}|i} = R(s_{\overline{4}|i} + a_{\overline{3}|i})$$

(*d*) At 12, X is the accumulated value of an annuity due.

$$X = R\, s_{\overline{7}|i}(1+i)$$

(*e*) At 15, $X = R\, s_{\overline{7}|i}(1+i)^4$

Fig. 5-21

5.33 A couple deposited \$100 monthly in a fund paying interest at $j_{12} = 12\%$. The first deposit was made on June 1, 1979; the last deposit, on November 1, 1989. Find the value of the fund on (*a*) September 1, 1984; (*b*) December 1, 1991.

We arrange the data as in Fig. 5-22.

(*a*) The value X of the fund on September 1, 1984, is the accumulated value of an ordinary annuity with $R = 100$, $n = 64$, and $i = 0.01$.

$$X = 100s_{\overline{64}|.01} = \$8904.62$$

(*b*) The value Y of the fund on December 1, 1991 is

$$Y = 100s_{\overline{126}|.01}(1.01)^{25} = \$32\ 104.75$$

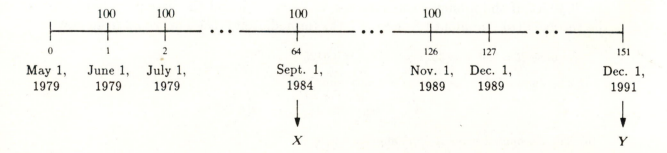

Fig. 5-22

5.34 If the couple in Problem 5.33 want to make equal monthly withdrawals from January 1, 1994, to December 1, 1999, how much will they get each month?

　　　Let R denote the monthly withdrawal. We arrange the data as in Fig. 5-23. Using December 1, 1993, as the focal date, we write an equation of value for R:

$$R\,a_{\overline{72}|.01} \;=\; 100s_{\overline{126}|.01}(1.01)^{49}$$
$$R \;=\; \frac{100s_{\overline{126}|.01}(1.01)^{49}}{a_{\overline{72}|.01}} = \$796.95$$

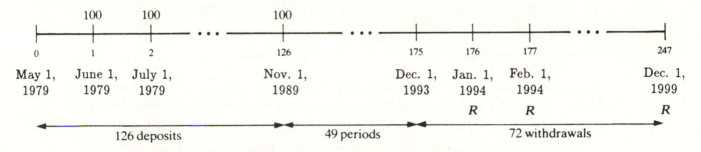

Fig. 5-23

5.5　FINDING THE TERM OF AN ANNUITY

　　　Formulas (5.1) and (5.2) may be solved for n by the use of logarithms. Normally, for given values of S (or A), R, and i, you will find n to be nonintegral. It is then necessary to make the concluding payment different from R in order to have equivalence. One of the following procedures is followed in practice, usually (ii).

　(i) The last regular payment is increased by a sum which will make the payments equivalent to S (or A).

　(ii) A smaller concluding payment is made one period after the last full payment.

SOLVED PROBLEMS

5.35 A fund of \$8000 is to be accumulated by semiannual payments of \$200. If the fund earns interest at $j_2 = 12\%$, find the number of full deposits required and the final deposits under the two alternatives.

　　　With $S = 8000$, $R = 200$, and $i = 0.06$, solve (5.1) for n, using logarithms.

$$200s_{\overline{n}|.06} \;=\; 8000$$
$$s_{\overline{n}|.06} \;=\; 40$$
$$\frac{(1.06)^n - 1}{0.06} \;=\; 40$$
$$(1.06)^n - 1 \;=\; 2.4$$
$$(1.06)^n \;=\; 3.4$$
$$n \log 1.06 \;=\; \log 3.4$$
$$n \;=\; \frac{\log 3.4}{\log 1.06} = 21.00220291$$

Thus, an annuity of 21 payments has an accumulated value slightly less than $8000, while one of 22 payments has an accumulated value greater than $8000.

Under alternative (i) the 21st deposit will be increased by X, as shown in Fig. 5-24. Using 21 as the focal date, we obtain the equation of value for X:

$$200s_{\overline{21}|.06} + X = 8000 \qquad \text{or} \qquad 7998.55 + X = 8000 \qquad \text{or} \qquad X = \$1.45$$

Thus the 21st deposit will be $201.45.

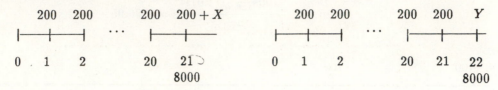

Fig. 5-24   Fig. 5-25

Under alternative (ii) let Y be the concluding deposit made 6 months after the 21st deposit of $200, as shown in Fig. 5-25. Using 22 as the focal date, we obtain the equation of value for Y:

$$200s_{\overline{21}|.06}(1.06) + Y = 8000 \qquad \text{or} \qquad 8478.46 + Y = 8000 \qquad \text{or} \qquad Y = -\$478.46$$

The negative value of Y indicates that no concluding deposit is required; because of the interest accumulated after the 21st deposit of $200, the fund will exceed the required amount by $478.46. **Check:** Carrying the accumulated value of 21 deposits forward for one-half year will result in the accumulated value

$$200s_{\overline{21}|.06}(1.06) = \$8478.46$$

5.36 A debt of $4000 bears interest at $j_2 = 12\%$. It is to be repaid by semiannual payments of $400. Find the number of full payments needed and the final payment under alternative (ii).

With $A = 4000$, $R = 400$, and $i = 0.06$, solve (5.2) for n by the use of logarithms.

$$400a_{\overline{n}|.06} = 4000$$
$$a_{\overline{n}|.06} = 10$$
$$\frac{1-(1.06)^{-n}}{0.06} = 10$$
$$1-(1.06)^{-n} = 0.6$$
$$(1.06)^{-n} = 0.4$$
$$-n\log 1.06 = 0.4$$
$$n = -\frac{\log 0.4}{\log 1.06} = 15.72520854$$

Thus, an annuity of 15 payments has a discounted value less than $4000, while one of 16 payments has a discounted value greater than $4000.

Under alternative (ii) the debt will be discharged by 15 semiannual payments of $400 each and a 16th payment of $$X$ six months later, as indicated in Fig. 5-26. Using 16 as the focal date, we obtain the equation of value for X:

$$400s_{\overline{15}|.06}(1.06) + X = 4000(1.06)^{16}$$
$$9869.01 + X = 10\,161.41$$
$$X = \$292.40$$

Using 0 as the focal date, we obtain the equation of value for X:

$$400a_{\overline{15}|.06} + X(1.06)^{-16} = 4000$$
$$3884.90 + X(1.06)^{-16} = 4000$$
$$X(1.06)^{-16} = 115.10$$
$$X = 115.10(1.06)^{16} = \$292.39$$

The one-cent difference is due to roundoff error.

Fig. 5-26

5.37 Robert is accumulating a \$10 000 fund by depositing \$100 at the end of each month, starting September 1, 1992. If the interest rate on the fund is $j_{12} = 12\%$ until May 1, 1995, and then drops to $j_{12} = 10\frac{1}{2}\%$, find the date and amount of the reduced final deposit.

Let n be the number of full \$100 deposits after May 1, 1995, and let X be the reduced final deposit. We arrange the data as shown in Fig. 5-27. The accumulated value of the first 33 deposits on May 1, 1995, is

$$100s_{\overline{33}|.01} = \$3886.90$$

Since n additional deposits of \$100 are required to bring this sum up to \$10 000 at $j_{12} = 10\frac{1}{2}\%$ ($i = 0.00875$), we have

$$3886.90(1.00875)^n + 100s_{\overline{n}|.00875} = 10\,000$$
$$3886.90(1.00875)^n + 100\frac{(1.00875)^n - 1}{0.00875} = 10\,000$$
$$34.010375(1.00875)^n + 100(1.00875)^n - 100 = 87.50$$
$$134.010375(1.00875)^n = 187.50$$
$$(1.00875)^n = 1.399145402$$
$$n\log 1.00875 = \log 1.399145402$$
$$n = \frac{\log 1.399145402}{\log 1.00875} = 38.55187247$$

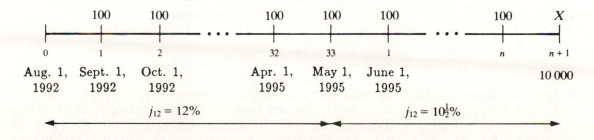

Fig. 5-27

Thus 38 additional deposits of $100 will be required. The last $100 will be made 38 months after May 1, 1995, that is, on July 1, 1998.

Using August 1, 1998 as the focal date, we obtain as the equation of value for X

$$3886.90(1.00875)^{39} + 100s_{\overline{38}|.00875}(1.00875) + X = 10\,000$$
$$5459.61 + 4524.21 + X = 10\,000$$
$$X = \$16.18$$

The final deposit of $16.18, on August 1, 1998, will bring the fund up to $10 000.

5.38 A parcel of land, valued at \$35 000, is sold for \$15 000 down. The buyer agrees to pay the balance with interest at $j_{12} = 12\%$ by paying \$500 monthly as long as necessary, the first payment due 2 years from now. Find the number of \$500 payments needed and the size of the concluding payment one month after the last \$500 payment.

We have $A = 35\,000 - 15\,000 = 20\,000$, $R = 500$, $i = 0.01$, $k = 23$. Solving (5.5) for n using logarithms,

$$500a_{\overline{n}|.01}(1.01)^{-23} = 20\,000$$
$$a_{\overline{n}|.01} = 40(1.01)^{23}$$
$$\frac{1 - (1.01)^{-n}}{0.01} = 40(1.01)^{23}$$
$$(1.01)^{-n} = 0.497134793$$
$$n = -\frac{\log 0.497134793}{\log 1.01} = 70.23827523$$

Thus, 70 full payments of \$500 will be needed, together with a smaller payment \$X one month after the last \$500 payment, as shown in Fig. 5-28. Using 94 as the focal date, we obtain the equation of value for X:

$$500s_{\overline{70}|.01}(1.01) + X = 20\,000(1.01)^{94}$$
$$50\,841.55 + X = 50\,961.14$$
$$X = \$119.59$$

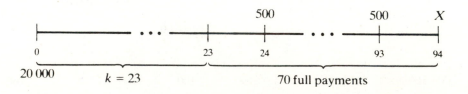

Fig. 5-28

5.39 A couple deposited \$200 monthly in a fund paying interest at $j_{12} = 4.5\%$. The first deposit was made on June 1, 1980; the last deposit, on November 1, 1990. They plan to make monthly withdrawals of \$1000, starting on May 1, 1995. Find (*a*) the number of \$100 withdrawals, (*b*) the date and the size of the smaller concluding withdrawal one month after the last \$1000 withdrawal.

(a) We arrange the data as in Fig. 5-29. With April 1, 1995, as the focal date, the equation of value is

$$1000a_{\overline{n}|.00375} = 200s_{\overline{126}|.00375}(1.00375)^{53}$$

$$a_{\overline{n}|.00375} = 39.189497$$

$$\frac{1 - (1.00375)^{-n}}{0.00375} = 39.18949744$$

$$(1.00375)^{-n} = 0.853039385$$

$$n = -\frac{\log 0.853039385}{\log 1.00375} = 42.46597471$$

There will be 42 full withdrawals.

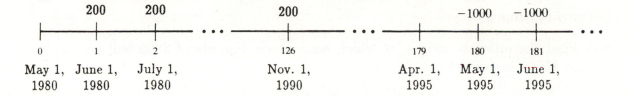

Fig. 5-29

(b) The accumulated value of the fund on April 1, 1995, is

$$200s_{\overline{126}|.00375}(1.00375)^{53} = \$39\ 189.50$$

Let X be the smaller concluding withdrawal; by (a), this withdrawal is made on November 1, 1998 (see Fig. 5-30). Using November 1, 1998, as the focal date, we write an equation of value for X:

$$\begin{aligned} X &= 39\ 189.50(1.00375)^{43} - 1000s_{\overline{42}|.00375}(1.00375) \\ &= 46\ 032.94 - 45\ 566.50 = \$466.44 \end{aligned}$$

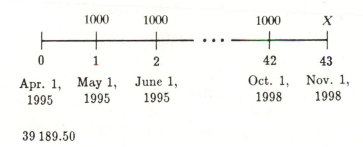

Fig. 5-30

5.6   FINDING THE INTEREST RATE

A very practical application of (5.1) and (5.2) is in finding the interest rate. In many business transactions the true interest rate is concealed in one way or another. To decide between different propositions (options, investments), it is necessary to know the true interest rate of each.

When R, n and either S or A are given, the nominal rate j_m, and with it the unknown rate $i = j_m/m$, may be determined approximately by linear interpolation. For most practical purposes, linear interpolation gives sufficient accuracy.

For fixed n, $s_{\overline{n}|i}$ increases when i increases, whereas $a_{\overline{n}|i}$ decreases. In general, the closer the bounds on the nominal rate j_m, the better the approximation of j_m, and hence of i, furnished by linear interpolation. We adopt the practice of interpolating between two nominal rates that are 1% apart and using factors $s_{\overline{n}|i}$ and $a_{\overline{n}|i}$ rounded off to 4 decimal places.

To obtain a starting value to solve an equation $s_{\overline{n}|i} = k$ by linear interpolation, we may use the formula $i = \dfrac{(k/n)^2 - 1}{k}$.

To obtain a starting value to solve the equation $a_{\overline{n}|i} = k$ by linear interpolation, we may use the formula $i = \dfrac{1 - (k/n)^2}{k}$.

SOLVED PROBLEMS

5.40 Find the interest rate, j_2, at which semiannual deposits of \$500 will accumulate to \$6000 in 5 years.

We have $S = 5000$, $R = 500$, $n = 16$. By (5.1),

$$500 s_{\overline{10}|i} = 6000 \qquad \text{or} \qquad s_{\overline{10}|i} = 12$$

To determine the rate $j_2 = 2i$ such that

$$s_{\overline{10}|i} = \frac{(1+i)^{10} - 1}{i} = 12$$

we find two factors $s_{\overline{10}|i}$, one greater than 12 and one less than 12, and corresponding to j_2-values that differ by 1%. These j_2-values provide upper and lower bounds on the unknown rate j_2, which is then approximated by linear interpolation.

A starting value to solve $s_{\overline{10}|i} = 12$ is $i = \dfrac{(12/10)^2 - 1}{12} = 0.03666667$ or $j_2 = 2i = 7.33\%$.

$$\text{for } j_2 = 7\%; \quad s_{\overline{10}|.035} = 11.7314$$
$$\text{for } j_2 = 8\%; \quad s_{\overline{10}|.04} = 12.0061$$

Arranging our data in an interpolation table, we have:

$$
0.2747 \left\{ 0.2686 \left\{ \begin{array}{c|c} s_{\overline{10}|i} & j_2 \\ \hline 11.7314 & 7\% \\ 12.0000 & j_2 \\ 12.0061 & 8\% \end{array} \right\} x \right\} 1\%
\qquad
\begin{array}{l} \dfrac{x}{1\%} = \dfrac{0.2686}{0.2747} \\[2mm] x = 0.98\% \\[2mm] j_2 = 7.98\% \end{array}
$$

We may check the accuracy of our answer by calculating the accumulated value of the deposits at $i = j_2/2 = 0.0399$:

$$S = 500 s_{\overline{10}|.0399} = \$6000.27$$

5.41 A dealer sells an article for \$600. He will allow you to buy it for \$240 down, with the balance to be paid in \$30 monthly installments for a year. If you pay cash, he will give you a 10% discount. (*a*) What is the interest rate j_{12} paid by the purchaser who uses the installment plan? (*b*) What is the annual effective rate being charged on the installment plan?

(a) For any installment plan the following equation of value must hold:

$$\text{cash price} = (\text{down payment}) + (\text{discounted value of installments})$$

Here, the cash price is $600 - 60 = \$540$, and so

$$540 = 240 + 30a_{\overline{12}|i} \qquad \text{or} \qquad a_{\overline{12}|i} = \frac{540 - 240}{30} = 10$$

A starting value to solve $a_{\overline{12}|i}$ is $i = \dfrac{1 - (10/12)^2}{10} = 0.030555556$ or $j_{12} = 12i = 36.67\%$

$$\text{for } j_{12} = 35\%; \quad a_{\overline{12}|i} = 10.0037$$
$$\text{for } j_{12} = 36\%; \quad a_{\overline{12}|i} = 9.9540$$

Arranging our data in an interpolation table, we have:

$a_{\overline{12}|i}$	j_{12}
10.0037	35%
10.0000	j_{12}
9.9540	36%

$$-0.0497 \begin{cases} -0.0037 \begin{cases} \\ \\ \end{cases} \end{cases} \quad \begin{cases} \begin{cases} \\ x \\ \end{cases} 1\% \end{cases}$$

$$\frac{x}{1\%} = \frac{-0.0037}{-0.0497}$$

$$x = 0.07\%$$

$$j_{12} = 35.07\%$$

We may check the accuracy of our answer by calculating the discounted value of the installment plan at $i = j_{12}/12 = 0.029225$:

$$240 + 30a_{\overline{12}|.029225} = \$540.01$$

(b) Let j denote the annual effective rate; then

$$1 + j = \left(1 + \frac{0.3507}{12}\right)^{12} = 1.4129407 \qquad \text{and} \qquad j = 41.29\%$$

5.42 You are offered a loan of $10\ 000 with no payments for 6 months, then $600 per month for 1 year, and $500 per month for the following year. What annual effective rate of interest does this loan charge?

Let i be the rate per month and $f(i)$ be the present value of all payments at i, that is,

$$f(i) = 600a_{\overline{12}|i}(1 + i)^{-6} + 500a_{\overline{12}|i}(1 + i)^{-18}.$$

We want to solve for i the equation $f(i) = 10\ 000$.

By trial and error we obtain two j_{12}-values that differ by 1% and provide lower and upper bounds on the unknown rate $j_{12} = 12i$. Arranging our data in an interpolation table, we have:

$f(i)$	j_{12}
10 014.23	19%
10 000.00	j_{12}
9 874.04	20%

$$-140.19 \begin{cases} -14.23 \begin{cases} \\ \\ \end{cases} \end{cases} \quad \begin{cases} \begin{cases} \\ x \\ \end{cases} 1\% \end{cases}$$

$$\frac{x}{1\%} = \frac{-14.23}{-140.19}$$

$$x = 0.10\%$$

$$j_{12} = 19.10\%$$

Annual effective rate $j = (1 + \frac{0.1910}{12})^{12} - 1 = 20.86\%$.

Supplementary Problems

ACCUMULATED VALUE

5.43 Find the accumulated value of an ordinary simple annuity of \$3000 per year for 7 years, if money is worth (a) $j_1 = 8\%$, (b) $j_1 = 10\frac{3}{4}\%$, (c) $j_1 = 17.29\%$.
Ans. (a) \$26 768.41; (b) \$29 125.07; (c) \$35 633.96

5.44 Find the accumulated value of: (a) \$500 a month for 4 years 3 months at 10% compounded monthly; (b) \$800 per quarter for 6 years 3 months at $14\frac{1}{4}\%$ compounded quarterly; (c) \$1000 per half-year for 10 years at 12.23% compounded semiannually.
Ans. (a) \$31 613.95; (b) \$31 420.22; (c) \$37 243.60

5.45 Lucy deposits \$100 every 3 months into a savings account that pays interest at $j_4 = 6\%$. If she makes her first deposit on July 1, 1993, find her total savings just after she makes her deposit in January 1996. *Ans.* \$1186.33

5.46 One thousand dollars at the end of each year for 5 years is equivalent to what single payment at the end of 5 years, if interest is at (a) $j_1 = 5\frac{1}{4}\%$, (b) $j_1 = 10\%$? *Ans.* (a) \$5553.29; (b) \$6105.10

5.47 Find the accumulated value of quarterly deposits of \$300 each, immediately after the 32nd deposit, if the deposits earned interest at $j_4 = 10\%$ for the first 5 years and at $j_4 = 12\%$ for the last 3 years. *Ans.* \$15 183.78

5.48 Jack makes regular deposits of \$500 at the end of each half-year for 5 years and then \$800 for the next 3 years. Find the accumulated value of his deposits if interest is at $j_2 = 11\%$.
Ans. \$14 386.98

5.49 Paul has deposited \$1000 in a savings account paying interest at $j_1 = 10\%$ and now finds that his deposit has accumulated to \$1610.51. If he had been able to invest the \$1000 over the same period in a guaranteed investment certificate paying interest at $j_2 = 13\frac{1}{4}\%$ and had deposited this interest in his savings account, to what sum would his \$1000 now have accumulated?
Ans. \$1808.93

5.50 Beginning June 30, 1992, and every three months until December 31, 1996, Laura deposits \$300 into her savings account. Starting September 30, 1997, she makes quarterly withdrawals of \$500. What is the balance in her account after the withdrawal of June 30, 1999, if interest is at $j_4 = 8\%$ until March 31, 1995, and $j_4 = 6\%$ afterward? *Ans.* \$3515.68

5.51 Prove:

$$(a) \quad (1+i)s_{\overline{n}|i} = s_{\overline{n+1}|i} - 1 \qquad (b) \quad s_{\overline{m+n}|i} = s_{\overline{m}|i} + (1+i)^m s_{\overline{n}|i}$$

Illustrate both (a) and (b) using a time diagram.

5.52 If $s_{\overline{n}|i} = 12.3$ and $i = 6\%$, find $s_{\overline{n+2}|i}$ and $s_{\overline{2n}|i}$. *Ans.* 15.88028, 33.6774

5.53 It is desired to check a column of values of $s_{\overline{n}|i}$ from $n = 20$ through $n = 40$ by verifying their sum by means of an independent formula. Derive an expression for the sum of these values.
Ans. $[(1+i)^{20} s_{\overline{21}|i} - 21]/i$

5.54 Find an expression for an accumulation factor for n equal payments of \$1 assuming simple interest at rate i per payment period. *Ans.* $n + n(n-1)i/2$

5.55 Prove that $\dfrac{s_{\overline{2n}|}}{s_{\overline{n}|}} + \dfrac{s_{\overline{n}|}}{s_{\overline{2n}|}} - \dfrac{s_{\overline{3n}|}}{s_{\overline{2n}|}} = 1$

5.56 (a) Expand $(1+i)^n$ by the binomial theorem to show that

$$s_{\overline{n}|i} = n + \frac{n(n-1)}{1\cdot 2}i + \frac{n(n-1)(n-2)}{1\cdot 2\cdot 3}i^2 + \cdots$$

(b) Use the first four terms of the expansion in (a) to find the value of $s_{\overline{20}|.01}$ and compare it with the directly calculated value.

5.57 What quarterly deposits should be made into a savings account paying $j_4 = 4\%$ to accumulate $10 000 at the end of 10 years? *Ans.* $215.98

5.58 A man wants to accumulate a $200 000 retirement fund. He plans to make the first deposit on March 1, 1984, and the last on September 1, 2005. Find the size of each deposit, if he makes the deposits (a) semiannually in a fund that pays $12\frac{1}{2}\%$ per annum compounded semiannually; (b) monthly in a fund that pays $12\frac{1}{2}\%$ per annum compounded monthly. *Ans.* (a) $932.58; (b) $152.70

5.59 (a) Barbara wants to accumulate $10 000 by the end of 10 years. She starts making quarterly deposits in her savings account, which pays $j_4 = 8\%$. Find the size of these deposits. (b) After 4 years, the bank changes the rate to $j_4 = 6\%$. Find the size of the quarterly deposits now required if the $10 000 goal is to be met. *Ans.* (a) $165.56; (b) $195.18

5.60 John wants to accumulate $5000 in a fund at the end of 2 years. He deposits $150 at the end of each month during the first year, and $150 + X$ at the end of each month of the following year. Find X if $j_{12} = 12\%$. *Ans.* $75.22

5.61 To prepare for early retirement, a self-employed consultant deposits $5500 into a retirement savings plan each year for 20 years, starting on her 31st birthday. When she is 51, she wishes to draw out 30 equal annual payments. What is the size of each withdrawal, if $j_1 = 12\%$ for the first 10-year period, $j_1 = 10\%$ for the next 10-year period, and $j_1 = 11\%$ for the 30-year retirement period? *Ans.* $38 878.17

5.62 Anne wants to accumulate $100 000 at the end of 20 years. She deposits $1000 at the end of each of the first 10 years, and $1000 + X$ at the end of each of the second 10 years. Interest is at $j_1 = 10\frac{1}{4}\%$. (a) Find X. (b) If the last four payments of $1000 were missed, what would be the value of X? *Ans.* (a) $2546.43; (b) $3312.68

DISCOUNTED VALUE

5.63 Derive

$$a_{\overline{n}|i} = \frac{1 - (1 + i)^{-n}}{i}$$

as the sum of a geometric progression.

5.64 Find the discounted values of the following ordinary annuities: (a) $500 a month for 4 years 3 months at $j_{12} = 10\%$; (b) $800 per quarter for 6 years 3 months at $j_4 = 14\frac{1}{4}\%$; (c) $1000 per half-year for 10 years at $j_2 = 12.23\%$. *Ans.* (a) $20 704.67; (b) $13 096.22; (c) $11 363.60

5.65 How much money is needed now to provide for $6000 at the end of each half-year (first payment one half-year from now) for 4 years, if money is worth (a) $j_2 = 8\%$? (b) $j_2 = 12\%$? (c) $j_2 = 16\%$? *Ans.* (a) $40 396.47; (b) $37 258.76; (c) $34 479.83

5.66 Mr. Jones wants to save enough money to send his two children to college. They are 4 years apart in age, so he wants to have a sum of money that will provide $5000 a year for 8 years. Find the sum required 1 year before the first withdrawal, if interest is at $j_1 = 7\%$. *Ans.* $29 856.49

5.67 An annuity pays R per month starting February 1, 1992, and ending January 1, 1995. If the value of this annuity on January 1, 1995, is $8000 and $j_{12} = 11\%$, what was its value on January 1, 1992? *Ans.* $5760.04

5.68 An heir is to receive an inheritance of $1000 each half-year for 10 years, the first payment to be made in 6 months. If money is worth $j_2 = 10\%$, what is the cash value of this inheritance? *Ans.* $12 462.21

5.69 Find the present value of annual payments of $1000 each over 10 years, if interest is $j_1 = 11\%$ for the first 4 years and $j_1 = 13\%$ for the last 6 years. *Ans.* $5735.76

5.70 A contract calls for payments of $250 a month for 10 years and an additional payment of $2000 at the end of that time. At the beginning of the 5th year (just after the 48th payment is made) the contract is sold to a buyer at a price that will yield $j_{12} = 14\%$. What does the buyer pay? *Ans.* $13 000.17

5.71 An annuity provides $60 at the end of each month for 3 years and $80 at the end of each month for the following 2 years. If $j_{12} = 11\%$, find the present value of the annuity. *Ans.* $3068.54

5.72 An oil company requires an arctic drilling machine and is deciding whether to purchase it for $1 000 000 cash or to lease it for $240 000, payable at the end of each half-year. The salvage value is $100 000 at the end of the machine's 6-year life. Maintenance costs are $10 000 each 6 months, but payable by the lessor, if the machine is leased. If the company can earn 10% on its capital, compounded semiannually, advise the company whether to lease or to buy. *Ans.* Buy

5.73 The Ace Manufacturing Company is considering the purchase of one of two machines. Machine A costs $200 000 and machine B costs $400 000. The machines are projected to have a life of 5 years and to yield revenues as tabulated below.

	Cash Revenue	
End of Year	Machine A	Machine B
1	none	$ 90 000
2	$100 000	90 000
3	100 000	90 000
4	100 000	90 000
5	100 000	300 000

If the market rate of interest is 14% per annum, compounded yearly, which machine should the company purchase? *Ans.* machine A

5.74 The Northeast Mining Company is considering the exploitation of a mining property. If the company goes ahead, the estimated cash flows are as follows:

End of Year	Cash Inflow	Cash Outflow
0 (now)	0	$3 000 000
1	$1 000 000	2 000 000
2	1 000 000	0
3	1 000 000	0
4	1 000 000	0
5	1 000 000	0
6	1 000 000	0
7	1 000 000	0
8	1 000 000	0

The project would be financed out of working capital, on which Northeast expects to earn at least $j_1 = 16\%$. Advise whether Northeast should proceed. *Ans.* No; NPV is negative.

5.75 Prove   (a) $\dfrac{1}{s_{\overline{n}|i}} + i = \dfrac{1}{a_{\overline{n}|i}}$

(b) $a_{\overline{m+n}|i} = a_{\overline{m}|i} + (1+i)^{-m} a_{\overline{n}|i} = (1+i)^{-n} a_{\overline{m}|i} + a_{\overline{n}|i}$

Illustrate (b) on a time diagram.

5.76 If $a_{\overline{2n}|i} = 1.6 a_{\overline{n}|i}$ and $i = 10\%$, find $s_{\overline{2n}|i}$. *Ans.* 17.7

5.77 Prove

$$(a) \quad s_{\overline{k-l}|i} = s_{\overline{k}|i} - (1+i)^k a_{\overline{l}|i} \quad (k > l)$$

$$(b) \quad a_{\overline{k-l}|i} = a_{\overline{k}|i} - (1+i)^{-k} s_{\overline{l}|i} \quad (k > l)$$

5.78 If money doubles itself in n years at rate i, show that $a_{\overline{n}|i}$, $a_{\overline{2n}|i}$, and $s_{\overline{n}|i}$ are in arithmetic progression.

5.79 Prove that $(1 - i a_{\overline{n}|i})$, 1, $(i s_{\overline{n}|i} + 1)$ are in geometric progression.

5.80 If $X (s_{\overline{2n}|i} + a_{\overline{2n}|i}) = (s_{\overline{3n}|i} + a_{\overline{n}|i})$, what is the value of X? *Ans.* $(1+i)^n$

5.81 A car is purchased for \$2000 down and \$200 a month for 6 years. Interest is at $j_{12} = 10\%$. (*a*) Determine the price of the car. (*b*) If the first 4 payments are missed, what payment at the time of the 5th payment will update the payments? (*c*) Assuming no payments are missed, what single payment at the end of 2 years will completely pay off the debt? (*d*) After 27 payments have been made, the contract is sold to a buyer who wishes to yield $j_{12} = 12\%$. Determine the sale price.
Ans. (*a*) \$12 795.73; (*b*) \$1061.81; (*c*) \$8085.63; (*d*) \$7218.90

5.82 A person buys a boat with a cash price of \$4500. He pays \$500 down and the balance is financed at $j_{12} = 14.79\%$. If he is to make 24 equal monthly payments, what will be the size of each payment? *Ans.* \$193.55

5.83 With the death of the insured on September 1, 1992, a life insurance policy pays out \$80 000 as a death benefit. The beneficiary is to receive monthly payments, with the first payment on October 1, 1992. Find the size of the monthly payment if interest is earned at $j_{12} = 11\%$ and the beneficiary is to receive 120 payments. *Ans.* \$1102.00

5.84 A car selling for \$8500 may be purchased for \$1500 down and the balance in equal monthly payments for 3 years. Find the monthly payment at $j_{12} = 18\%$. *Ans.* \$253.07

5.85 At age 65, Mr. Jones takes his life savings of \$120 000 and buys a 15-year annuity certain with monthly payments. Find the size of these payments (*a*) at $j_{12} = 12\%$, (*b*) at $j_{12} = 9\%$.
Ans. (*a*) \$1440.20; (*b*) \$1217.12

5.86 On the birth of their first child, a couple put \$1500 in a special savings account paying interest at $j_1 = 11\%$. This fund, used to pay college fees, allows for four equal withdrawals, corresponding to the 18th through 21st birthdays. Find the size of the withdrawals. *Ans.* \$2850.22

5.87 A family needs to borrow \$5000 for home renovations. The loan is to be repaid with monthly payments over 5 years. (*a*) If they go to a finance company, the interest rate will be $j_{12} = 21\%$; (*b*) if they use their credit card, the interest will be at $j_{12} = 18\%$; (*c*) if they go to the bank, the rate will be $j_{12} = 15\%$. Find the respective monthly payments and the total of all payments for each loan.
Ans. (*a*) \$135.27, \$8116.20; (*b*) \$126.97, \$7618.20; (*c*) \$118.95, \$7137.00

5.88 Mr. Horvath has been accumulating a retirement fund at an annual effective rate of 9% which will provide him with an income of \$12 000 per year for 20 years, the first payment on his 65th birthday. If he now wishes to reduce the number of income payments to 15 (for 15 years), what should he receive annually? *Ans.* \$13 589.73

OTHER SIMPLE ANNUITIES

5.89 Deposits of \$500 are made at the beginning of each half-year for 5 years into an account paying $j_2 = 6\%$. How much is in the account (*a*) at the end of 5 years? (*b*) Just before the 6th deposit? *Ans.* (*a*) \$5903.90; (*b*) \$2734.20

5.90 The monthly rent for a town house is \$520 payable at the beginning of each month. If money is worth $j_{12} = 9\%$, (*a*) what is the equivalent yearly rental payable in advance? (*b*) What is the cash equivalent of 5 years of rent? *Ans.* (*a*) \$5990.75; (*b*) \$25 238.03

5.91 A debt of \$1000 with interest at $j_{12} = 18\%$ is to be paid off over 18 months by equal monthly payments, the first due today. Find the monthly payment. *Ans.* \$62.86

5.92 Five years from now, a company will need \$150 000 to replace worn-out equipment. Starting now, what monthly deposits must be made in a fund paying $j_{12} = 8\%$ for 5 years to accumulate this sum? *Ans.* \$2027.94

5.93 The premium on a life insurance policy can be paid either yearly in advance or monthly in advance. If the annual premium is \$120, what monthly premium would be equivalent at $j_{12} = 11\%$? *Ans.* \$10.51

5.94 An insurance policy pays a death benefit of \$100 000 either in a lump sum or in equal amounts at the beginning of each month for 10 years. What size would these monthly payments be, if $j_{12} = 10\%$? *Ans.* \$1310.59

5.95 A realtor rents office space for \$5800 every three months, payable in advance. He immediately invests half of each payment in a fund paying 13% compounded quarterly. How much is in the fund at the end of 5 years? *Ans.* \$82 534.24

5.96 A refrigerator is bought for \$60 down and \$60 a month for 15 months. If interest is charged at $j_{12} = 18\frac{1}{2}\%$, what is the cash price of the refrigerator? *Ans.* \$858.06

5.97 Jacques signed a contract that calls for payments of \$500 at the beginning of each 6 months for 10 years. If money is worth $j_2 = 13\%$, find the value of the remaining payments (*a*) just after he makes the 4th payment, (*b*) just before he makes the 6th payment. (*c*) If after making the first 3 payments he failed to make the next 3 payments, what would he have to pay when the next (seventh) payment is due to put himself back on schedule?
Ans. (*a*) \$4883.88; (*b*) \$5006.92; (*c*) \$2203.59

5.98 A mutual fund promises a rate of growth of 10% a year on funds left with it. How much would an investor who makes deposits of \$100 at the beginning of each year have on deposit by the time the first deposit has grown to \$259? (As an approximation, assume that time is an integral number of years and that the investor is about to make, but has not made, an annual deposit at that time.) *Ans.* \$1753.12

5.99 Find the discounted value of an ordinary annuity deferred 3 years 6 months and paying \$500 semiannually for 7 years, if interest is (*a*) 17% per annum compounded semiannually, (*b*) 7% per annum compounded semiannually. *Ans.* (*a*) \$2262.56; (*b*) \$4291.72

5.100 Find the value on July 1, 1989, of semiannual payments of \$500 for 6 years, if the first payment is on January 1, 1993, and interest is (*a*) $11\frac{1}{4}\%$ per annum payable semiannually, (*b*) 9% per annum payable semiannually. *Ans.* (*a*) \$3081.69; (*b*) \$3501.06

5.101 On Mr. Smith's 55th birthday, the Smiths decide to sell their house and move into an apartment. They realize \$80 000 on the sale of the house and invest this money in a fund paying $j_1 = 9\%$. On Mr. Smith's 65th birthday, they make the first of 15 equal withdrawals that will exhaust the fund over 15 years. What is the amount of each withdrawal? *Ans.* \$21 555.41

5.102 The XYZ Furniture Store sells a davenport for \$950. It can be purchased for \$50 down and no payments for 3 months; at the end of the third month, you make your first payment and continue until a total of 18 payments are made. Find the size of each payment, if interest is at $j_{12} = 18\%$. *Ans.* \$59.16

5.103 An 8-year-old child is bequeathed \$1 000 000. The law requires that this money be set aside in a trust fund until the child reaches age 18. The child's parents decide that the money should be paid out in 20 equal annual installments, with the first payment at age 18. Find these payments if the trust fund pays interest at $j_1 = 10\%$. *Ans.* \$276 963.65

5.104 Doreen bought a car on September 1 by paying \$2000 down and agreeing to make 36 monthly payments of \$350, the first due on December 1. If interest is at 18% compounded monthly, find the equivalent cash price. *Ans.* \$11 397.21

5.105 A woman wins \$100 000 in the New York State Lottery. She takes only \$20 000 in cash and invests the balance at $j_{12} = 8\%$, with the understanding that she will receive 180 equal monthly payments, the first one to be made in 4 years. Find the size of the payments. *Ans.* \$1044.76

5.106 Derive formula (5.3) using the sum of a geometric progression and show that it is equivalent to $S = R(s_{\overline{n+1}|i} - 1)$.

5.107 Derive formula (5.4) using the sum of a geometric progression and show that it is equivalent to $A = R(a_{\overline{n-1}|i} + 1)$.

5.108 Show that (5.5) is equivalent to $A = R(a_{\overline{k+n}|i} - a_{\overline{k}|i})$.

5.109 Show that $s_{\overline{n+1}|i} - \ddot{a}_{\overline{n+1}|i} = \ddot{s}_{\overline{n}|i} - a_{\overline{n}|i}$.

5.110 Starting on his 45th birthday, a man deposits \$1000 a year in a savings account that pays interest at 13% per annum. He makes his last deposit on his 64th birthday. On his 65th birthday he transfers his total savings to a special retirement fund that pays $14\frac{1}{2}\%$ per annum. From this fund he will receive \$X on his birthday for 15 years, starting immediately. Find X. *Ans.* \$13 332.71

5.111 Deposits are \$100 per month for 3 years, nothing for 2 years and then \$200 per month for 3 years. Interest rates start at $j_{12} = 8\%$ and fall to $j_{12} = 6\%$ on the date of the first \$200 deposit. Find the accumulated value at the time of the final \$200 deposit. *Ans.* \$13 566.11

5.112 How much must you deposit at the end of each month for 25 years to fund an annuity of \$2500 per month for 20 years? The first annuity payment is made 5 years after the last deposit, and interest changes from $j_{12} = 8\%$ to $j_{12} = 7\%$ at the time of the first annuity payment. *Ans.* \$228.91

5.113 The present value of an annuity of \$1000 payable for n years commencing one year from now is \$6053. The annual effective rate of interest is 12.5%. Find the present value of an annuity of \$1000 commencing one year from now and payable for $n + 2$ years. *Ans.* \$6461.63

5.114 A person deposits \$100 at the beginning of each year for 20 years. Simple interest at an annual rate of i is credited to each deposit from the date of deposit to the end of the 20-year period. The total amount thus accumulated is \$2840. If, instead, compound interest had been credited at an annual effective rate of i, what would the accumulated value of these deposits have been at the end of 20 years? *Ans.* \$3096.92

FINDING THE TERM OF AN ANNUITY

5.115 A couple wants to accumulate \$10 000 by making deposits of \$800 at the end of each half-year into a savings account that earns interest at $j_2 = 9\%$. (a) How many full deposits must they make? (b) What additional deposit made at the time of the last full deposit will be necessary to bring the savings account to \$10 000? (c) What deposit made 6 months after the last full deposit will bring the account to \$10 000?
Ans. (a) 10; (b) \$169.43; (c) no deposit necessary (account has \$10 272.94)

5.116 A man dies and leaves his wife an estate of \$100 000. The money is invested at $j_{12} = 12\%$. (a) How many monthly payments of \$1500 would the widow receive? (b) What would be the size of the concluding payment, made one month after the last \$1500 payment?
Ans. (a) 110; (b) \$616.24

5.117 A firm buys a machine for \$30 000 by paying \$5000 down and \$5000 at the end of each year. If interest is at $j_1 = 10\%$, (a) how many full payments must be made? (b) What final payment, one year after the last full payment, will be necessary? *Ans.* (a) 7; (b) \$1410.28

5.118 On July 1, 1993, Sharon has $10 000 in an account paying interest at $j_4 = 12\frac{1}{2}\%$. She plans to withdraw $500 every three months, with the first withdrawal on October 1, 1993. (*a*) How many full withdrawals can she make? (*b*) What will be the size and the date of the concluding withdrawal, made 3 months after the last full withdrawal?
Ans. (*a*) 31; (*b*) $438.06 on July 1, 2001

5.119 On his 25th birthday, Juan deposited $2000 in a fund paying $j_1 = 10\%$, and continued to make such deposits each year, the last on his 49th birthday. Beginning on his 50th birthday, he plans to make equal annual withdrawals of $10 000. (*a*) How many such withdrawals can he make? (*b*) What additional sum paid with the last full withdrawal will exhaust the fund? (*c*) What sum paid one year after the last full withdrawal will exhaust the fund?
Ans. (*a*) 43; (*b*) $853.53; (*c*) $938.88

5.120 A couple bought land worth $30 000. They paid $5000 down and signed a note agreeing to repay the balance with interest at $j_1 = 12\%$, by annual payments of $5000 as long as necessary and a smaller concluding payment one year later. The note was sold just after the 4th annual payment to an investor who wants to earn a yield of $j_1 = 13\%$. Find the selling price.
Ans. $15 115.89

5.121 From July 1, 1986, to January 1, 1991, a couple made semiannual deposits of $500 in a special savings account paying $j_2 = 11\%$. Starting July 1, 1995, they make semiannual withdrawals of $800. (*a*) How many full withdrawals can they make? (*b*) Give the size and date of the concluding withdrawal.
Ans. (*a*) 21; (*b*) $193.63 on January 1, 2006

5.122 If the couple in Problem 5.121 decide to exhaust their savings by equal semiannual withdrawals from July 1, 1995, to July 1, 2005, inclusive, how much will they get each half-year?
Ans. $804.86

5.123 In October 1987, an industrialist gives your school $30 000 to be used to award a $4000 scholarship at each fall convocation, starting the year the industrialist dies. If the industrialist died May 1990 and the money earns interest at $j_1 = 8\%$, for how many years will full scholarships be awarded? *Ans.* 15

5.124 Solve for n (*a*) $S = R\,s_{\overline{n}|i}$, (*b*) $A = R\,a_{\overline{n}|i}$.
Ans. (*a*) $n = \log(\dfrac{Si}{R} + 1)/\log(1 + i)$; (*b*) $n = -\log(1 - \dfrac{Ai}{R})/\log(1 + i)$

5.125 A car loan of $10 000 at $j_{12} = 12\%$ is being paid off by n payments. The first $(n-1)$ payments are $263.34 per month. The final monthly payment is $263.24. Find n. *Ans.* 48

5.126 A widow, as beneficiary of a $50 000 insurance policy, will receive $15 000 immediately and $1800 every three months thereafter. The company pays interest at $j_4 = 9\%$; after 3 years, the rate is increased to $j_4 = 11\%$. (*a*) How many full payments of $1800 will she receive? (*b*) What additional sum paid with the last full payment will exhaust her benefits? (*c*) What payment 3 months after the last full payment will exhaust her benefits?
Ans. (*a*) 26; (*b*) $795.22; (*c*) $817.09

5.127 A loan of $20 000 is to be repaid by annual payments of $4000 per annum for the first 5 years and payments of $4500 per year thereafter for as long as necessary. Find the total number of payments and the amount of the smaller final payment made one year after the last regular payment. Assume an annual effective rate of 18%. *Ans.* 12, $4456.81

5.128 A friend agrees to lend you $2000 on September 1 each year for 4 years to help with education costs. One year after the last payment you are expected to start annual repayment of $800 for as long as necessary. Interest on the loan is at $j_1 = 6.5\%$. Find the number of repayments needed, and the amount of the reduced final payment. *Ans.* 20, $798.39

5.129 On November 10, 1993, I. M. Broke obtained a bank loan of $4000 at 10% compounded monthly. I. M. Broke will repay the loan by making monthly payments of $250 beginning on December 10, 1994. Determine the number of full payments, the date and the size of the smaller concluding payment required. *Ans.* 19; $50.31 on July 10, 1996

5.130 Erika deposited $100 monthly in a fund paying interest at $j_{12} = 12\%$. The first deposit was made on June 1, 1980, and the last deposit on November 1, 1990. (a) Find the value of the fund on (i) September 1, 1985 (after the payment is made); (ii) December 1, 1992. (b) From May 1, 1995, she plans to draw down the fund with monthly withdrawals of $1000. Find the date and the size of the smaller concluding withdrawal one month after the last $1000 withdrawal. *Ans.* (a) (i) $8904.62, (ii) $32 104.75; (b) December 1, 1999, $475.78

FINDING THE INTEREST RATE

5.131 Find the interest rate j_4 at which deposits of $250 at the end of every 3 months will accumulate to $5000 in 4 years. *Ans.* 11.60%

5.132 A used car sells for $600 cash, or $100 down and $90 a month for 6 months. Find the interest rate j_{12} if the purchaser buys the car on the installment plan. *Ans.* 26.93%

5.133 An insurance company will pay $80 000 to a beneficiary, or monthly payments of $1000 for 10 years. What rate j_{12} is the insurance company using? *Ans.* 8.70%

5.134 You borrow $1600 from a loan company and agree to pay $160 a month for 12 months. Find the nominal rate compounded monthly and the effective rate charged. *Ans.* 35.07%, 41.29%

5.135 Goods worth $1000 are purchased using the following carrying-charge plan: A down payment of $100 is required, after which 18% of the unpaid balance is added on and the amount is then divided into 12 equal monthly installments. What annual effective rate of interest does the plan include? *Ans.* 36.76%

5.136 A finance company charges 15% "interest in advance" and allows the client to repay the loan in 12 equal monthly payments. The monthly payment is obtained by dividing the total of principal and interest by 12. Find the nominal rate compounded monthly and the annual effective rate charged. *Ans.* 26.62%, 30.12%

5.137 To buy a car costing $13 600 you can pay $1600 down and the balance in 36 monthly payments of $450 each. You can also borrow the money from a loan company and repay $13 600 by making quarterly payments of $1060 over 5 years, first payment to be made in 3 months. By comparing the annual effective rates of interest charged, determine which option is better. *Ans.* Borrow

5.138 On February 1, 1975, Allan made the first of a sequence of regular annual deposits of $1000 into a savings account. The last deposit was made February 1, 1991. If the account earned $j_1 = 10\frac{1}{2}\%$, the balance after the last deposit would have been $42 472.13; whereas it would have been $44 500.84 at $j_1 = 11\%$. In fact, the balance was $43 500. What annual effective rate of interest did the account earn? *Ans.* 10.75%

5.139 A television set sells for $700. Sales tax of 7% is added to that. The T.V. may be purchased for $100 down and monthly payments of $60 for one year. (a) What is the interest rate j_{12}? (b) What is the annual effective interest rate? *Ans.* (a) 19.61%; (b) 21.47%

5.140 A store offers to sell a watch for $55 cash or $5 a month for 12 months. What nominal rate j_{12} is the store actually charging on the installment plan, if the first payment is made immediately? *Ans.* 19.48%

5.141 The "Fly by Night" Used Car Lot uses the following to illustrate their 12% finance plan on a car paid for over 3 years.

Cost of car	12 000.00	
12% finance charge	4 320.00	(12% of 12 000 × 3 years)
Total cost	16 320.00	

Monthly payment $= \dfrac{16\ 320}{36} = \$453.33$

What is the true interest rate j_{12} being charged? *Ans.* 21.20%

Chapter 6

General and Other Annuities

6.1 GENERAL ANNUITIES

So far, we have assumed that the periodic payments of an annuity are made at the times the interest is compounded. In this chapter, we will consider *general annuities*, for which periodic payments are made more or less frequently than the interest is compounded.

A general annuity may be transformed into an equivalent simple annuity in two ways:

  (i) By changing the given interest rate to an equivalent one (see Chapter 4) for which the new interest conversion period is the same as the payment period

  (ii) By replacing the given payments, W, by equivalent payments, R, made at the ends of the interest periods (see Problems 6.11 and 6.12)

We shall employ (i), which is preferred when calculators are used. Problems 6.13 and 6.14 will illustrate (ii).

SOLVED PROBLEMS

ORDINARY GENERAL ANNUITIES

6.1 Find the accumulated value of an annuity of $300 at the end of each month for 5 years at 6% compounded quarterly.

First we find the rate i per month equivalent to $0.06/4$ per quarter-year:

$$
\begin{aligned}
(1+i)^{12} &= \left(1 + \frac{0.06}{4}\right)^4 \\
1 + i &= (1.015)^{1/3} \\
i &= (1.015)^{1/3} - 1 = 0.004975206
\end{aligned}
$$

Now we calculate the accumulated value S of an ordinary *simple* annuity with $R = 300$, $n = 60$, and $i = 0.004975206$ (Fig. 6-1). From (5.1),

$$S = R\, s_{\overline{n}|i} = 300 s_{\overline{60}|.004975206} = \$20\ 915.01$$

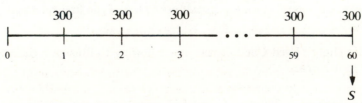

Fig. 6-1

6.2 Solve Problem 6.1 using a rate of 9% compounded daily.

First find the rate i per month such that

$$(1+i)^{12} = \left(1 + \frac{0.09}{365}\right)^{365}$$

$$i = \left(1 + \frac{0.09}{365}\right)^{365/12} - 1 = 0.007527264$$

Now calculate the accumulated value S of an ordinary simple annuity with $R = 300$, $n = 60$, and $i = 0.007527264$:

$$S = R\, s_{\overline{n}|i} = 300 s_{\overline{60}|.007527264} = \$22\ 646.68$$

6.3 A contract calls for the payment of \$5000 at the end of each half-year for 10 years. Find the present value of the contract at $j_1 = 8\%$.

First we find the rate i per half-year such that

$$(1+i)^2 = 1.08$$

$$i = (1.08)^{1/2} - 1 = 0.039230485$$

Then we calculate the discounted value A of an ordinary simple annuity with $R = 5000$, $n = 20$, and $i = 0.039230485$ (Fig. 6-2). From (5.2),

$$A = R\, a_{\overline{n}|i} = 5000 a_{\overline{20}|.039230485} = \$68\ 417.01$$

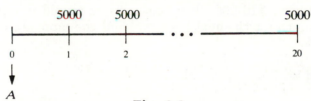

Fig. 6-2

6.4 Solve Problem 6.3 using the rate 8% compounded continuously.

First we find the rate i per half-year such that

$$(1+i)^2 = e^{0.08}$$

$$i = e^{0.04} - 1 = 0.040810774$$

Now we calculate the discounted value A of an ordinary simple annuity with $R = 5000$, $n = 20$, and $i = 0.040810774$

$$A = 5000 a_{\overline{20}|.040810774} = \$67\ 466.38$$

6.5 Deposits of \$100 are made at the end of each month to a savings account paying interest at 5% compounded semiannually. How much money will be accumulated in the account at the end of 3 years if (a) simple interest, (b) compound interest, is paid for a fractional part of a conversion period?

(a) First we find an equivalent payment R per half-year by accumulating monthly payments to the end of a 6-month period at simple interest 5% per annum (see Fig. 6-3). The first payment accumulates 5 months' interest; the second, 4 month's interest; and so on. Hence, the total interest is equal to $0 + 1 + 2 + \cdots + 5 = 15$ months' simple interest on \$100, that is,

$$(100)(0.05)\left(\frac{15}{12}\right) = \$6.25$$

Thus the semiannual payment is $R = 600 + 6.25 = \$606.25$, and the accumulated value S is

$$S = R\, s_{\overline{n}|i} = 606.25 s_{\overline{6}|.025} = \$3872.57$$

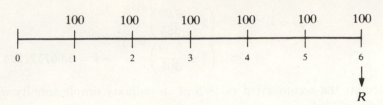

Fig. 6-3

(b) First we find the rate i per month such that

$$
\begin{aligned}
(1+i)^{12} &= (1.025)^2 \\
i &= (1.025)^{1/6} - 1 = 0.004123915
\end{aligned}
$$

Then we calculate the accumulated value S of an ordinary simple annuity with $R = 100$, $n = 36$, and $i = 0.004123915$:

$$S = R\, s_{\overline{n}|i} = 100 s_{\overline{36}|.004123915} = \$3872.37$$

Note that the accumulated value at simple interest is larger than the one at compound interest.

6.6 Mr. Colby borrows \$10 000. The loan is to be repaid in equal installments at the end of each month for the next 5 years. Find the monthly payments if the interest is at $j_1 = 10\%$.

First we find the rate i per month such that

$$
\begin{aligned}
(1+i)^{12} &= 1.1 \\
i &= (1.1)^{1/12} - 1 = 0.00797414
\end{aligned}
$$

Then we calculate the monthly payment R of an ordinary simple annuity with $A = 10\,000$, $n = 60$, and $i = 0.00797414$ (see Fig. 6-4).

$$R = \frac{A}{a_{\overline{n}|i}} = \frac{10\,000}{a_{\overline{60}|.00797414}} = \$210.36$$

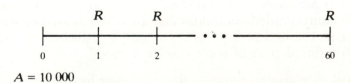

$A = 10\,000$

Fig. 6-4

6.7 A company wishes to have \$150 000 in a fund at the end of 8 years. What deposit at the end of each month must they make, if the fund pays interest at 5%

compounded daily?

First find the rate i per month such that

$$(1+i)^{12} = \left(1+\frac{0.05}{365}\right)^{365}$$
$$i = \left(1+\frac{0.05}{365}\right)^{365/12} - 1 = 0.004175073$$

Now calculate the monthly payment R of an ordinary simple annuity with $S = 150\,000$, $n = 96$, and $i = 0.004175073$ (see Fig. 6-5).

$$R = \frac{S}{s_{\overline{n}|i}} = \frac{150\,000}{s_{\overline{96}|.004175073}} = \$1273.45$$

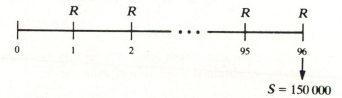

Fig. 6-5

6.8 How much a month, for 3 years at $j_\infty = 11\%$, would you have to save in order to receive \$500 a month for 3 years afterward?

First we find the rate i per month such that

$$(1+i)^{12} = e^{0.11}$$
$$i = e^{0.11/12} - 1 = 0.009208809$$

Then we solve an equation of value for the monthly deposit X, using the end of 3 years as the focal date. (See Fig. 6-6.) At 36:

$$X\, s_{\overline{36}|.009208809} = 500 a_{\overline{36}|.009208809}$$
$$X = \frac{500 a_{\overline{36}|.009208809}}{s_{\overline{36}|.009208809}} = \$359.46$$

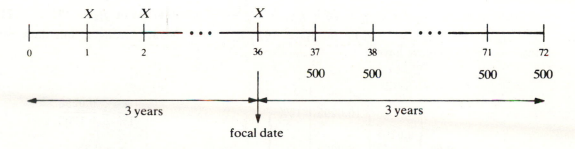

Fig. 6-6

6.9 Steve buys a used car worth $7500. He pays $1500 down and agrees to pay $250 at the end of each month for as long as necessary. Find the number of full payments and the final payment one month later, if interest is at $j_1 = 14.2\%$.

First we find the rate i per month such that

$$(1+i)^{12} = 1.142$$
$$i = (1.142)^{1/12} - 1 = 0.011126537$$

Then, with $A = 7500 - 1500 = 6000$, $R = 250$, and $i = 0.011126537$, we solve (5.2) for n by use of logarithms.

$$250a_{\overline{n}|i} = 6000$$
$$\frac{1 - (1+i)^{-n}}{i} = 24$$
$$(1.011126537)^{-n} = 0.732963108$$
$$n = -\frac{\log 0.732963108}{\log 1.011126537} = 28.0756718$$

Thus 28 full payments of $250 will be required.

Let X be the final payment one month later. We arrange the data as shown in Fig. 6-7. Using 29 as the focal date, we obtain the equation of value for X:

$$250s_{\overline{28}|i}(1+i) + X = 6000(1+i)^{29}$$
$$8251.09 + X = 8270.10$$
$$X = \$19.01$$

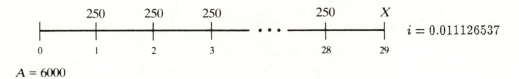

Fig. 6-7

6.10 If it takes $50 per month for 18 months to repay a loan of $800, what nominal rate compounded semiannually is being charged?

First we find the rate j_{12} by the method of Chapter 5. We have $A = 800$, $R = 50$, $n = 18$; using (5.2),

$$50a_{\overline{18}|i} = 800 \qquad \text{or} \qquad a_{\overline{18}|i} = 16$$

A starting value to solve $a_{\overline{18}|i} = 16$ is $i = \dfrac{1 - \left(\frac{16}{18}\right)^2}{16} = 0.013117284$ or $j_{12} = 12i \doteq 15.74\%$. For $j_{12} = 15\%$, $a_{\overline{18}|i} = 16.0295$; for $j_{12} = 16\%$, $a_{\overline{18}|i} = 15.9093$. Arranging our data in an interpolation table, we have

		$a_{\overline{18}|i}$	j_{12}		
		16.0295	15%		
0.1202	0.0295	16.0000	j_{12}	x / 1%	
		15.9093	16%		

$$\frac{x}{1\%} = \frac{0.0295}{0.1202}$$
$$x = 0.25\%$$
$$j_{12} = 15.25\%$$

Now we change $j_{12} = 15.25\%$ into an equivalent rate $j_2 = 2i$ such that

$$(1+i)^2 = \left(1 + \frac{0.1525}{12}\right)^{12}$$

$$i = \left(1 + \frac{0.1525}{12}\right)^{6} - 1$$

$$j_2 = 2\left[\left(1 + \frac{0.1525}{12}\right)^{6} - 1\right] = 15.74\%$$

6.11 Derive a formula whereby an ordinary general annuity with payments W made p times a year may be replaced by an equivalent ordinary simple annuity with payments R made m times per year (see Fig. 6-8).

For the general annuity, let i' be the interest rate per payment period that is equivalent to i, the rate per conversion period (which will be the payment period of the simple annuity). Thus,

$$(1+i')^p = (1+i)^m \tag{1}$$

The values of the two annuities at the end of 1 year must be equal, that is,

$$R\, s_{\overline{m}|i} = W\, s_{\overline{p}|i'} \qquad \text{or} \qquad R\frac{(1+i)^m - 1}{i} = W\frac{(1+i')^p - 1}{i'} \tag{2}$$

Together, (1) and (2) imply

$$\frac{R}{i} = \frac{W}{i'} \tag{3}$$

Solving (1) for i' and substituting in (3) gives

$$R = W\frac{i}{(1+i)^{m/p} - 1} = \frac{W}{s_{\overline{m/p}|i}} \tag{6.1}$$

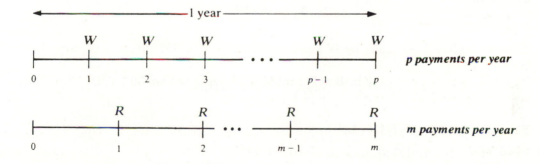

Fig. 6-8

6.12 Convert an annuity with semiannual payments of \$500 into an equivalent annuity with (a) annual payments, if money is worth $j_1 = 8\%$; (b) quarterly payments, if money is worth 10% per annum compounded semiannually.

Under (a), the given annuity is general; under (b), it is simple.

(a) Figure 6-8 and formula (6.1) apply, with $W = 500$, $m = 1$, $p = 2$, $i = 0.18$.

$$R = \frac{W}{s_{\overline{m/p}|i}} = \frac{500}{s_{\overline{1/2}|.08}} = \$1019.62$$

(b) Figure 6-8 and formula (6.1) apply, with $R = 500$, $m = 2$, $p = 4$, $i = 0.05$.

$$W = R\, s_{\overline{m/p}|i} = 500 s_{\overline{1/2}|.05} = \$246.95$$

6.13 Use the replacement formula (6.1) of Problem 6.11 to solve (a) Problem 6.1, (b) Problem 6.3.

(a) First we replace W monthly by R quarterly. We have $W = 300$, $p = 12$, $m = 4$, and $i = 0.015$. From (6.1),

$$R = \frac{W}{s_{\overline{m/p}|i}} = \frac{300}{s_{\overline{1/3}|.015}} = \$904.49$$

Then we calculate the accumulated value S of an ordinary simple annuity with $R = 904.49$, $n = 20$, and $i = 0.0225$. From (5.1),

$$S = R\, s_{\overline{n}|i} = 904.49 s_{\overline{20}|.015} = \$20\ 915.13$$

This answer differs by 12 cents from the one in Problem 6.1, because of rounding off the payment R as given by (6.1).

(b) First we replace W semiannually by R annually. We have $W = 5000$, $p = 2$, $m = 1$, and $i = 0.08$. From (6.1),

$$R = \frac{W}{s_{\overline{m/p}|i}} = \frac{5000}{s_{\overline{1/2}|.08}} = \$10\ 196.15$$

Then we calculate the discounted value A of an ordinary simple annuity with $R = 10\ 196.15$, $n = 10$, and $i = 0.08$. From (5.2),

$$A = R\, a_{\overline{n}|i} = 10\ 196.15 a_{\overline{10}|.08} = \$68\ 417.00$$

In this case there is a roundoff error of 1 cent.

6.14 Use the replacement formula (6.1) of Problem 6.11 to solve Problem 6.6.

First we find the annual payment R of an ordinary simple annuity with $A = 10\ 000$, $n = 5$, and $i = 0.10$.

$$R = \frac{A}{a_{\overline{n}|i}} = \frac{10\ 000}{a_{\overline{5}|.10}} = \$2637.97$$

Then we replace R annually by W monthly. We have $R = 2637.97$, $p = 12$, $m = 1$, $i = 0.10$, from (6.1),

$$W = R\, s_{\overline{m/p}|i} = 2637.97 s_{\overline{1/12}|.10} = \$210.36$$

OTHER GENERAL ANNUITIES

6.15 Find the accumulated and discounted values of payments of \$100 made at the beginning of each month for 5 years at $j_4 = 14\%$.

First find the rate i per month such that

$$(1 + i)^{12} = (1.035)^4$$
$$i = (1.035)^{1/3} - 1 = 0.011533142$$

Now calculate the accumulated value S and the discounted value A of a simple annuity due, with $R = 100$, $n = 60$, and $i = 0.011533142$ (see Fig 6-9). From (5.3),

$$S = R\, s_{\overline{n}|i}(1 + i) = 100 s_{\overline{60}|.011533142}(1.011533142) = \$8681.11$$

From (5.4),

$$A = R\, a_{\overline{n}|i}(1 + i) = 100 a_{\overline{60}|.011533142}(1.011533142) = \$4362.83$$

Note that $A = S(1 + i)^{-60}$.

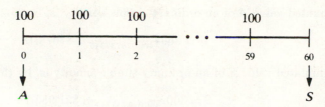

Fig. 6-9

6.16 A lot is sold for \$8000 down and six semiannual payments of \$3000, the first due at the end of 2 years. Find the cash value of the lot, if money is worth 6% compounded (*a*) daily, (*b*) continuously.

(*a*) First we find the rate i per half-year such that

$$(1+i)^2 = \left(1 + \frac{0.06}{365}\right)^{365}$$

$$i = \left(1 + \frac{0.06}{365}\right)^{365/2} - 1 = 0.030451993$$

We arrange the data as shown in Fig. 6-10. The cash value of the lot is the down payment plus the discounted value of a simple deferred annuity with $R = 3000$, $n = 6$, $k = 3$ and $i = 0.030451993$. From (5.5),

$$\text{cash value} = 8000 + 3000a_{\overline{6}|i}(1+i)^{-3} = \$22\,830.71$$

(*b*) Proceeding as in (*a*),

$$(1+i)^2 = e^{0.06}$$
$$i = e^{0.03} - 1 = 0.030454534$$

and

$$\text{cash value} = 8000 + 3000a_{\overline{6}|i}(1+i)^{-3} = \$22\,830.47$$

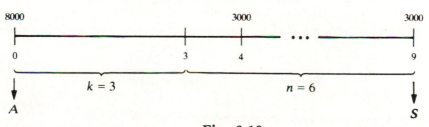

Fig. 6-10

6.17 An annuity consists of 40 payments of \$300 each, made at intervals of 3 months. Interest is at $j_1 = 12\%$. Find the value of this annuity (*a*) 3 months before the time of the first payment, (*b*) at the time of the last payment, (*c*) at the time of the first payment, (*d*) 3 months after the last payment, (*e*) 4 years 3 months before the first payment.

We calculate the rate i per quarter-year such that

$$(1+i)^4 = 1.12$$
$$i = (1.12)^{1/4} - 1 = 0.028737345$$

We then have $R = 300$, $n = 40$, and $i = 0.028737345$.

(a) The discounted value A of an ordinary simple annuity is, by (5.2),

$$A = 300a_{\overline{40}|.028737345} = \$7078.18$$

(b) The accumulated value S of an ordinary simple annuity is, by (5.1),

$$S = 300s_{\overline{40}|.028737345} = \$21\,983.75$$

(c) The discounted value A of a simple annuity due is, by (5.4),

$$A = 300a_{\overline{40}|.028737345}(1.028737345) = \$7281.59$$

(d) The accumulated value S of a simple annuity due is, by (5.3),

$$S = 300s_{\overline{40}|.028737345}(1.028737345) = \$22\,615.50$$

(e) The discounted value A of a simple deferred annuity with period of deferment $k = 16$ is, by (5.5),

$$A = 300a_{\overline{40}|.028737345}(1.028737345)^{-16} = \$4498.31$$

6.18 What equal monthly payments for 10 years will pay off a loan of \$20 000 with interest at 12% compounded semiannually, if the first payment is made (a) immediately? (b) 6 months from now?

First we find the rate i per month such that

$$(1+i)^{12} = (1.06)^2$$
$$i = (1.06)^{1/6} - 1 = 0.009758794$$

We then have $A = 20\,000$, $n = 120$, and $i = 0.009758794$.

(a) The monthly payment R of a simple annuity due is, by (5.4),

$$R = \frac{A}{a_{\overline{n}|i}(1+i)} = \frac{20\,000}{a_{\overline{120}|.009758794}(1.009758794)} = \$280.86$$

(b) The monthly payment R of a simple deferred annuity with period of deferment $k = 5$ is, by (5.5),

$$R = \frac{A}{a_{\overline{n}|i}(1+i)^{-k}} = \frac{20\,000}{a_{\overline{120}|.009758794}(1.009758794)^{-5}} = \$297.72$$

6.2   PERPETUITIES

A *perpetuity* is an annuity whose payments begin on a fixed date and continue forever. Examples of perpetuities are the interest payments from a sum of money invested permanently, scholarships paid from an endowment on a perpetual basis, and the dividends on a share of preferred stock (assuming that the company will never become bankrupt).

It is meaningless to speak about the accumulated value of a perpetuity, since there is no end to the term of a perpetuity. The present or discounted value, however, is well-defined, being the equivalent dated value of the set of payments at the beginning of the term of the perpetuity.

Let A be the discounted value of an *ordinary simple perpetuity*, which is an infinite series of equal payments, made at the ends of interest periods (Fig. 6-11). Let i be the interest rate per period, and R the periodic payment of the perpetuity. Then, as proved in Problem 6.19,

$$A = \frac{R}{i} \tag{6.2}$$

which is certainly consistent with the requirement that the equivalent fixed investment A be such as to yield a return of $Ai = R$ at the end of each interest period. (Some textbooks write $a_{\overline{\infty}|}$ for $1/i$.)

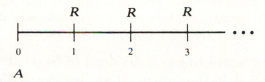

Fig. 6-11

SOLVED PROBLEMS

6.19 Derive (6.2).

A is the equivalent dated value of the infinite series of payments, shown in Fig. 6-11, at the beginning of the term of the perpetuity.

Solution 1:

$$A = R(1+i)^{-1} + R(1+i)^{-2} + R(1+i)^{-3} + \cdots = R[(1+i)^{-1} + (1+i)^{-2} + (1+i)^{-3} + \cdots]$$

Using (2.5) to sum the bracketed infinite geometric progression, we obtain

$$A = R\frac{(1+i)^{-1}}{1-(1+i)^{-1}} \times \frac{1+i}{1+i} = R\frac{1}{1+i-1} = \frac{R}{i}$$

Solution 2:

$$A = \lim_{n\to\infty} R\,a_{\overline{n}|i} = R \lim_{n\to\infty} \frac{1-(1+i)^{-n}}{i} = \frac{R}{i}[1 - \lim(1+i)^{-n}] = \frac{R}{i}$$

since $\displaystyle \lim_{n\to\infty}(1+i)^{-n} = \lim_{n\to\infty}\frac{1}{(1+i)^n} = 0$ for $i > 0$.

6.20 Find the discounted value of an ordinary simple perpetuity paying \$50 a month, if (a) $j_{12} = 9\%$, (b) $j_{12} = 12\%$, (c) $j_{12} = 15\%$.

(a) $R = 50$, $i = 0.0075$; from (6.2),

$$A = \frac{R}{i} = \frac{50}{0.0075} = \$6666.67$$

(b)
$$A = \frac{R}{i} = \frac{50}{0.01} = \$5000$$

(c)
$$A = \frac{R}{i} = \frac{50}{0.0125} = \$4000$$

6.21 How much money is needed to establish a scholarship fund paying \$1500 annually, if the fund will earn interest at $j_1 = 7\%$ and the first payment will be made (a) at the end of the first year? (b) Immediately? (c) Five years from now?

(a) The payments form an ordinary simple perpetuity, with $R = 1500$, $i = 0.07$. By (6.2),

$$A = \frac{1500}{0.07} = \$21\ 428.57$$

(b) An extra $1500 is needed immediately. Thus the required sum is

$$21\ 428.57 + 1500 = \$22\ 928.57$$

(c) If the first scholarship is awarded 5 years from now, the fund will have to contain \$22 928.57 at that time. Consequently, we have to find the discounted value of \$22 928.57 due in 5 years at $j_1 = 7\%$. This is $\$22\ 928.57(1.07)^{-5} = \$16\ 347.75$.

6.22 A certain stock is expected to pay a dividend of \$4.00 at the end of each quarter for an indefinite period in the future. If an investor wishes to realize an annual effective yield of 12%, how much should he pay for the stock?

We find the rate i per quarter-year such that

$$\begin{aligned}(1+i)^4 &= 1.12 \\ i &= (1.12)^{1/4} - 1 = 0.028737345\end{aligned}$$

The price to yield $j_1 = 12\%$, or $i = 0.028737345$ per quarter-year, is the discounted value A of an ordinary simple perpetuity with $R = 4$; from (6.2),

$$A = \frac{R}{i} = \frac{4}{0.028737345} = \$139.19$$

6.23 In 1992 a research foundation was established by a fund of \$250 000 invested at a rate that would provide \$30 000 payments at the end of each year, forever. (a) What interest rate was being earned on the fund? (b) After the payment in 1997, the foundation learned that the rate of interest earned on the fund was being changed to $j_1 = 9\%$. If the foundation wants to continue annual payments forever, what size will the new payments be? (c) If, instead, the foundation continues with the \$30 000 payments annually, how many full payments can be made at the new interest rate?

(a) We have $A = 250\ 000$, $R = 30\ 000$; from (6.2),

$$i = \frac{R}{A} = \frac{30\ 000}{250\ 000} = 0.12 = 12\%$$

(b) After the first five payments, the principal is still \$250 000. Thus, $A = 250\ 000$, $i = 0.09$; from (6.2),

$$R = Ai = 250\ 000(0.09) = \$22\ 500$$

(c) We have $A = 250\ 000$, $R = 30\ 000$, $i = 0.09$. Solving (5.2) for n,

$$\begin{aligned}250\ 000 &= 30\ 000 a_{\overline{n}|.09} \\ \frac{1 - (1.09)^{-n}}{0.09} &= 8.\dot{3} \\ (0.09)^{-n} &= 0.25 \\ n &= \frac{\log 0.25}{\log 1.09} = 16.08646345\end{aligned}$$

Sixteen full payments can be made (beyond the first five payments).

6.24 On September 1, 1993, a philanthropist gives a college a fund of $50 000, which is invested at $j_2 = 10\%$. If semiannual scholarships are to be awarded from this grant indefinitely, what is the size of each scholarship, assuming the first one is awarded on (*a*) September 1, 1993? (*b*) September 1, 1995?

(*a*) Using September 1, 1993, as the focal date, we write an equation of value for the payment R.

$$R + \frac{R}{0.05} = 50\ 000$$
$$R\left(1 + \frac{1}{0.05}\right) = 50\ 000$$
$$21R = 50\ 000$$
$$R = \$2380.95$$

(*b*) Using September 1, 1995, as the focal date, we write an equation of value for the payment R.

$$R + \frac{R}{0.05} = 50\ 000(1.05)^4$$
$$21R = 50\ 000(1.05)^4$$
$$R = \$2894.06$$

6.25 A scholarship fund is established by an endowment of $20 000 invested at 10% compounded continuously. What payment will this fund provide (*a*) at the end of each year, forever? (*b*) At the end of every 2 years, forever?

(*a*) We find the rate i per year such that

$$1 + i = e^{0.10}$$
$$i = e^{0.10} - 1 = 0.105170918$$

From (6.2), we calculate the payment R of an ordinary simple perpetuity with $A = 20\ 000$ and $i = 0.105170918$:

$$R = Ai = 20\ 000(0.105170918) = \$2103.42$$

(*b*) Similarly to (*a*),

$$1 + i = e^{2(0.10)}$$
$$i = e^{0.20} - 1 = 0.221402758$$
and
$$R = Ai = 20\ 000(0.221402758) = \$4428.06$$

6.3 ANNUITIES WHOSE PAYMENTS VARY

Thus far, all annuities considered have had a level series of payments. Unfortunately, this is not always the case in real life. In those situations where payments are in arithmetic or geometric progression, we can apply the methods of Chapter 2.

SOLVED PROBLEMS

6.26 Find the discounted value at rate i per interest period, of a simple increasing annuity whose n payments at the ends of the interest periods are R, $2R$, $3R$, ..., nR.

We arrange the data on a time diagram, Fig. 6-12. Using 0 as the focal date, we write the equation of value for the discounted value A of the annuity:

$$A = R(1+i)^{-1} + 2R(1+i)^{-2} + \cdots + (n-1)R(1+i)^{-(n-1)} + nR(1+i)^{-n} \tag{1}$$

If we multiply (1) by $1 + i$, we obtain

$$(1+i)A = R + 2R(1+i)^{-1} + \cdots + (n-1)R(1+i)^{-(n-2)} + nR(1+i)^{-(n-1)} \qquad (2)$$

Subtracting (1) from (2),

$$iA = R[1 + (1+i)^{-1} + (1+i)^{-2} + \cdots + (1+i)^{-(n-1)}] - nR(1+i)^{-n} \qquad (3)$$

The expression in brackets on the right-hand side of (3) is the sum of a geometric progression having n terms, with first term $t_1 = 1$ and common ratio $r = (1+i)^{-1}$. By (2.4),

$$
\begin{aligned}
1 + (1+i)^{-1} + (1+i)^{-2} + \cdots + (1+i)^{-(n-1)} &= 1\frac{1 - (1+i)^{-n}}{1 - (1+i)^{-1}} \\
&= \frac{1 - (1+i)^{-n}}{1 - (1+i)^{-1}} \times \frac{1+i}{1+i} = (1+i)\frac{1 - (1+i)^{-n}}{i} \\
&= (1+i)a_{\overline{n}|i}
\end{aligned}
$$

Hence (3) becomes

$$iA = R(1+i)a_{\overline{n}|i} - nR(1+i)^{-n} \qquad \text{or} \qquad A = \frac{R}{i}[(1+i)a_{\overline{n}|i} - n(1+i)^{-n}] = \frac{R}{i}(\ddot{a}_{\overline{n}|i} - nv^n)$$

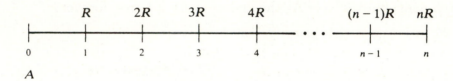

Fig. 6-12

6.27 Find the accumulated value S of the simple increasing annuity of Problem 6.26.

$$
\begin{aligned}
S &= A(1+i)^n = \frac{R}{i}[(1+i)a_{\overline{n}|i} - n(1+i)^{-n}](1+i)^n \\
&= \frac{R}{i}[(1+i)a_{\overline{n}|i}(1+i)^n - n] = \frac{R}{i}[(1+i)s_{\overline{n}|i} - n] = \frac{R}{i}(\ddot{s}_{\overline{n}|i} - n)
\end{aligned}
$$

6.28 Mrs. Jones invests \$1000 at the end of each year for 10 years in an investment fund which pays $j_1 = 13\%$. The fund pays the interest out in check form at the end of each year, and does not allow deposits of less than \$1000. Mrs. Jones deposits her annual interest payment from the fund into her bank account, which pays interest at $j_1 = 10\%$. How much money does she have at the end of 10 years?

Mrs. Jones will have exactly \$10 000 in the investment fund at the end of 10 years, since all interest is paid out. Interest deposits to her bank account will be at the end of each year, starting at the end of the 2nd year, as shown in Fig. 6-13.

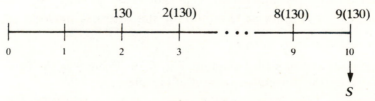

Fig. 6-13

Note that there are a total of 9 deposits. The accumulated value S of the interest deposits at $i = 0.10$ may be calculated using the equation of Problem 6.27:

$$S = \frac{R}{i}[(1+i)s_{\overline{n}|i} - n] = \frac{130}{0.10}[(1.10)s_{\overline{9}|.10} - 9] = \$7718.65$$

giving a total accumulated value $10\,000 + 7718.65 = \$17\,718.65$.

6.29 For Problem 6.28, show that if Mrs. Jones can reinvest her annual interest payments at $j_1 = 13\%$, the answer is $\$1000\ s_{\overline{10}|.13}$, as would be expected.

With $X = \$1000$, $i = 0.13$, $R = Xi$, and $n = 9$, the equation of Problem 6.27 gives as the accumulated value of the interest deposits

$$S = X[(1+i)s_{\overline{n}|i} - n]$$

and the total accumulated value is

$$(n+1)X + S = X[(1+i)s_{\overline{n}|i} + 1] = X\ s_{\overline{n+1}|i} = \$1000\ s_{\overline{10}|.13}$$

where the second equality follows from Problem 5.51(a).

6.30 A court is trying to determine the discounted value of the future income of a man paralyzed in a car accident. At the time of the accident, the man was earning $\$45\,000$ a year, and he anticipated getting a 4% raise each year. He is 30 years away from retirement. If money is worth $j_1 = 5\%$, what is the discounted value of his future income? (Assume that the payments are at the end of each year, with the first payment one year hence.)

We arrange the data as in Fig. 6-14. The discounted value A of the future income at $i = 0.10$ is

$$A = 45\,000(1.04)(1.05)^{-1} + 45\,000(1.04)^2(1.05)^{-2} + \cdots + 45\,000(1.04)^{30}(1.05)^{-30}$$

The expression on the right-hand side is the sum of a geometric progression having $n = 30$ terms, first term $t_1 = 45\,000(1.04)(1.05)^{-1}$, and common ratio $r = (1.04)(1.05)^{-1}$. Thus, applying (2.4), we obtain

$$A = t_1\frac{1-r^n}{1-r} = 45\,000(1.04)(1.05)^{-1}\frac{1-(1.04)^{30}(1.05)^{-30}}{1-(1.04)(1.05)^{-1}} = \$1\,167\,898.49$$

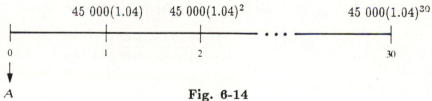

Fig. 6-14

Alternate Solution

Rewrite the equation for A as

$$A = 45\,000\left[\left(\frac{1.05}{1.04}\right)^{-1} + \left(\frac{1.05}{1.04}\right)^{-2} + \cdots + \left(\frac{1.05}{1.04}\right)^{-30}\right]$$

Let i_e be a rate of interest such that

$$1 + i_e = \frac{1.05}{1.04} \qquad \text{or} \qquad i_e = \frac{1.05}{1.04} - 1 = 0.009615385$$

Then $\quad A = 45\,000[(1+i_e)^{-1} + (1+i_e)^{-2} + \cdots + (1+i_e)^{-30}] = 45\,000a_{\overline{30}|i_e} = \$1\,167\,898.49$

6.31 The Smiths need to have their house painted immediately. Painting the house will cost \$1200, so they are trying to decide if they should put on aluminum siding instead. If it is assumed that the house must be painted every 5th year (forever) and that the cost of painting will rise by 2% per year (forever), how much should the Smiths be willing to pay for aluminum siding, given that they can earn 8% per annum on their money?

The discounted value A of the future painting expenses is

$$A = 1200 + 1200(1.02)^5(1.08)^{-5} + 1200(1.02)^{10}(1.08)^{-10} + \cdots$$

The expression on the right-hand side is the sum of an infinite geometric progression, with $t_1 = 1200$ and

$$0 < r = (1.02)^5(1.08)^{-5} < 1$$

Thus, applying (2.5),

$$A = \frac{t_1}{1-r} = \frac{1200}{1 - (1.02)^5(1.08)^{-5}} = \$4827.40$$

The Smiths should be willing to pay up to \$4827.40 for aluminum siding.

6.32 Find the discounted value of a series of payments that start at \$18 000 at the end of Year One and then increase by \$2000 each year, forever. Interest is $j_1 = 10\%$.

The discounted value A of this increasing perpetuity is

$$A = 18\ 000(1.10)^{-1} + 20\ 000(1.10)^{-2} + 22\ 000(1.10)^{-3} + \cdots \tag{1}$$

If we multiply (1) by $1 + i = 1.10$, we obtain

$$1.10A = 18\ 000 + 20\ 000(1.10)^{-1} + 22\ 000(1.10)^{-2} + \cdots \tag{2}$$

Subtracting (1) from (2),

$$0.1A = 18\ 000 + 2000[(1.10)^{-1} + (1.10)^{-2} + \cdots] \tag{3}$$

The expression in brackets on the right-hand side of (3) is the sum of an infinite geometric progression, with $t_1 = (1.10)^{-1}$ and $r = (1.10)^{-1}$. Applying (2.5), we obtain

$$(1.10)^{-1} + (1.10)^{-2} + \cdots = \frac{(1.10)^{-1}}{1 - (1.10)^{-1}} \times \frac{1.10}{1.10} = \frac{1}{1.10 - 1} = \frac{1}{0.10} = 10$$

and substituting into (3),

$$0.1A = 18\ 000 + 2000(10) \quad \text{or} \quad A = \$380\ 000$$

Supplementary Problems

ORDINARY GENERAL ANNUITIES

6.33 Find the accumulated and the discounted value of each of the following annuities:

	Payment	Payment Interval	Term	Interest Rate
(a)	100	3 months	10 years	18% comp. monthly
(b)	2000	12 months	5 years	16% comp. quarterly
(c)	250	1 month	3 years	10% comp. daily
(d)	300	3 months	10 years	9% comp. continuously

Ans. (a) \$10 878.94, \$1822.47; (b) \$14 024.88, \$6400.77; (c) \$10 451.87, \$7743.26; (d) \$19 243.26, \$7823.72

6.34 A contract calls for payments of \$100 at the end of each month for 5 years and an additional payment of \$2000 at the end of 5 years. What is the present worth of the contract, if money is worth (a) 15% compounded semiannually, (b) 8% compounded continuously. *Ans.* (a) \$5215.73; (b) \$6269.37

6.35 Deposits of \$100 are made at the end of each quarter to a bank account for 5 years. Find the accumulated value of these payments, given (a) $j_{12} = 6\%$, (b) $j_4 = 6\%$, (c) $j_1 = 6\%$, (d) $j_\infty = 6\%$. *Ans.* (a) \$2314.08; (b) \$2312.37; (c) \$2304.96; (d) \$2314.94

6.36 Payments of \$1000 are to be made at the end of each half-year for the next 10 years. Find their discounted value, if interest is at 12% per annum compounded (a) half-yearly, (b) quarterly, (c) annually, (d) continuously. *Ans.* (a) \$11 469.92; (b) \$11 386.59; (c) \$11 629.86; (d) \$11 300.85

6.37 An annuity pays \$200 at the end of each month for 5 years, and \$300 at the end of each month for the next 5 years. Find the discounted value of the annuity at 12% compounded continuously. *Ans.* \$16 370.16

6.38 Deposits of \$200 are made at the end of each month for 5 years into a bank account where interest is paid at $j_4 = 8\%$. Find the accumulated value of the account at the end of 5 years, if (a) no interest, (b) simple interest, (c) compound interest is paid for the fractional part of a period. *Ans.* (a) \$14 578.42; (b) \$14 675.61; (c) \$14 675.18

6.39 Is it cheaper to buy a car for \$9800 and after 3 years trade it in for \$2000, or to rent a car for \$250 a month payable at the end of each month for 3 years? Assume that maintenance and license costs are identical for both alternatives and that money is worth $j_1 = 8\%$. *Ans.* Renting is cheaper.

6.40 An experimental forest is to be grown by planting 1000 trees each year on the same day of the year. The trees each contain 0.003 cubic meter of wood when they are planted and increase in volume at the rate of 5% per quarter-year thereafter. What will be the total volume of all the trees in the forest after 10 years? (Assume that no trees die.) *Ans.* 102.2009 m^3

6.41 A property worth \$80 0000 is sold for \$10 000 down and equal monthly payments for the next 20 years. Find the monthly payment if the interest rate is $j_1 = 8.14\%$. *Ans.* \$579.04

6.42 A couple would like to accumulate \$20 000 in 3 years as a down payment on a house, by making deposits at the end of each week in an account paying interest at $j_{12} = 6\%$. Find the weekly deposit, assuming that compound interest is paid for any part of a conversion period. *Ans.* \$117.11

6.43 Upon her husband's death, a woman finds that he held a \$30 000 life insurance policy. If the insurance company pays 10% per annum, what monthly income (paid at the end of each month) can she expect over a ten-year period? *Ans.* \$389.33

6.44 A city wants to accumulate $500 000 over the next 20 years to redeem an issue of bonds. What payment will be required at the end of each 6 months, if interest is earned at 7% per annum compounded monthly? *Ans.* $5843.61

6.45 To prepare for an early retirement, a self-employed businessman makes deposits of $11 500 each year for 20 years, starting on his 32nd birthday. When he is 55, he wishes to make 30 equal annual withdrawals. What is the size of each withdrawal if interest is 10% compounded (a) semiannually, (b) continuously? *Ans.* (a) $98 347.98; (b) $104 376.31

6.46 A condominium owners' association will need $100 000 three years from now, to be used for major renovations. For the last two years they made deposits of $4000 at the end of each quarter into a fund that earns interest at $j_{365} = 12\%$. What quarterly deposits to the fund will be needed to reach the goal of $100 000 three years from now? *Ans.* $3439.55

6.47 A deposit of $2000 is made to open an account on April 1, 1992. Quarterly deposits of $300 are then made for 5 years, the first deposit on July 1, 1992. Starting on October 1, 1998, the first of a sequence of $1000 quarterly withdrawals is made. Assuming interest at 10% compounded continuously, find the balance in the account (a) on October 1, 1995; (b) on October 1, 2001. *Ans.* (a) $7804.35; (b) $2058.21

6.48 The ABC Development Company recently sued the XYZ Trust Company. ABC had borrowed $8.2 million from XYZ "at 13%." The loan was paid off over 3 years in monthly installments. ABC thought the rate of interest was $j_1 = 13\%$, but XYZ was using $j_{365} = 13\%$. The court sided with ABC and awarded the total dollar difference (without interest) to them. Find the size of the award. *Ans.* $111 185.28

6.49 How many monthly deposits of $100 each, and what final deposit one month later, will be necessary to accumulate $3000, if interest is (a) at $6\frac{1}{2}\%$ compounded semiannually? (b) At 15% compounded quarterly? *Ans.* (a) 27, $88.29; (b) 25, $56.12

6.50 A couple requires a $60 000 mortgage loan at $j_2 = 10.25\%$ to buy a new house. Find their monthly mortgage payment over (a) 30 years, (b) 20 years, (c) 10 years. (d) If the couple can afford to pay $950 monthly, how many full payments will be required and what will be the smaller concluding payment?
Ans. (a) $528.22; (b) $580.51; (c) $794.16; (d) 90, $196.47

6.51 The Andersons can buy a certain house listed at $100 000 and pay $35 000 down. They can get a mortgage from the bank for $65 000 at $j_2 = 12\%$, payable over 25 years in monthly installments. The seller of the house offers the house for $105 000; he will give them a $70 000 mortgage at $j_2 = 10\%$, payable monthly, and will assume a 25-year amortization period. If the two interest rates are guaranteed for 25 years, what should the Andersons do? *Ans.* Take the seller's offer.

6.52 A $40 000 mortgage is to be paid off by monthly payments over 25 years. Find the monthly payment if (a) $j_2 = 9\%$, (b) $j_2 = 11\%$, (c) $j_2 = 13\%$. *Ans.* (a) $331.19; (b) $385.01; (c) $440.96

6.53 A $50 000 mortgage is obtained at a rate $j_2 = 10\frac{1}{4}\%$. The mortgage can be paid over 20, 25 or 30 years. Find the monthly payments necessary under each of these three options.
Ans. $483.76; $455.68; $440.18

6.54 For each loan listed below, find the unknown quantity.

	Loan	Interest Rate	Term	Monthly Payment
(a)	?	$j_2 = 11\frac{3}{4}\%$	25 years	$575
(b)	$57 000	$j_2 = ?$	20 years	$832
(c)	$48 000	$j_2 = 16\frac{1}{2}\%$	?	$750
(d)	$80 000	$j_2 = 10\frac{3}{8}\%$	30 years	?

Ans. (*a*) $56 681.35; (*b*) 17.52%; (*c*) 144 full payments plus $144.77; (*d*) $711.39

6.55 A finance company advertises loans of $1000 to be repaid in 36 monthly installments of $40 each. Find the annual effective rate of interest charged. *Ans.* 28.65%

6.56 A used car valued at $2500 is sold for $500 down and $200 a month for the next 12 months. Find the annual effective rate of interest charged. *Ans.* 41.29%

6.57 Convert an annuity with quarterly payments of $1000 into an equivalent annuity with (*a*) semi-annual payments, if money is worth 5% compounded semiannually; (*b*) monthly payments, if money is worth 7% compounded quarterly. *Ans.* (*a*) $2012.42; (*b*) $331.41

6.58 The Whites pay $300 at the end of each month on their home mortgage. Find the equivalent semiannual payments at 10% per annum compounded semiannually. *Ans.* $1837.14

6.59 At $j_2 = 10\%$, replace an annuity of $500 payable at the end of each half-year by an equivalent annuity payable (*a*) at the end of each month, (*b*) at the end of each year. *Ans.* (*a*) $81.65; (*b*) $1025

6.60 The Friedmans have a $50 000 mortgage with monthly payments over 25 years at $j_2 = 10\%$. Because they each get paid weekly, they decide to switch to weekly payments (still at $j_2 = 10\%$ and paid over 25 years). Compare their weekly payments to their monthly payments. *Ans.* $102.89 weekly, $447.24 monthly

6.61 You want to take a $105 000 mortgage at $j_2 = 9\%$ and can afford to pay up to $300 per week. What repayment period in years should you request and what will be your weekly payment? *Ans.* 11 years, $286.81

6.62 An ordinary annuity consists of $n = pt$ payments of $R each, there being p payments per year over t years at rate j_∞. Determine (*a*) the accumulated value S, (*b*) the discounted value A.
Ans. (*a*) $S = R\,s_{\overline{n}|j_\infty} \equiv R\dfrac{e^{j_\infty t} - 1}{e^{j_\infty/p} - 1}$; (*b*) $A = R\,a_{\overline{n}|j_\infty} \equiv R\dfrac{1 - e^{-j_\infty t}}{e^{j_\infty/p} - 1}$

6.63 Use Problem 6.11 to obtain the accumulated value S and the discounted value A of an ordinary general annuity with payments W, p times per year for k years at rate $j_m = mi$.
Ans. $S = \dfrac{W}{s_{\overline{m/p}|i}}$, $A = \dfrac{W}{s_{\overline{m/p}|i}}a_{\overline{km}|i}$

OTHER GENERAL ANNUITIES

6.64 Find the accumulated value of payments of $200 made at the beginning of each month for 3 years at $j_2 = 13\%$. *Ans.* $8795.06

6.65 An insurance policy requires premium payments of $15 at the beginning of each month for 20 years. Find the discounted value of these payments at $j_4 = 11\%$. *Ans.* $1476.08

6.66 Find the discounted value at $j_2 = 10\%$ of 20 annual payments of $200 each, the first one due five years hence. *Ans.* $1133.07

6.67 An annuity consists of 60 payments of $200 at the end of each month. Interest is at $j_1 = 11.05\%$. Find the value of this annuity (*a*) at the time of the first payment, (*b*) two years before the first payment, (*c*) at the time of the last payment. *Ans.* (*a*) $9380.77; (*b*) $7606.79; (*c*) $15 705

6.68 A father has saved a fund to provide for his son's 4-year university program. The fund will pay $300 at the beginning of each month for eight months (September through April), plus an extra $2000 each September 1st, for four years. If $j_4 = 8\%$, what is the value of the fund on the first day of university (before any withdrawals)? *Ans.* $15 495.16

6.69 On November 1, 1993, a research fund of $300 000 was established to provide for equal annual grants for 10 years. What will be the size of each grant if (*a*) the fund earns interest at 8% compounded daily and the first grant is awarded on November 1, 1993; (*b*) the fund earns interest at 10% compounded monthly and the first grant is awarded on November 1, 1996? *Ans.* (*a*) $41 884.02; (*b*) $60 795.57

6.70 A college graduate must repay the government $4200 for outstanding student loans. The graduate can afford to pay $150 monthly. If the first payment is made at the end of the first year and money is worth 11% compounded quarterly, find the number of full payments required and the size of the smaller concluding payment. *Ans.* 36, $71.19

6.71 A $5000 loan is repaid by 24 monthly payments of $175 each, followed by 24 monthly payments of $160 each. What interest rate j_1 is being charged? *Ans.* 29.70%

6.72 A used car may be purchased for $7600 cash or $600 down and 20 monthly payments of $400 each, the first payment to be made in 6 months. What annual effective rate of interest does the installment plan use? *Ans.* 11%

6.73 Which is cheaper?
(*a*) Buy a car for $13 600 and after three years trade it in for $3600.
(*b*) Lease a car for $318 a month payable at the beginning of each month for three years.
Assume maintenance costs are identical and that money is worth $j_1 = 8\%$. *Ans.* Lease

6.74 What sum of money should be set aside to provide an income of $500 a month for a period of 3 years if the money earns interest at $j_{12} = 6\%$ and the first payment is to be received (*a*) 1 month from now, (*b*) immediately, (*c*) 2 years from now?
Ans. (*a*) $16 435.51; (*b*) $16 517.69; (*c*) $14 654.25

6.75 The proceeds of $100 000 death benefit are left on deposit with an insurance company for seven years at an annual effective interest rate of 5%. The balance at the end of seven years is paid to the beneficiary in 120 equal monthly payments of X, with the first payment made immediately. During the payout period, interest is credited at an annual effective interest rate of 3%. Calculate X. *Ans.* $1352.74

6.76 Obtain the accumulated value S and the discounted value A of a general annuity due, having payments W, p times per year for k years at rate $j_m = mi$. (*Hint:* Use Problem 6.63.)
$$Ans.\ S = W\frac{s_{\overline{km}|i}}{s_{\overline{m/p}|i}}(1+i)^{m/p},\ A = W\frac{a_{\overline{km}|i}}{s_{\overline{m/p}|i}}(1+i)^{m/p}$$

PERPETUITIES

6.77 Find the discounted value of an ordinary simple perpetuity paying $400 a year, if interest is (*a*) $j_1 = 8\%$, (*b*) $j_1 = 12.48\%$. *Ans.* (*a*) $5000; (*b*) $3205.13

6.78 How much money is needed to establish a fund that pays a research scholarship of $10 000 each half-year, if the endowment can be invested at $j_2 = 10\%$ and if the first scholarship will be provided (*a*) a half-year from now? (*b*) immediately? (*c*) Four years from now?
Ans. (*a*) $200 000; (*b*) $210 000; (*c*) $142 136.27

6.79 It costs a railroad company $500 at the end of each month to maintain a level-crossing gate system. How much can the company contribute toward the cost of an underpass that will eliminate the level-crossing system, if money is worth 15% per annum compounded monthly? *Ans.* $40 000

6.80 The XYZ Company has a stock that pays a semiannual dividend of $4. If the stock sells for $64, (*a*) what yield j_2 did the investor desire? (*b*) What is the equivalent rate j_1?
Ans. (*a*) 12.5%; (*b*) 12.89%

6.81 A college estimates that its new campus center will require $3000 for upkeep at the end of each year for the next 5 years and $5000 at the end of each year thereafter, indefinitely. If money is worth 12% effective, how large an endowment is necessary for the future upkeep of the campus center? *Ans.* $34 457.12

6.82 Deposits of $1000 are put into a fund at the beginning of each year for the next 20 years. At the end of the 20th year, annual payments from the fund commence, and continue forever. If $j_1 = 12\%$, find the size of these payments. *Ans.* $8646.29

6.83 Derive

$$a_{\overline{n}|i} = \frac{1 - (1+i)^{-n}}{i}$$

as the difference between the discounted value of an ordinary simple perpetuity of $1 per period and the discounted value of an ordinary simple perpetuity of $1 per period deferred for n periods.

6.84 How much money is needed to endow a perpetual series of lectures costing $2000 semiannually (first lecture at the end of 6 months), if money is worth (a) $12\frac{1}{2}\%$ per annum compounded monthly? (b) $12\frac{1}{2}\%$ per annum compounded semiannually? (c) 9% per annum compounded monthly? *Ans.* (a) $32 970.56; (b) $32 000; (c) $43 618.37

6.85 On the assumption that a farm will net $15 000 annually indefinitely, what is a fair price for it, if money is worth (a) $j_1 = 8\%$? (b) $j_{12} = 15\%$? *Ans.* (a) $187 500; (b) $93 309.97

6.86 How much money is needed to establish a research fund paying $10 000 a year indefinitely, if the fund earns interest (a) at $j_\infty = 12\%$ and the first payment is provided at the end of 3 years? (b) At $j_{12} = 12\%$ and the first payment is provided immediately? *Ans.* (a) $61 697.83; (b) $88 848.80

6.87 A sum of $20 000 is invested in a fund that yields 8% compounded semiannually. What payment will this fund provide, forever, (a) At the end of each month? (b) At the beginning of each year? *Ans.* (a) $131.16; (b) $1508.88

6.88 A perpetuity paying $1000 at the end of each year is replaced with an annuity paying X at the end of each month for 10 years. Find X if $j_4 = 15\%$. *Ans.* $100.98

6.89 Use an infinite geometric progression to derive the discounted value A of an ordinary general perpetuity with payments W, p times per year at rate $j_m = mi$.
Ans. $A = \dfrac{W}{i} \dfrac{1}{s_{\overline{m/p}|i}}$

6.90 A family is considering putting aluminum siding on their house as it needs painting immediately. Painting the house costs $4200 and must be done every 4 years. What price can the family afford for the aluminum siding if they earn interest at $j_1 = 8.75\%$ on their savings? *Ans.* $14 734.88

6.91 You take out a loan for L at $j_{12} = 18\%$ and repay $300 at the end of each month for as long as necessary. This loan is invested at $j_1 = 10\%$ and provides for a perpetuity due which pays the prize in the annual "Liar's Contest." The prize is $200 for the 1st year and increases by $150 each year until it reaches $500. From then on the prize remains $500. Find the time and amount of the final repayment on the loan. *Ans.* $182.88 at the end of 20 months.

6.92 A university receives a certain sum as a bequest and invests it to earn $j_1 = 8\%$. The fund can be used to pay for a lecturer at $60 000 payable at the end of each year forever, or the money can be used to pay for a new building that the university is planning to erect. The building will be paid for with 25 equal annual payments, the first of which is due 4 years from today when the building will be occupied. Find the amount of each building payment. *Ans.* $88 506.21

6.93 Find the monthly deposit needed for 5 years to provide for a perpetuity of $400 per month. The 1st perpetuity payment is made 2 years after the last deposit, and interest changes from $j_{12} = 8\%$ to $j_{12} = 9\%$ on that date. *Ans.* $623.50

6.94 You deposit $1000 per year for 10 years at $j_1 = 8\%$. This fund then provides for a perpetuity of $3000 per year, with the first payment made n years after the final deposit. At the time of the first perpetuity payment, interest rates fall to $j_4 = 7\%$. Find n. *Ans.* 14.65461368 years

6.95 A perpetuity pays $4000 per year, as follows:
> in odd-numbered years, a payment of $4000 is made at the end of the year.
> in even-numbered years, a payment of $1000 is made at the end of each quarter.

Interest is at annual effective rate of 8%. At the beginning of an odd-numbered year, this perpetuity is exchanged for another of equal value which provides semiannual payments, the first payment due six months hence. What is the semiannual payment of the new perpetuity? *Ans.* $1989.36

ANNUITIES WHOSE PAYMENTS VARY

6.96 Find the discounted value of a series of 15 payments made at the end of each year at $j_1 = 6\%$, if the first payment is $300, the second is $600, the third is $900, and so on. *Ans.* $20 180.04

6.97 At rate i per year, find the discounted value A of a simple decreasing annuity whose n payments at the end of each year are nR, $(n-1)R$, $(n-2)R$, ..., $2R$, R.

Ans. $A = \dfrac{R}{i}(n - a_{\overline{n}|i})$

6.98 Find (a) the discounted, (b) the accumulated, value of a simple decreasing annuity of 20 payments at the end of each year at $j_1 = 12\%$, if the first payment is $2000, the second is $1900, and so on, the last payment being $100. *Ans.* (a) $10 442.13; (b) $100 727.85

6.99 Find the discounted value of 11 payments made at the end of each year at rate $j_1 = i$, if the successive payments are $1, $2, $3, $4, $5, $6, $5, $4, $3, $2, $1. *Ans.* $a_{\overline{6}|i}(1 + a_{\overline{5}|i})$

6.100 Find the present value of an annuity of which the payments are $200 per month at the end of each month during the first year, $195 per month during the second year, $190 per month during the third year, etc. ($5 monthly will be paid at the end of each month during the 40th year and nothing thereafter). Interest is at $j_{12} = 12\%$. *Ans.* $16 090.80

6.101 An investor deposits $1000 at the beginning of each year in a special fund paying interest at $j_1 = 10\%$. The interest payments are then deposited in a bank account paying interest at $j_1 = 6\%$. How much money has been accumulated at the end of 6 years? *Ans.* $8323.06

6.102 Find the discounted value of payments made at the beginning of each year, indefinitely, at $j_1 = 12\%$, if the first payment is $100, the second is $200, the third is $300, and so on. *Ans.* $8711.11

6.103 A certain site is returning an annual rent of $5000 per year payable at the beginning of the year. It is expected that the rent will increase 6% per year, on the average, forever. Calculate the present value of the site at $j_1 = 10\%$. *Ans.* $137 500

6.104 Consider the perpetuity whose year-end payments are R, $R + p$, $R + 2p$, ..., $R + (n-1)p$, $R + np$, $R + np$, ... The payments increase by a constant amount p until they reach $R + np$, after which they continue without change. Find the discounted value A of this perpetuity at rate i per annum.

Ans. $A = \dfrac{R + pa_{\overline{n}|i}}{i}$

6.105 An investor deposits \$$X$ at the end of each interest period for n periods in a fund that pays interest at rate i_1 per period. The interest payments are reinvested at rate i_2 per period. (a) Determine the total accumulated value at the end of n periods. (b) Letting $i_1 = i_2 = i$ in the result of (a), check the answer to Problem 6.29.

Ans. (a) $nX + \dfrac{i_1 X}{i_2}[(1+i_2)s_{\overline{n-1}|i_2} - (n-1)]$

6.106 Mrs. Tong invests \$10 000 in a preferred stock that pays an annual dividend at 12%. (That is, \$1200 at the end of each year.) Mrs. Tong invests the dividend payment in her bank account, which pays annual interest at $j_1 = 10\%$. (a) What is the total accumulated value of her assets at the end of 5 years? (b) Show that if the two rates of interest (12% and 10%) had been equal (at rate i) the answer would have been \$10 000$(1+i)^5$. *Ans.* (a) \$17 326.12

6.107 For the same lump sum payment, an insurance company will provide a choice of an ordinary annuity of \$1000 per year for 20 years or an ordinary increasing annuity of \$$X$ per year initially, inflating at 10% per year, for 20 years. (a) Find X if interest is at $j_1 = 8\%$. (b) What is the final payment of the increasing annuity? *Ans.* (a) \$442.88; (b) \$2708.61

6.108 Jeanne has won a lottery that pays \$1000 per month in the first year, \$1100 per month in the second year, \$1200 per month in the third year, etc. Payments are made at the end of each month for 10 years. Using an annual effective interest rate of 3%, calculate the present value of this prize. *Ans.* \$147 928.85

6.109 Find the accumulated value of quarterly payments of \$100, \$110, \$120, ... for 20 years if interest is at $j_{12} = 12\%$. *Ans.* \$113 990.19

6.110 What is the present value of quarterly payments of \$100, \$110, \$120, \$130, ..., for 20 years, if $j_2 = 8\%$? *Ans.* \$15 770.38

6.111 An annuity provides for 30 annual payments. The first payment of \$100 is made immediately and the remaining payments increase by 8% per annum. Interest is calculated at $j_1 = 13.4\%$. Calculate the present value of this annuity. *Ans.* \$1614.11

6.112 The principal on a loan of \$5000 is to be repaid in equal annual installments of \$1000 each at the end of each of the next 5 years. Interest will be paid at the end of each year at $j_1 = 4\frac{1}{2}\%$ on the entire outstanding principal, including the \$1000 payment then due. Find the purchase price of the loan to yield $j_1 = 5\%$. *Ans.* \$4932.95

6.113 Find the discounted value one year before the first payment of a series of 20 annual payments the first of which is \$500. The payments inflate at $j_1 = 6\%$ and interest is at $j_1 = 8\%$. *Ans.* \$7797.87

6.114 A man deposits \$100 at the beginning of each year into a fund paying interest to him in cash at $j_1 = 6\%$ each year on the principal in the fund. At the end of each year he deposits the interest payments into a second fund earning interest at $j_1 = 4\%$. At the end of which year will the interest fund first exceed the principal fund? *Ans.* 24

6.115 Consider an annuity where payments vary as represented in the following diagram ($0 < i_1 < 1$):

$$R(1+i_1) \quad R(1+i_1)^2 \quad R(1+i_1)^3 \qquad\qquad R(1+i_1)^n$$

| | | | | ... | |
|0|1|2|3| |n|

Further, assume the interest rate is i_2 per interest period. Let i be the rate such that

$$(1+i) = \frac{1+i_2}{1+i_1}$$

Show that the discounted value A of the above annuity at rate i_2 is

$$A = R\, a_{\overline{n}|i}$$

6.116 Derive an algebraic proof of the following identities and interpret by means of time diagrams.

(a) $a_{\overline{1}|i} + a_{\overline{2}|i} + a_{\overline{3}|i} + \ldots + a_{\overline{n}|i} = \dfrac{n - a_{\overline{n}|i}}{i}$

(b) $s_{\overline{1}|i} + s_{\overline{2}|i} + s_{\overline{3}|i} + \ldots + s_{\overline{n-1}|i} = \dfrac{s_{\overline{n}|i} - n}{i}$

6.117 Consider an annuity with a term of n periods in which payments begin at P at the end of the first period and change by Q each period thereafter. Assuming an interest rate i per period, and $P > 0$, $P + (n-1)Q > 0$ (to avoid negative payments if $Q < 0$) show that

(a) The discounted value of the annuity is given by

$$Pa_{\overline{n}|i} + Q\frac{a_{\overline{n}|i} - n(1+i)^{-n}}{i}$$

(b) The accumulated value of the annuity is given by

$$Ps_{\overline{n}|i} + Q\frac{s_{\overline{n}|i} - n}{i}$$

6.118 Consider the increasing \$1 annuity in the diagram below:

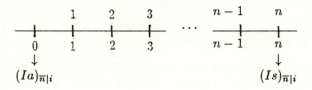

where $(Ia)_{\overline{n}|i}$ denotes the discounted value, $(Is)_{\overline{n}|i}$ denotes the accumulated value of this annuity and i is the interest rate per period. Show that

$$(Ia)_{\overline{n}|i} = a_{\overline{n}|i} + \frac{a_{\overline{n}|i} - n(1+i)^{-n}}{i} = \frac{\ddot{a}_{\overline{n}|i} - nv^n}{i}$$

$$(Is)_{\overline{n}|i} = s_{\overline{n}|i} + \frac{s_{\overline{n}|i} - n}{i} = \frac{\ddot{s}_{\overline{n}|i} - n}{i}$$

6.119 Consider the decreasing \$1 annuity in the diagram below:

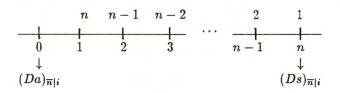

where $(Da)_{\overline{n}|i}$ denotes the discounted value, $(Ds)_{\overline{n}|i}$ denotes the accumulated value of this annuity, and i is the interest rate per period. Show that

$$(Da)_{\overline{n}|i} = na_{\overline{n}|i} - \frac{a_{\overline{n}|i} - n(1+i)^{-n}}{i} = \frac{n - a_{\overline{n}|i}}{i}$$

$$(Ds)_{\overline{n}|i} = ns_{\overline{n}|i} - \frac{s_{\overline{n}|i} - n}{i} = \frac{n(1+i)^n - s_{\overline{n}|i}}{i}$$

6.120 Consider a perpetuity in which payments begin at P at the end of the first period and increase by Q per period thereafter. Assuming the interest rate i per period and $P > 0$, $Q > 0$ (to avoid negative payments), show that the discounted value of this perpetuity is given by

$$\frac{P}{i} + \frac{Q}{i^2}$$

[*Hint*: Take the limit of the formulas in Problem 6.117(a) as n approaches infinity.]

6.121 Using the result of Problem 6.120:

(*a*) Show that the discounted value of a \$1 increasing perpetuity, denoted by $(Ia)_{\overline{\infty}|i}$, is

$$(Ia)_{\overline{\infty}|i} = \frac{1}{i} + \frac{1}{i^2}$$

(*b*) Calculate the present value of a perpetuity whose payments start at \$100 at the end of the first month and increase by \$2 per month thereafter, assuming $j_{12} = 6\%$. *Ans.* (*b*) \$100 000

In the following exercises use the formulas developed in Problems 6.117 through 6.119.

6.122 What is the accumulated value of quarterly payments of \$100, \$110, \$120, ... for 15 years if interest is $j_2 = 10\%$? *Ans.* \$43 627.25

6.123 A decreasing annuity will pay \$800 at the end of 6 months, \$750 at the end of 1 year, etc., until a final payment of \$350 is made at the end of 5 years. Find the present value of the payments if money is worth 8% compounded monthly. *Ans.* \$4780.77

6.124 Find the accumulated value of deposits of \$1, \$2, \$3, ..., \$98, \$99, \$100, \$99, \$98, ..., \$3, \$2, \$1 if interest is 2% per period. *Ans.* \$97 489.02

6.125 It is desired to accumulate a fund of \$18 000 at the end of 3 years by equal deposits at the beginning of each month. If the deposits earn interest at $j_{12} = 9\%$ but the interest can be reinvested only at $j_{12} = 6\%$, find the size of the necessary deposit. *Ans.* \$435.84

Chapter 7

Amortization and Sinking Funds

7.1 AMORTIZATION OF A DEBT

The most common method of repaying an interest-bearing loan is by *amortization*. In this method, a series of periodic payments is made; when these payments are equal— the usual case — their size can be determined by the procedures of Chapters 5 and 6. (See Problem 7.1.)

The indebtedness at any time is called the *outstanding balance* or *outstanding principal*; it is just the discounted value of all unmade payments. Each sequential payment pays the interest on the unpaid balance and also repays a part of the outstanding principal. Over the term of the loan, as the outstanding principal decreases, the interest portion of each payment decreases and the principal portion increases. This shifting distribution is shown in an *amortization schedule* (Problems 7.4 and 7.5).

The common commercial practice is to round the payment up to the nearest cent; we shall follow this practice unless specified otherwise. However, instead of rounding up to the nearest cent, the lender may round up to the nearest dime or to the nearest dollar. Rounding up of the payment amount results in a reduced last payment, as may be determined from an equation of value at the time of the last payment. (See Problems 7.2 and 7.3.)

SOLVED PROBLEMS

7.1 A debt of $6000 with interest at 16% compounded semiannually is to be amortized by equal semiannual payments of R over the next 3 years, the first payment due in 6 months. Find the payment rounded up to the nearest cent.

The six payments of R form an ordinary simple annuity with $A = 6000$, $n = 6$, and $i = 0.08$. From (5.2),

$$R = \frac{A}{a_{\overline{n}|i}} = \frac{6000}{a_{\overline{6}|.08}} = \$1297.892317 \approx \$1297.90$$

7.2 Find the concluding payment in Problem 7.1.

Let X be the concluding payment. We arrange our data on a time diagram, Fig. 7-1, and set up an equation of value for X at time 6.

$$
\begin{aligned}
X + 1297.90 s_{\overline{5}|.08}(1.08) &= 6000(1.08)^6 \\
X + 8223.40 &= 9521.25 \\
X &= \$1297.85
\end{aligned}
$$

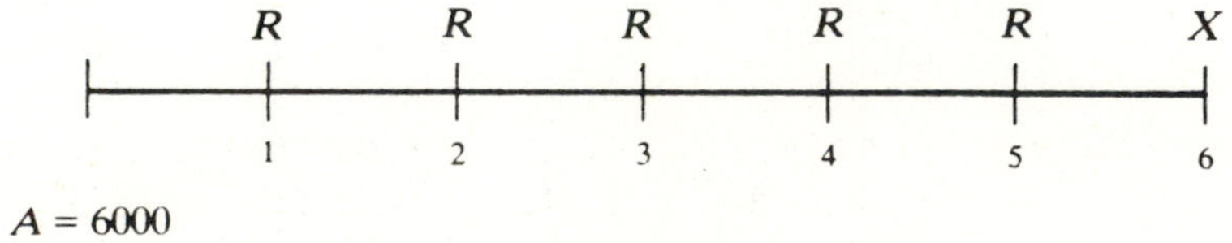

Fig. 7-1

124

7.3 Find the concluding payment X, if the payment in Problem 7.1 is rounded up to the nearest dollar.

Following Problem 7.2, with $R = \$1298$,

$$
\begin{aligned}
X + 1298 s_{\overline{5}|.08}(1.08) &= 6000(1.08)^6 \\
X + 8224.04 &= 9521.25 \\
X &= \$1297.21
\end{aligned}
$$

7.4 Construct a complete amortization schedule for the debt of Problems 7.1 and 7.3.

Table 7-1

Payment Number	Periodic Payment	Interest at 8%	Principal Repaid	Outstanding Principal
				6000.00
1	1298.00	480.00	818.00	5182.00
2	1298.00	414.56	883.44	4298.56
3	1298.00	343.88	954.12	3344.44
4	1298.00	267.56	1030.44	2314.00
5	1298.00	185.12	1112.88	1201.12
6	1297.21	96.09	1201.12	0
TOTALS	7787.21	1787.21	6000.00	

See Table 7-1. Note that the interest due at the time of the first payment is 8% of $6000, or $480. The first payment of $1298 will pay this interest and will also reduce the outstanding principal by $1298 - 480 = \$818$, bringing it to $5182. The interest due at the time of the second payment is 8% of $5182, or $414.56; the procedure is repeated until the loan is repaid in full.

7.5 Mr. Adams borrows $2000 to be repaid with quarterly payments over 2 years at $j_{12} = 24\%$. Construct a complete amortization schedule.

Find the rate i per quarter-year such that

$$
\begin{aligned}
(1 + i)^4 &= (1.02)^{12} \\
i &= (1.02)^3 - 1 = 0.061208
\end{aligned}
$$

The quarterly payment, R, is then

$$
R = \frac{2000}{a_{\overline{8}|.061208}} = \$323.6134121 \approx \$323.62
$$

Now we can construct the complete amortization schedule, Table 7-2.

Table 7-2

Payment Number	Periodic Payment	Interest at 6.1208%	Principal Repaid	Outstanding Principal
				2000.00
1	323.62	122.42	201.20	1798.80
2	323.62	110.10	213.52	1585.28
3	323.62	97.03	226.59	1358.69
4	323.62	83.16	240.46	1118.23
5	323.62	68.44	255.18	863.05
6	323.62	52.83	270.79	592.26
7	323.62	36.25	287.37	304.89
8	323.55	18.66	304.89	0
TOTALS	2588.89	588.89	2000.00	

In Tables 7-1 and 7-2, it should be noted that the total amount of principal repaid equals the original debt. The total of all periodic payments equals the total interest plus the total principal repaid. Finally, the entries in the Principal Repaid column (except for the final smaller payment) are in the ratio $1 + i$. For example, in Table 7-2,

$$\frac{213.52}{201.20} \cong \frac{226.59}{213.52} \cong \cdots \cong 1.061208$$

7.6 A loan is being repaid in equal installments at the end of each year for 10 years. Interest is at $j_1 = 10\%$. If the amount of principal repaid in the fifth payment is \$200, find (a) the amount of principal repaid in the eighth payment, (b) the amount of the loan (assume no rounding of the payments).

(a) Since the entries in the Principal Repaid column of the amortization schedule are in the ratio $1 + i$, the principal repaid in the eighth payment is $200(1.10)^3 = \$266.20$.

(b) The sum of the 10 repayments of principal is

$$200(1.1)^{-4} + \cdots + 200 + \cdots + 200(1.1)^5 = 200(1.1)^{-4}s_{\overline{10}|.1} = \$2177.10$$

and this must equal the amount of the loan.

7.7 Mrs. Smith borrows \$15 000 to be repaid in equal monthly installments over 4 years at $j_{12} = 9\%$. Find the total amount of interest she will pay in the lifetime of the loan.

The monthly payment, R, is

$$R = \frac{15\ 000}{a_{\overline{48}|.0075}} = 373.2756356$$

total value of all payments $= 48 \times 373.2756356 = \$17\ 917.23$
total principal repaid $= \$15\ 000$
total interest paid $= 17\ 917.23 - 15\ 000 = \$2917.23$

7.8 Consider a loan of \$$A$ to be repaid with level payments of \$$R$ at the end of each period for n periods, at rate i per period. Show that in the kth line of the amortization schedule $(1 \le k \le n)$

(a) Outstanding principal $= R\ a_{\overline{n-k}|i}$

(b) Interest $= iR\ a_{\overline{n-k+1}|i} = R[1 - (1 + i)^{-(n-k+1)}] = R[1 - v^{n-k+1}]$

(c) Principal repaid $= R(1 + i)^{-(n-k+1)} = Rv^{n-k+1}$

(a) Outstanding principal after the $(k-1)$st payment is the discounted value of the remaining $n - (k - 1) = n - k + 1$ payments, that is, $R\ a_{\overline{n-k+1}|i}$.

(b) Interest paid in the kth payment is

$$iR\ a_{\overline{n-k+1}|i} = iR\frac{1 - (1 + i)^{-(n-k+1)}}{i} = R[1 - (1 + i)^{-(n-k+1)}] = R[1 - v^{n-k+1}]$$

(c) Principal paid in the kth payment is

$$R - R[1 - (1 + i)^{-(n-k+1)}] = R(1 + i)^{-(n-k+1)} = Rv^{n-k+1}$$

7.9 A loan is being repaid with 20 annual installments at $j_1 = 9\%$. In what installment are the principal and interest portions most nearly equal to each other?

From Problem 7.8, in the kth payment:

$$\text{Interest} = R[1 - (1 + i)^{-(n-k+1)}] = R[1 - (1.09)^{-(20-k+1)}]$$
$$\text{Principal} = R(1 + i)^{-(n-k+1)} = R(1.09)^{-(20-k+1)}$$

We want to find k such that

$$
\begin{aligned}
R[1 - (1.09)^{-(20-k+1)}] &= R(1.09)^{-(20-k+1)} \\
(1.09)^{-(20-k+1)} &= \tfrac{1}{2} \\
(1.09)^{20-k+1} &= 2 \\
(21 - k)\log(1.09) &= \log 2 \\
-k \log 1.09 &= \log 2 - 21 \log 1.09 \\
k &= \frac{21 \log 1.09 - \log 2}{\log 1.09} \\
k &= 12.95676827
\end{aligned}
$$

In the 13th payment, the principal and interest portions are most nearly equal.

7.2 OUTSTANDING PRINCIPAL

There are two methods that can be used to find the outstanding principal P on a debt A, being amortized by equal payments R over n periods at rate i per period (see Fig. 7-2).

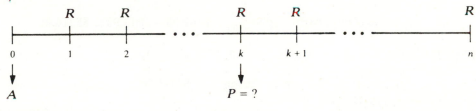

Fig. 7-2

Retrospective Method: Looking back in time, the outstanding principal P just after the kth payment is equal to the accumulated value of the debt less the accumulated value of the k payments made to date:

$$P = A(1 + i)^k - R\, s_{\overline{k}|i} \tag{7.1}$$

Prospective Method: Looking ahead, the outstanding principal P just after the kth payment is equal to the discounted value of the $n - k$ payments yet to be made. If all payments, including the last one, are equal,

$$P = R\, a_{\overline{n-k}|i} \tag{7.2}$$

While (7.1) and (7.2) are algebraically equivalent (see Problem 7.8), (7.2) can only be used if all remaining payments are equal. We have seen that most loans have a concluding irregular (usually smaller) payment. In such situations, the Prospective Method is still applicable provided (7.2) is suitably modified. (See Problem 7.11.)

The amortization method is quite often used to pay off loans incurred in purchasing a property. In such cases, the outstanding principal is called the *seller's equity*. The amount of principal that has been paid already, plus the down payment, is called the *buyer's equity* or the *owner's equity*. Clearly,

$$\text{buyer's equity} + \text{seller's equity} = \text{selling price}$$

(See Problem 7.14.) This formula does not account for any change in the value of the property.

SOLVED PROBLEMS

7.10 Show that (7.1) and (7.2) are equivalent if all payments are equal.

Substituting $A = R\, a_{\overline{n}|i}$ in (7.1),

$$
\begin{aligned}
P &= R\, a_{\overline{n}|i}(1+i)^k - R\, s_{\overline{k}|i} = R\left[\frac{1-(1+i)^{-n}}{i}\right](1+i)^k - R\left[\frac{(1+i)^k-1}{i}\right] \\
&= R\frac{(1+i)^k - (1+i)^{-(n-k)} - (1+i)^k + 1}{i} = R\frac{1-(1+i)^{-(n-k)}}{i} \\
&= R\, a_{\overline{n-k}|i}
\end{aligned}
$$

7.11 The amortization schedule in Problem 7.4 shows an outstanding balance of \$2314.00 after four payments (2 years). Confirm this balance, using (*a*) the retrospective, (*b*) the prospective, method.

(*a*) With $A = 6000$, $R = 1298$, $k = 4$, and $i = 0.08$, (7.1) gives

$$P = 6000(1.08)^4 - 1298 s_{\overline{4}|.08} = 8162.93 - 5843.93 = \$2314.00$$

(*b*) With $R = 1298$, $k = 4$, $n = 6$, and $i = 0.08$, (7.2) gives

$$P = 1298 a_{\overline{2}|.08} = \$2314.68$$

This is an incorrect answer. From Problem 7.3, the concluding payment is \$1297.21; hence, the discounted value of the remaining two payments is actually

$$P = 1298(1.08)^{-1} + 1297.21(1.08)^{-2} = 1201.85 + 1112.15 = \$2314.00$$

7.12 A loan of \$8000 is to be amortized with equal monthly payments over 2 years at $j_{12} = 15\%$. Find the outstanding principal after 7 months and split the eighth payment into principal and interest portions.

The monthly payment is

$$R = \frac{8000}{a_{\overline{24}|.0125}} = \$387.8931844 \approx \$387.90$$

The outstanding principal P after 7 payments is

$$P = 8000(1.0125)^7 - 387.90 s_{\overline{7}|.0125} = \$5907.53$$

The interest portion of the eighth payment is

$$(5907.53)(0.0125) = \$73.84$$

And the principal portion is $387.90 - 73.84 = \$314.06$.

7.13 Solve Problem 7.12 assuming $j_\infty = 16\%$.

Find the rate i per month such that

$$(1+i)^{12} = e^{0.16}$$
$$i = 0.013422619$$

The monthly payment R is then

$$R = \frac{8000}{a_{\overline{24}|.013422619}} = \$392.1145563 \approx \$392.12$$

$$P(\text{after 7 months}) = 8000(1.013422619)^7 - 392.12s_{\overline{7}|.013422619} = \$5924.75$$

The interest portion of the eighth payment is

$$(5924.75)(0.013422619) = \$79.53$$

and the principal portion is $392.12 - 79.53 = \$312.59$.

7.14 Harry buys a cottage worth \$42 000 by paying \$7000 down and the balance with interest at $j_{12} = 9\%$ in monthly installments of \$600 for as long as necessary. Find Harry's equity at the end of 5 years.

Using the retrospective method, we calculate the seller's equity at the end of 5 years:

$$P = 35\,000(1.0075)^{60} - 600s_{\overline{60}|.0075}$$
$$= 54\,798.84 - 45\,254.48 = \$9544.36$$

Then, buyer's equity $= 42\,000 - 9544.36 = \$32\,455.64$

7.15 The Andersons borrow \$15 000 to buy a car. The loan will be repaid over three years with monthly payments at $j_{12} = 6\%$. Find the total interest paid in the 12 payments of the second year.

The monthly payment is

$$R = \frac{15\,000}{a_{\overline{36}|.005}} = \$456.3290618 \approx \$456.33$$

Using (7.1), the outstanding principal after 1 year is

$$P = 15\,000(1.005)^{12} - 456.33s_{\overline{12}|.005} = 15\,925.17 - 5629.09 = \$10\,296.08$$

and the outstanding principal after 2 years is

$$P = 15\,000(1.005)^{24} - 456.33s_{\overline{24}|.005} = 16\,907.40 - 11\,605.36 = \$5302.04$$

The total principal repaid in the second year is therefore

$$10\,296.08 - 5302.04 = \$4994.04$$

The balance of the 12 payments made in the second year represents interest. Hence,

$$\text{interest paid} = (12)(456.33) - 4994.04 = \$481.92$$

7.3 MORTGAGES

Buying a house is likely to be the most expensive purchase an individual or couple ever makes. Usually, a large loan, or *mortgage*, is taken out when one buys a home, which is ordinarily repaid in monthly installments over a long period (25 to 30 years). The lender holds the legal right to repossess the property upon failure to repay.

Prevailing practice in the U.S. is to have monthly payments on mortgage loans, with interest compounded monthly. Because of today's unpredictable rates of interest, traditional fixed-rate mortgages have been replaced by a variety of creative financing arrangements—it is important to understand the costs of these complex contracts.

Canadian mortgage regulations require that the interest can be compounded, at most, semiannually, whereas mortgage payments are usually made monthly. Thus mortgage amortizations in Canada are, in effect, general annuities. (See Problem 7.21.)

SOLVED PROBLEMS

7.16 A couple buys a condominium on May 1, 1994, for \$65 000. They make a 20% down payment and get a 29-year mortgage loan at $j_{12} = 10\%$ for the balance; the loan is to be amortized by equal monthly payments rounded up to the nearest dime. If they make the first payment on June 1, 1994, how much interest can they deduct when they prepare their income tax return for 1994? Show the first 3 lines and last 3 lines of the amortization schedule.

As the down payment is 20% of \$65 000, or \$13 000, the couple borrows \$52 000. Thus, $A = 52\ 000$, $n = 29 \times 12 = 348$, $i = \frac{5}{6}\%$; and we calculate the monthly payment as

$$R = \frac{52\ 000}{a_{\overline{348}|5/600}} = \$458.90 \text{ (rounded up to the nearest dime)}$$

Figure 7-3 shows the payments made during 1994. Outstanding principal on December 1, 1994, is

$$P_7 = 52\ 000 \left(\frac{605}{600}\right)^7 - 458.90 s_{\overline{7}|5/600} = 55\ 110.23 - 3293.73 = \$51\ 816.50$$

so that total principal repaid during 1994 is $52\ 000 - 51\ 816.50 = \$183.50$. Consequently, total interest paid during 1994 is

$$(7)(458.90) - 183.50 = \$3028.80$$

They can list \$3028.80 on their income tax return as a mortgage interest deduction.

	458.90	458.90	458.90	458.90	458.90	458.90	458.90
0	1	2	3	4	5	6	7
May 1	June 1	July 1	Aug. 1	Sept. 1	Oct. 1	Nov. 1	Dec. 1

$A = 52\ 000$

Fig. 7-3

The beginning and the end of the amortization schedule are given in Table 7-3. To compute the last 3 lines, we first find the outstanding principal after the 345th payment:

$$P_{345} = 52\,000 \left(\frac{605}{600}\right)^{345} - 458.90 s_{\overline{345}|5/600} = 910\,806.68 - 909\,476.27 = \$1330.41$$

The last monthly payment is smaller, owing to the rounding of the regular monthly payment. Table 7-3 shows that at the beginning the payments go mainly toward interest due, and at the end mainly toward principal.

Table 7-3

Payment Number	Monthly Payment	Interest at $\frac{5}{6}\%$	Principal Repaid	Outstanding Principal
				52 000.00
1	458.90	433.33	25.57	51 974.43
2	458.90	433.12	25.78	51 948.65
3	458.90	432.91	25.99	51 922.66
....	....	....	....	....
345				1 330.41
346	458.90	11.09	447.81	882.60
347	458.90	7.36	451.54	431.06
348	434.65	3.59	431.06	0

7.17 In an effort to advertise low rates of interest, but still achieve high rates of return, lenders sometimes charge *points*. Each point is a 1% discount from the face value of the loan. Suppose that a cottage is being sold for \$55 000 and that the buyer pays \$15 000 down and gets a \$40 000, 15-year mortgage at $j_{12} = 9\%$ from a lender who charges 5 points. What is the true interest rate on the loan?

The lender will advance 95% of \$40 000, or \$38 000. However, the monthly payment is calculated on the basis of a \$40 000 loan:

$$R = \frac{40\,000}{a_{\overline{180}|.0075}} = \$405.71$$

Thus, the true monthly interest rate i is the solution of the equation

$$405.71 a_{\overline{180}|i} = 38\,000 \quad \text{or} \quad a_{\overline{180}|i} = 93.6630$$

Applying the method of interpolation (Section 5.6), we have:

		$a_{\overline{180}|i}$	j_{12}	
		98.5934	9%	
5.5360	4.9304	93.6630	j_{12}	x ... 1%
		93.0574	10%	

$$\frac{x}{1\%} = \frac{4.9304}{5.5360}$$

$$x = 0.89\%$$

$$j_{12} = 9.89\%$$

The true interest rate on the mortgage loan is $j_{12} = 9.89\%$.

Points are charged at the beginning of the mortgage loan. Borrowers who decide to pay off their mortgage loan ahead of time will not recover any part of this charge. As a result, they will actually pay an even higher interest rate, as illustrated in Problem 7.18.

7.18 The borrower in Problem 7.17 sells his cottage at the end of 5 years and pays off the mortgage. What rate did the borrower pay?

The outstanding principal at the end of 5 years, *just before* the 60th payment, is

$$P = 40\,000(1.0075)^{60} - 405.71s_{\overline{59}|.0075}(1.0075)$$
$$= 62\,627.24 - 30\,194.62 = \$32\,432.62$$

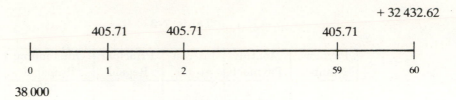

Fig. 7-4

We arrange our data as shown in Fig. 7-4. We want to find the rate j_{12} which makes the discounted value, A', of the payments, including the last payment of \$32 432.62, equal to \$38 000. Choosing the trial values $j_{12} = 10\%$ and $j_{12} = 11\%$, we calculate:

$$A'_{10\%} = 405.71a_{\overline{59}|5/600} + 32\,432.62\left(\frac{605}{600}\right)^{-60} = 18\,848.31 + 19\,712.18 = \$38\,560.49$$

$$A'_{11\%} = 405.71a_{\overline{59}|11/1200} + 32\,432.62\left(\frac{1211}{1200}\right)^{-60} = 18\,425.17 + 18\,758.94 = \$37\,184.11$$

Then, using linear interpolation, we get

$$1376.38\left\{560.49\left\{\begin{array}{c|c} A' & j_{12} \\ \hline 38\ 560.49 & 10\% \\ 38\ 000.00 & j_{12} \\ 37\ 184.11 & 11\% \end{array}\right\}x\right\}1\%$$

$$\frac{x}{1\%} = \frac{560.49}{1376.38}$$

$$x = 0.41\%$$

$$j_{12} = 10.41\%$$

The true rate is $j_{12} = 10.41\%$ if the borrower pays off the mortgage at the end of 5 years.

7.19 With mortgage rates at $j_{12} = 12\%$, the ABC Savings and Loans Company makes a special offer to its customers. It will lend mortgage money and determine the monthly payment for the next 5 years as if $j_{12} = 9\%$. Over the 5-year period, the mortgage will be carried at $j_{12} = 12\%$ and any deficiency that results will be added to the outstanding balance to be refinanced in 5 years' time. If the Browns are taking out a \$90 000 mortgage to be repaid over 25 years under this scheme, find their outstanding balance at the end of 5 years.

The monthly payment R, assuming $j_{12} = 9\%$, is

$$R = \frac{90\,000}{a_{\overline{300}|.0075}} = \$755.2767273 \approx \$755.28$$

That is, they owe $11 819.23 more than they originally borrowed. This balance will be refinanced at the new prevailing rate of interest.

7.20 XYZ Savings and Loans issues mortgages where payments are determined by the rate of interest that prevails on the day the loan is made. After that, the rate of interest varies according to market forces but the monthly payments do not change in dollar size. Instead, the length of time to full repayment is either lengthened (if interest rates rise) or shortened (if interest rates fall). Mr. Adams takes out a 20-year, $70 000 mortgage at $j_{12} = 9\%$. After exactly 2 years (24 payments) interest rates change. Find the duration of the loan and the final smaller payment if the new interest rate stays fixed at (a) $j_{12} = 10\%$, (b) $j_{12} = 8\%$.

The monthly amount R to repay a $70 000 loan over 20 years at $j_{12} = 9\%$ ($i = 0.0075$) is

$$R = \frac{70\ 000}{a_{\overline{240}|.0075}} = \$629.8081691 \approx \$629.81$$

The outstanding principal after 2 years is, by (7.1),

$$
\begin{aligned}
P &= 70\ 000(1.0075)^{24} - 629.81 s_{\overline{24}|.0075} \\
&= 83\ 748.95 - 16\ 493.76 = \$67\ 255.19
\end{aligned}
$$

Payments remains at $629.81, but the duration of the loan will change.

(a) If $j_{12} = 10\%$ ($i = .1/12$), we find n such that

$$
\begin{aligned}
67\ 255.19 &= 629.81 a_{\overline{n}|i} \\
a_{\overline{n}|i} &= 106.7864753 \\
\frac{1 - (1+i)^{-n}}{i} &= 106.7864753 \\
(1+i)^{-n} &= 0.110112706 \\
n &= -\frac{\log 0.110112706}{\log(1+i)} = 265.8517003
\end{aligned}
$$

Thus there are 265 more payments of $629.81, plus a final smaller payment of

$$
\begin{aligned}
X &= 67\ 255.19(1+i)^{266} - 629.81 s_{\overline{265}|i}(1+i) \\
&= 611\ 537.17 - 611\ 000.43 = \$536.74
\end{aligned}
$$

The total duration of the loan is $24 + 265 + 1 = 290$ months, or 24 years 2 months.

(b) If $j_{12} = 8\%$ ($i = .08/12$), we find n such that

$$
\begin{aligned}
67\ 255.19 &= 629.81 a_{\overline{n}|i} \\
a_{\overline{n}|i} &= 106.7864753 \\
(1+i)^{-n} &= 0.288090165 \\
n &= -\frac{\log 0.288090165}{\log(1+i)} = 187.2938181
\end{aligned}
$$

Thus there are 187 more payments of $629.81, plus a final smaller payment of

$$
\begin{aligned}
X &= 67\ 255.19(1+i)^{188} - 629.81 s_{\overline{187}|i}(1+i) \\
&= 234\ 549.87 - 234\ 364.38 = \$185.49
\end{aligned}
$$

The total duration of this loan is $24 + 187 + 1 = 212$ months, or 17 years 8 months.

7.21 An advertisement by the Royal Trust in Canada said:

"Our new Double-Up mortgage can be paid off faster and that dramatically reduces the interest you pay. You can double your payment any or every month, with no penalty. Then your payment reverts automatically to its normal amount the next month.

What this means to you is simple. You will pay your mortgage off sooner. And that's good news because you can save thousands of dollars in interest as a result."

Consider a $120 000 mortgage at 8% per annum compounded semiannually, amortized over 25 years with no anniversary prepayments.

(*a*) Find the required monthly payment for the above mortgage.

(*b*) What is the total amount of interest paid over the full amortization period (assuming $j_2 = 8\%$)?

(*c*) Suppose that payments were "Doubled-Up," according to the advertisement, every sixth and twelfth month.

   (i) How many years and months would be required to pay off the mortgage?

  (ii) What would be the total amount of interest paid over the full amortization period?

 (iii) How much of the loan would still be outstanding at the end of 3 years, just after the Doubled-Up payment then due?

 (iv) How much principal is repaid in the payment due in 37 months?

(*a*) First we find the rate i per month such that

$$(1+i)^{12} = (1.04)^2$$
$$i = (1.04)^{1/6} - 1 = 0.006558197$$

The monthly payment R is

$$R = \frac{120\ 000}{a_{\overline{300}|i}} = 915.8561463 \approx \$915.86$$

(*b*) The concluding payment $= 120\ 000(1+i)^{300} - 915.86 s_{\overline{299}|i}(1+i)$
$$= 852\ 802.00 - 851\ 889.73 = \$912.27$$

The total amount of interest $= (299 \times 915.86 + 912.27) - 120\ 000$
$$= \$154\ 754.41$$

(*c*) (i) We calculate an equivalent semiannual payment

$$915.86 s_{\overline{6}|i} + 915.86 = \$6501.91$$

We find n such that

$$6501.91 a_{\overline{n}|.04} = 120\ 000$$
$$\frac{1 - (1.04)^{-n}}{.04} = \frac{120\ 000}{6501.91}$$
$$(1.04)^{-n} = 0.261755392$$
$$n = -\frac{\log 0.261755392}{\log 1.04} = 34.17441251$$

It would require 34.17441251 half-years, that is, 17 years and 2 months to pay off the mortgage.

(ii) The concluding payment at the end of 17 years and 2 months (206th payment) is

$$120\,000(1+i)^{206} - 6501.91 s_{\overline{34}|.04}(1+i)^2 - 915.86(1+i)$$
$$= 461\,309.67 - 460\,186.96 - 921.87 = \$200.84$$

The total amount of interest $= [(205+34) \times 915.86 + 200.84] - 120\,000$
$= \$99\,091.38$

(iii) The outstanding principal at the end of 3 years is

$$120\,000(1.04)^6 - 6501.91 s_{\overline{6}|.04} = \$108\,711.27$$

(iv) The interest paid in the 37th payment $= 108\,711.27i = \$712.95$
The principal paid in the 37th payment $= 915.86 - 712.95 = \$202.91$

7.4 REFINANCING A LOAN

Amortization Method

Often borrowers will want to renegotiate long-term loans, especially if interest rates have dropped. Most contracts stipulate penalties at the time of any early renegotiation.

If refinancing a loan at a lower rate is being contemplated, the savings due to the lower interest charges must be compared with the cost of refinancing to decide whether the refinancing would be profitable.

Sum-of-Digits Method

This method, which has a built-in prepayment penalty, was initially used as a close approximation to the amortization method. (See Problem 7.27.) At high rates of interest, however, the outstanding principal as determined by the sum-of-digits method leads to a significant penalty against the borrower. (See Problem 7.31.)

SOLVED PROBLEMS

7.22 A borrower has an $8000 loan with the Easy-Credit Finance Company. The loan is to be repaid over 4 years at $j_{12} = 18\%$. The contract stipulates a penalty in case of early repayment, equal to 3 months' payments. Just after the 20th payment, the borrower determines that his local bank would lend him the money at $j_{12} = 13.5\%$. Should he refinance?

The original monthly payment, R_1, at $j_{12} = 18\%$ is

$$R_1 = \frac{8000}{a_{\overline{48}|.015}} = \$234.9999969 \approx \$235$$

The outstanding balance P after 20 payments is

$$P = 8000(1.015)^{20} - 235 s_{\overline{20}|.015} = 10\,774.84 - 5434.06 = \$5340.78$$

Adding the penalty, we obtain as the total sum to be refinanced $5340.78 + 3(235) = \$6045.78$

The new monthly payment, R_2, at $j_{12} = 13\frac{1}{2}\%$ would be

$$R_2 = \frac{6045.78}{a_{\overline{28}|.01125}} = 252.9130458 \approx \$252.92$$

Since R_2 exceeds the original monthly payment of $235, the borrower should not refinance.

7.23 A couple purchased a home and signed a mortgage contract for \$105 000 to be paid in monthly installments over 25 years at $j_{12} = 10\frac{1}{2}\%$. The contract stipulates that after 5 years the mortgage will be renegotiated at the new prevailing rate of interest. Find (*a*) the monthly payment for the initial 5-year period, (*b*) the outstanding principal after 5 years, (*c*) the new monthly payment after 5 years at $j_{12} = 9\%$.

 (*a*) With $A = 105\,000$, $n = 300$, and $j_{12} = 10\frac{1}{2}\%$ ($i = 0.00875$), we calculate the monthly payment for the initial 5-year period:

$$R_1 = \frac{105\,000}{a_{\overline{300}|.00875}} = \$991.3907904 \approx \$991.40$$

 (*b*)
$$\begin{aligned} P &= 105\,000(1.00875)^{60} - 991.40 s_{\overline{60}|.00875} \\ &= 177\,093.31 - 77\,794.08 = \$99\,299.23 \end{aligned}$$

Note that in 5 years, with payments totaling $(60)(991.40) = \$59\,484$, only

$$105\,000 - 99\,299.23 = \$5700.77$$

of principal has been repaid.

 (*c*) Using $A = 99\,299.23$, $n = 240$, $j_{12} = 9\%$ ($i = 0.0075$), we calculate the new monthly payment after 5 years:

$$R_2 = \frac{99\,299.23}{a_{\overline{240}|.0075}} = \$893.4209463 \approx \$893.43$$

Hence, their monthly payment will fall by \$97.97.

7.24 Solve Problem 7.23(*c*) assuming $j_{12} = 12\%$.

 Now $i = 0.01$, and

$$R_2 = \frac{99\,299.23}{a_{\overline{240}|.01}} = \$1093.370052 \approx \$1093.38$$

Hence, their monthly payment would rise by \$101.98.

7.25 Elizabeth is repaying a debt of \$5000 with monthly payments over 3 years at $j_{12} = 16\frac{1}{2}\%$. At the end of the first year she makes an extra single payment of \$500. She then shortens the repayment period by 1 year and renegotiates the loan without an interest rate change. Find the new monthly payment and the amount of interest she saves by the refinancing.

 We have $A = 5000$, $n = 36$, $j_{12} = 16\frac{1}{2}\%$ ($i = 0.01375$); and so the original monthly payment is

$$R_1 = \frac{5000}{a_{\overline{36}|.01375}} = \$177.0219127 \approx \$177.03$$

and the total interest in the original loan is $(36)(177.0219127) - 5000 = \$1372.79$.

 The outstanding principal after 1 year is

$$5000(1.01375)^{12} - 177.03 s_{\overline{12}|.01375} = 5890.34 - 2292.61 = \$3597.73$$

After the extra payment, we have $A = 3597.73 - 500 = \$3097.73$, $n = 12$, $j_{12} = 16\frac{1}{2}\%$ ($i = 0.01375$); the new monthly payment is

$$R_2 = \frac{3097.73}{a_{\overline{12}|.01375}} = \$281.7931766 \approx \$281.80$$

and the total interest in the revised loan is

$$(12)(177.03) + 500 + (12)(281.7931766) - 5000 = \$1005.88$$

Therefore, her savings in interest is $1372.79 - 1005.88 = \$366.91$.

7.26 A consumer borrows \$10 000 to be repaid in monthly installments over 1 year at $j_{12} = 12\%$. Construct a complete amortization schedule.

Having calculated

$$R = \frac{10\ 000}{a_{\overline{12}|.01}} = \$888.4878868 \approx \$888.49$$

we construct Table 7-4.

Table 7-4

Payment Number	Periodic Payment	Interest at 1%	Principal Repaid	Outstanding Principal
				\$10 000.00
1	888.49	100.00	788.49	9 211.51
2	888.49	92.12	796.37	8 415.14
3	888.49	84.15	804.34	7 610.80
4	888.49	76.11	812.38	6 798.42
5	888.49	67.98	820.51	5 977.91
6	888.49	59.78	828.71	5 149.20
7	888.49	51.49	837.00	4 312.20
8	888.49	43.12	845.37	3 466.83
9	888.49	34.67	853.82	2 613.01
10	888.49	26.13	862.36	1 750.65
11	888.49	17.51	870.98	879.67
12	888.47	8.80	879.67	0
TOTALS	10 661.86	661.86	10 000.00	

7.27 For Problem 7.26, construct a complete repayment schedule using the sum-of-digits method.

By (2.2), the sum of digits from 1 to 12 is

$$\frac{12}{2}(1 + 12) = 78$$

Under the sum-of-digits method, the total interest in the loan, \$661.86, is allocated 12/78 to the first payment, 11/78 to the second payment, ..., 1/78 to the twelfth payment. This gives the schedule of Table 7-5.

Table 7-5

Payment Number	Periodic Payment	Interest Allocated	Principal Repaid	Outstanding Principal
				$10 000.00
1	888.49	101.82	786.67	9 213.33
2	888.49	93.34	795.15	8 418.18
3	888.49	84.85	803.64	7 614.54
4	888.49	76.37	812.12	6 802.42
5	888.49	67.88	820.61	5 981.81
6	888.49	59.40	829.09	5 152.72
7	888.49	50.91	837.58	4 315.14
8	888.49	42.43	846.06	3 469.08
9	888.49	33.94	854.55	2 614.53
10	888.49	25.46	863.03	1 751.50
11	888.49	16.97	871.52	879.98
12	888.47	8.49	879.98	0
TOTALS	10 661.86	661.86	10 000.00	

A comparison of Tables 7-4 and 7-5 leads to the following conclusions:

(i) Each outstanding principal balance under the sum-of-digits method exceeds the true balance under the amortization method. Hence, if the loan is refinanced before the end of its term, there is a built-in penalty for the borrower.

(ii) If the loan is paid off full-term, the total amount of interest is the same as under the amortization method, and no penalty is incurred.

7.28 For Problem 7.27, calculate the outstanding principal after five payments, without constructing the sum-of-digits schedule.

A banking institution would normally use the following format:

$$\text{Total Original Debt} = 12 \times 888.4878868 \qquad\qquad = \quad \$10\ 661.85$$

$$\text{Less: Interest Not Yet Due} \quad = \left(\frac{1+2+3+\cdots+7}{78}\right)(661.85)$$

$$= \left(\tfrac{28}{78}\right)(661.85) \qquad\qquad\qquad = \qquad\quad 237.59$$

$$\text{Less: Payments Already Made} = 5 \times 888.49 \qquad\qquad = \qquad 4\ 442.45$$

$$\text{Outstanding Principal} \qquad\qquad\qquad\qquad\qquad\quad = \qquad \overline{\$5\ 981.81}$$

7.29 A $15 000 loan is to be repaid over 10 years at $j_{12} = 18\%$. Construct the first three lines of the repayment schedule, using (a) the amortization method, (b) the sum-of-digits method.

$$R = \frac{15\ 000}{a_{\overline{120}|.015}} = \$270.2777986 \approx \$270.28$$

(a) The beginning of the amortization schedule is given in Table 7-6.

Table 7-6

Payment Number	Periodic Payment	Interest at $1\frac{1}{2}\%$	Principal Repaid	Outstanding Principal
				$15 000.00
1	270.28	225.00	45.28	14 954.72
2	270.28	224.32	45.96	14 908.76
3	270.28	223.63	46.65	14 862.11

(b) The total interest is $(120)(270.2777986) - 15\,000 = \$17\,433.34$, and the sum of digits from 1 to 120 is

$$\frac{120}{2}(1 + 120) = 7260$$

Thus, the interest portion of the first payment is

$$\left(\frac{120}{7260}\right)(17\,433.34) = \$288.15$$

and so on. This first apportionment of interest exceeds the size of the first payment, leading to a repayment schedule that begins as in Table 7-7.

Table 7-7

Payment Number	Periodic Payment	Interest Allocated	Principal Repaid	Outstanding Principal
				$15\,000.00
1	270.28	288.15	−17.87	15\,017.87
2	270.28	285.75	−15.47	15\,033.34
3	270.28	283.35	−13.07	15\,046.41

7.30 For the loan of Problem 7.29, find the outstanding principal after 2 years (24 payments), using (a) the amortization method, (b) the sum-of-digits method.

(a) $P = 15\,000(1.015)^{24} - 270.28 s_{\overline{24}|.015} = 21\,442.54 - 7739.07 = \$13\,703.47$

(b)

Total Debt = 120×270.2777986	=	\$32\,433.34
Less: Interest Not Yet Due = $\left(\frac{4656}{7260}\right)(17\,433.34)$	=	11\,180.39
Less: Payments to Date = 24×270.28	=	6\,486.72
Outstanding Principal	=	\$14\,766.23

7.31 For the loan in Problem 7.29, the borrower determines after 2 years of payments that he can renegotiate the loan at $j_{12} = 15\%$ over the remaining 8 years. Should he renegotiate the loan, if the lending institution uses the sum-of-digits method to determine the outstanding principal?

If he renegotiates, he will have to pay off the outstanding principal of \$14\,766.23 [Problem 7.30 (b)] over 8 years, with new monthly payments R at $j_{12} = 15\%$.

$$R = \frac{14\,766.23}{a_{\overline{96}|.0125}} = \$264.9859823 \approx \$264.99$$

Since the new monthly payment is less than \$270.28, he should renegotiate.

7.32 In Problem 7.31, what would the new monthly payment have been under the amortization method at $j_{12} = 15\%$?

Using the outstanding principal calculated in Problem 7.30(a), the new monthly payment is

$$R = \frac{13\,703.47}{a_{\overline{96}|.0125}} = \$245.9143233 \approx \$245.92$$

7.5   SINKING FUNDS

A specified sum of money can be accumulated by a specified future date through periodic deposits into a *sinking fund*. A schedule showing how a sinking fund accumulates to the desired amount is called a *sinking-fund schedule*.

The Sinking-Fund Method of Retiring a Debt

A common method of paying off a long-term loan is for the borrower to pay the interest on the loan as it falls due and to create a sinking fund to accumulate the principal at the end of the term of the loan. The sum of the interest payment and the sinking-fund deposit is called the *periodic expense* or *periodic cost of the debt*. The *book value* of the debt at any time is the original principal less the amount in the sinking fund at that time.

SOLVED PROBLEMS

7.33 A company wants to save $100 000 over the next 5 years so that they can expand their plant facility. How much must be deposited at the end of each year if their money earns interest at $j_1 = 6\%$? Construct a complete sinking-fund schedule.

The sinking-fund deposits form an ordinary simple annuity, with $S = 100\,000$, $i = 0.06$, and $n = 5$. From (5.1),

$$R = \frac{100\,000}{s_{\overline{5}|.06}} = \$17\,739.64$$

which leads to the sinking-fund schedule of Table 7-8. There is a 1¢ roundoff error in the final amount.

Table 7-8

Deposit Number	Deposit	Interest on Fund at 6%	Increase in Fund	Amount in Fund at End of Period
1	17 739.64	0	17 739.64	17 739.64
2	17 739.64	1064.38	18 804.02	36 543.66
3	17 739.64	2192.62	19 932.26	56 475.92
4	17 739.64	3388.56	21 128.20	77 604.12
5	17 739.64	4656.25	22 395.89	100 000.01

7.34 The Smiths want to save $12 000 for a down payment on a house. If they save $500 a month in an account paying interest at $j_{12} = 4.5\%$, how many deposits will be required and what will be the size of the final smaller deposit?

Find n such that $500s_{\overline{n}|.00375} = 12\,000$ or $s_{\overline{n}|.00375} = 24$, whence

$$\frac{(1.00375)^n - 1}{.00375} = 24$$
$$(1.00375)^n = 1.09$$
$$n = \frac{\ln 1.09}{\ln 1.00375} = 23.02378096$$

There will be 23 deposits of $500, plus a final deposit of

$$X = 12\,000 - 500s_{\overline{23}|.00375}(1.00375) = 12\,000 - 12\,032.02 = -\$32.02$$

The negative sign on X tells us (as in the second part of Problem 5.35) that it takes 24 months to save the $12\,000$, but that no 24th deposit is required.

7.35 A city needs to have $\$200\,000$ at the end of 15 years to retire a bond issue. What annual deposits are necessary if their money earns interest at $j_\infty = 12\frac{1}{2}\%$?

Find the rate i per year such that

$$1 + i = e^{0.125} \qquad \text{or} \qquad i = 0.133148453$$

With $S = 200\,000$, $n = 15$, $i = 0.133148453$, (5.1) gives

$$R = \frac{200\,000}{s_{\overline{15}|.133148453}} = \$4823.50$$

7.36 Construct the first three and last three lines of the sinking-fund schedule for Problem 7.35.

In order to complete the last three lines of the schedule, without doing the full schedule, we determine the amount in the sinking fund at the end of 12 years:

$$4823.50 s_{\overline{12}|i} = \$126\,129.35$$

and complete the schedule from that point. See Table 7-9. There is a $13\cent$ roundoff error in the final amount.

Table 7-9

Deposit Number	Deposit	Interest on Fund at i	Increase in Fund	Amount in Fund at End of Period
1	4823.50	0	4 823.50	4 823.50
2	4823.50	642.24	5 465.74	10 289.24
3	4823.50	1 370.00	6 193.50	16 482.74
....	....	....	....	....
12				126 129.35
13	4823.50	16 793.93	21 617.43	147 746.78
14	4823.50	19 672.26	24 495.76	172 242.54
15	4823.50	22 933.83	27 757.33	199 999.87

7.37 A city borrows $\$500\,000$ and agrees to pay interest semiannually at $j_2 = 10\%$. (*a*) What semiannual deposits must be made into a sinking fund earning interest at $j_2 = 6\%$ in order to repay the debt in 15 years? (*b*) What is the total semiannual expense of the debt?

(*a*) To find the semiannual sinking-fund deposit R, apply (5.1), with $S = 500\,000$, $i = 0.03$, $n = 30$:

$$R = \frac{500\,000}{s_{\overline{30}|.03}} = \$10\,509.63$$

(*b*) Interest on the debt is $500\,000(0.05) = \$25\,000$ semiannually. The total semiannual expense is then

$$25\,000 + 10\,509.63 = \$35\,509.63$$

7.38 For Problem 7.37, find the book value of the city's indebtedness after 10 years.

The amount in the sinking fund after 10 years is

$$10\,509.63 s_{\overline{20}|.03} = \$282\,397.69$$

and so the book value is $500\,000 - 282\,397.69 = \$217\,602.31$.

**7.6 COMPARISON OF AMORTIZATION AND
         SINKING-FUND METHODS**

One can compare the amortization and sinking-fund methods of repaying a debt by comparing the periodic expenses under the two methods.

Let i be the interest rate per period on the debt for both the sinking-fund and the amortization method, and let r be the interest rate for the same period on the sinking fund. Let the term of the loan be n interest conversion periods and let A be the principal on the loan.

The periodic expense, E_1, under the amortization method is, using Problem 5.75(a),

$$E_1 = \frac{A}{a_{\overline{n}|i}} = Ai + \frac{A}{s_{\overline{n}|i}}$$

The periodic expense, E_2, under the sinking-fund method is

$$E_2 = Ai + \frac{A}{s_{\overline{n}|r}}$$

Thus, if $i > r$, then $s_{\overline{n}|i} > s_{\overline{n}|r}$ and $E_1 < E_2$ (the amortization method is preferable); if $i = r$, then $E_1 = E_2$; and if $i < r$, then $E_1 > E_2$ (the sinking-fund method is preferable).

When the interest rate on the debt under the amortization method is different from the interest rate on the debt under the sinking-fund method, the above conclusions do not apply. Instead we must calculate and directly compare E_1 and E_2.

SOLVED PROBLEMS

7.39 A company can borrow \$200 000 for 15 years. They can amortize the debt at $j_1 = 11\%$ or they can pay interest on the loan at $j_1 = 10.5\%$ and set up a sinking fund at $j_1 = 7\frac{1}{2}\%$ to repay the principal in 15 years. Which plan is cheaper, and by how much per annum?

Under the amortization method:

$$E_1 = \frac{200\ 000}{a_{\overline{15}|.11}} = \$27\ 813.05$$

Under the sinking-fund method, the annual interest expense is $200\ 000(0.105) = \$21\ 000$ and the annual sinking-fund deposit required is

$$\frac{200\ 000}{s_{\overline{15}|.075}} = \$7657.45$$

for a total annual expense $E_2 = \$28\ 657.45$. Thus the amortization method is better by \$844.40 a year.

7.40 A company can borrow \$100 000 for 10 years by paying interest as it falls due at $j_2 = 14\%$ and by setting up a sinking fund at $j_{12} = 12\%$ that would require semiannual deposits. At what rate j_2 would an amortization method have the same semiannual expense?

Under the sinking-fund method, the semiannual interest payment is $100\ 000(0.07) = \$7000$. To find the semiannual sinking-fund deposit, first find the rate i per half-year such that

$$(1+i)^2 = (1.01)^{12}$$
$$i = (1.01)^6 - 1 = 0.061520151$$

Hence, the semiannual sinking fund deposit is

$$\frac{100\ 000}{s_{\overline{20}|.061520151}} = \$2674.34$$

for a total semiannual expense of $9674.34.

Now we find the rate of interest $j_2 = 2i$ such that

$$100\ 000 - 9674.34a_{\overline{20}|i} \qquad \text{or} \qquad a_{\overline{20}|i} = 10.3366$$

Using linear interpolation to find the rate j_2 (see Section 5.6), we calculate:

$a_{\overline{20}|i}$	j_2
10.5940	14%
10.3366	j_2
10.1945	15%

$$0.3995 \left\{ 0.2574 \left\{ \right. \right.$$

$$\left. \left. \right\} x \right\} 1\%$$

$$\frac{x}{1\%} = \frac{0.2574}{0.3995}$$

$$x = 0.64\%$$

$$j_2 = 14.64\%$$

7.41 A firm can borrow \$350 000 at $j_{12} = 12\%$ and amortize the debt with annual payments over 10 years. From a second source, the money can be borrowed at $j_1 = 11\frac{3}{4}\%$ if the interest is paid annually and annual deposits are made into a sinking fund to repay the \$350 000 in 10 years. What rate, j_4, must the sinking fund earn for the annual expense to be the same under the two options?

To find the annual expense E_1 of the amortization method, we find the rate i per year such that

$$1 + i = (1.01)^{12} \qquad \text{or} \qquad i = 0.12682503$$

Then

$$E_1 = \frac{350\ 000}{a_{\overline{10}|.12682503}} = \$63\ 684.98$$

Under the sinking-fund method, the annual interest payment is $350\ 000(0.1175) = \$41\ 125$, leaving

$$63\ 684.98 - 41\ 125.00 = \$22\ 559.98$$

for the annual sinking-fund deposit. Now we want to find the sinking-fund rate i per half-year such that

$$22\ 559.98 s_{\overline{10}|i} = 350\ 000$$
$$s_{\overline{10}|i} = 15.5142$$

Solving by linear interpolation,

$s_{\overline{10}|i}$	i
15.1929	9%
15.5142	i
15.9374	10%

$$0.7445 \left\{ 0.3213 \left\{ \right. \right.$$

$$\left. \left. \right\} x \right\} 1\%$$

$$\frac{x}{1\%} = \frac{0.3213}{0.7445}$$

$$x = 0.43\%$$

$$i = 9.43\%$$

Finally, we determine j_4 such that

$$\left(1 + \frac{j_4}{4}\right)^4 = 1.0943$$

$$j_4 = 4[(1.0943)^{1/4} - 1] = 9.11\%$$

Supplementary Problems

AMORTIZATION OF A DEBT

7.42 A loan of $20 000 is to be amortized with equal monthly payments over a 3-year period at $j_{12} = 8\%$. Find the concluding payment, if the monthly payment is rounded up to (a) the cent, (b) the dollar. *Ans.* (a) $626.62; (b) $623.85

7.43 A $4000 loan is to be amortized with eight equal quarterly payments over 2 years. If interest is at $j_4 = 10\%$, (a) find the quarterly payment, and (b) construct an amortization schedule. *Ans.* (a) $557.87

7.44 A debt of $1000, bearing interest at $j_{12} = 13\frac{1}{2}\%$, is amortized by monthly payments of $200 for as long as necessary. Construct the amortization schedule.

7.45 A debt of $2000 will be repaid by monthly payments of $500 for as long as necessary, the first payment to be made at the end of 6 months. If interest is at $j_{12} = 9\%$, find the size of the debt at the end of 5 months and make out the complete schedule starting at that time. *Ans.* $2076.13

7.46 A recreational vehicle worth $46 000 is purchased with a down payment of $6000 and monthly payments for 15 years. If interest is $j_2 = 10\%$, (a) find the monthly payment required, and (b) complete the first six lines of the amortization schedule. *Ans.* (a) $424.91

7.47 A couple purchases a house worth $116 000 by paying $16 000 down and then taking out a mortgage at $j_{12} = 8\%$ to be amortized over 25 years with equal monthly payments. (a) Find the monthly payment. (b) Make a partial amortization schedule showing the distribution of the first 6 payments as to principal and interest. *Ans.* (a) $771.82

7.48 Redo Problem 7.47, using (i) $j_{12} = 6\%$, (ii) $j_{12} = 10\%$. *Ans.* (i) $644.31, (ii) $908.71

7.49 A loan is being repaid over 10 years in equal annual installments. If the amount of principal in the third payment is $350, find the principal portion of the eighth payment, given (a) $j_1 = 8\frac{1}{2}\%$; (b) $j_{12} = 8\frac{1}{2}\%$. *Ans.* (a) $526.28; (b) $534.56

7.50 The Andersons borrow $8000 to be repaid in monthly installments over 4 years at $j_{12} = 13\frac{1}{2}\%$. Find the total amount of interest paid over the 4 years. *Ans.* $2397.31

7.51 A loan is being repaid with 10 annual installments. The principal portion of the seventh payment is $110.25 and the interest portion is $39.75. What annual effective rate of interest is being charged? *Ans.* 8%

7.52 A loan at $j_1 = 9\%$ is being repaid by monthly payments of $750 each. The total principal repaid in the 12 monthly installments of the 8th year is $400. What is the total interest paid in the 12 installments of the 10th year? *Ans.* $8524.76

7.53 You lend a friend $15 000 to be amortized by semiannual payments for 8 years, with interest at $j_2 = 9\%$. You deposit each payment in an account paying $j_{12} = 7\%$. What annual effective rate of interest have you earned over the entire 8-year period? *Ans.* 8.17%

7.54 A loan of $10 000 is being repaid by semiannual payments of $1000 on account of principal. Interest on the outstanding balance at j_2 is paid in addition to the principal repayments. The total of all payments is $12 200. Find j_2. *Ans.* 8%

7.55 A loan of A is to be repaid by 16 equal semiannual installments, including principal and interest, at rate i per half year. The principal in the first installment (6 months hence) is $30.83. The principal in the last is $100.00. Find the annual effective rate of interest. *Ans.* 16.99%

7.56 A loan is to be repaid by 16 quarterly payments of $50, $100, $150, ..., $800, the first payment due 3 months after the loan is made. Interest is at a nominal annual rate of 8% compounded quarterly. Find the total amount of interest contained in the payments. *Ans.* $1314.67

7.57 A loan of \$20 000 with interest at $j_{12} = 15\%$ is amortized by equal monthly payments over 15 years. In which payment will the interest portion be less than the principal portion, for the first time? *Ans.* 126th

OUTSTANDING PRINCIPAL

7.58 A loan is being repaid by monthly installments of \$250 at $j_{12} = 8\%$. If the loan balance after the fourth monthly payment is \$2800, find the original loan value. *Ans.* \$3710.11

7.59 Below is part of an amortization schedule based on monthly payments:

Payment	Distribution of Payment	
Number	Interest	Principal
k	39.64	92.03
$k+1$	38.72	92.95

Determine (a) the monthly payment, (b) the effective rate of interest per month (closest $\frac{1}{8}\%$), (c) the nominal rate of interest j_{12} (closest $\frac{1}{8}\%$). (d) Using the rate calculated in (b), find the outstanding balance just after payment k and find the remaining period of the loan beyond the date of payment $k+1$. *Ans.* (a) \$131.67; (b) 1%; (c) 12%; (d) \$3872.00, 34 months

7.60 To pay off the purchase of a car, Chantal got a \$15 000, 4-year bank loan at $j_{12} = 9\%$. She makes monthly payments. How much does she still owe on the loan at the end of 2 years (24 payments)? Use both the retrospective and prospective methods. *Ans.* \$8170.57

7.61 On July 1, 1995, Brian borrowed \$30 000 to be repaid over 3 years with monthly payments at $j_{12} = 8\%$ (first payment August 1, 1995). How much principal was repaid in 1995? How much interest? *Ans.* \$3750.17, \$950.33

7.62 A doctor buys a house worth \$380 000 by paying \$125 000 down and then taking a mortgage loan out for \$255 000. The mortgage is at $j_{12} = 7\frac{1}{2}\%$ and will be repaid over 20 years in monthly installments. How much of the debt does the doctor pay off in the first year? *Ans.* \$5720.22

7.63 To pay off the purchase of home furnishings, a couple takes out a bank loan of \$3000 to be retired with monthly payments over 2 years at $j_\infty = 8\%$. (a) What is the outstanding debt just after the tenth payment? (b) What is the principal portion of the eleventh payment? *Ans.* (a) \$1808.03; (b) \$123.63

7.64 Martha buys a piece of land worth \$40 000 by paying \$10 000 down and then taking out a loan for \$30 000. The loan will be retired with quarterly payments over 15 years at $j_4 = 14\%$. Find her equity at the end of 9 years. *Ans.* \$20 687.36

7.65 A loan of \$15 000 is being repaid by installments of \$350 at the end of each month for as long as necessary, plus a final smaller payment. If interest is at $j_4 = 10\%$, find the outstanding balance at the end of 2 years. *Ans.* \$9027.10

7.66 The Smiths borrow \$7500 to buy a car. The loan will be repaid over 4 years with monthly payments at $j_{12} = 15\%$. Find the total interest paid in the 12 payments of year 3. *Ans.* \$512.40

7.67 A loan of \$2000 is to be repaid by annual payments of \$400 per annum for the first 5 years and payments of \$450 per year thereafter for as long as necessary. Find the total number of payments and the amount of the smaller final payment made 1 year after the last regular payment. Assume annual effective rate of 18%. *Ans.* 12 payments, \$445.69

7.68 Five years ago, Justin deposited \$1000 into a fund out of which he draws \$100 at the end of each year. The fund guarantees interest at 5% on the principal on deposit during the year. If the fund actually earns interest at a rate in excess of 5%, the excess interest earned during the year is paid to Justin at the end of the year in addition to the regular \$100 payment. The fund has been earning 8% each year for the past 5 years. What is the total payment Justin now receives? *Ans.* \$123.54

7.69 A 5-year loan is being repaid with level monthly installments at the end of each month, beginning with January 1993 and continuing through December 1997. A 12% nominal annual interest rate compounded monthly was used to determine the amount of each monthly installment. On which date will the outstanding principal of this loan first fall below one-half of the original amount of the loan? *Ans.* November 1, 1995

7.70 A debt is amortized at $j_4 = 10\%$ by payments of $300 per quarter. If the outstanding principal is $2853.17 just after the kth payment, what was it just after the $(k-1)$st payment? *Ans.* $3076.26

7.71 A loan is made on January 1, 1975, and is to be repaid by 25 level annual installments. These installments are in the amount of $3000 each and are payable on December 31 of the years 1975 through 1999. However, just after the December 31, 1979, installment has been paid, it is agreed that, instead of continuing the annual installments on the basis just described, henceforth installments will be payable quarterly with the first such quarterly installment being payable on March 31, 1980, and the last one on December 31, 1999. Interest is at an annual effective rate of 10%. By changing from the old repayment schedule to the new one, the borrower will reduce the total amount of payments made over the 25-year period. Find the amount of this reduction. *Ans.* $2126.40

MORTGAGES

7.72 Mrs. Adams buys a house, taking a $60 000, 20-year mortgage at $j_{12} = 13\frac{1}{2}\%$ from a lender who charges 5 points. Find the true rate of interest, j_{12}, being charged. *Ans.* 14.39%

7.73 If, in Problem 7.72, Mrs. Adams pays off the outstanding balance of the mortgage after 12 years, what is the true rate of interest, j_{12}? *Ans.* 14.42%

7.74 ABC Savings and Loans develops a special scheme to help their customers pay their mortgages off quickly. Instead of making payments of X once a month, mortgage borrowers are asked to pay $X/4$ once a week (52 times a year). The Cohens are buying a house and need a $95 000 mortgage at $j_{12} = 9\%$. Determine (*a*) the monthly payment required to amortize the debt over 25 years, (*b*) the weekly payment as suggested in the scheme, (*c*) the number of weeks it will take to pay off the loan using the weekly schedule, (*d*) the interest saved using the weekly schedule. *Ans.* (*a*) $797.24; (*b*) $199.31; (*c*) 1003; (*d*) $39 280.92

7.75 Mr. Ramsay can buy a certain house for $190 000 if he takes out a $160 000 mortgage from a bank at $j_{12} = 12\%$. The loan would be amortized over 25 years, but the rate of interest would be fixed only for 5 years, after which the loan would be renegotiated. The seller of the house is willing to give Mr. Ramsay a mortgage at $j_{12} = 10.5\%$. The monthly payment would be determined using a 25-year repayment schedule. The seller would guarantee the rate of interest for 5 years, at which time Mr. Ramsay would have to renegotiate the outstanding principal. If Mr. Ramsay accepts this offer, the seller will want $200 000 for the house, forcing Mr. Ramsay to borrow $170 000. If Mr. Ramsay can earn $j_{12} = 6\%$ on his savings, what should he do? *Ans.* Pay $190 000

7.76 The Hwangs buy a home and assume a $90 000 mortgage to be amortized with monthly payments over 20 years at $j_{12} = 9\%$. The mortgage contract allows the Hwangs to make extra payments of principal each month. If they can pay an extra $100 each month, how long will it take to repay the mortgage, and what will be the size of the final smaller payment? *Ans.* 182 months, $266.43

7.77 Tristar Corporation built a new plant in 1992 at a cost of $1 700 000. It paid $200 000 cash and assumed a mortgage for $1 500 000 to be repaid over 10 years by equal semiannual payments due each June 30 and December 31, the first payment being due on December 31, 1992. The mortgage interest rate is 11% per annum compounded semiannually and the original date of the loan was July 1, 1992.

(*a*) What will be the total of the payments made in 1994 on this mortgage?

(b) Mortgage interest paid in any year (for this mortgage) is an income tax deduction for that year. What will be the interest deduction on Tristar Corporation's 1994 tax form?

(c) Suppose the plant is sold on January 1, 1996. The buyer pays $650 000 cash and assumes the outstanding mortgage. What is Tristar Corporation's capital gain (amount realized less original price) on the investment in the building?

Ans. (a) $251 038; (b) $147 230.30; (c) $94 366.50

7.78 You are choosing between two mortgages for $60 000 with a 20-year amortization period. Both charge $j_{12} = 10.5\%$ and permit the mortgage to be paid off in less than 20 years. Mortgage A allows you to make weekly payments, with each payment being $\frac{1}{4}$ of the normal monthly payment. Mortgage B allows you to make double the usual monthly payment every 6 months. Assuming that you will take advantage of the mortgage provisions, calculate the total interest charges over the life of each mortgage to determine which mortgage costs less.
Ans. Mortgage B

7.79 The Smiths buy a home and take out an $160 000 mortgage on which the interest rate is allowed to float freely. At the time the mortgage is issued, interest rates are $j_2 = 10\%$ and the Smiths choose a 25-year amortization schedule. Six months into the mortgage, interest rates rise to $j_2 = 12\%$. Three years into the mortgage (after 36 payments) interest rates drop to $j_2 = 11\%$ and 4 years into the mortgage, interest rates drop to $j_2 = 9\frac{1}{2}\%$. Find the outstanding balance of the mortgage after 5 years. (The monthly payment is set at issue and does not change.)
Ans. $161 937.39

REFINANCING A LOAN

7.80 A borrower is repaying an $8000 loan at $j_{12} = 15\%$ with monthly payments over 3 years. Just after the twelfth payment (at the end of 1 year), he has the balance refinanced at $j_{12} = 12\%$. If the number of payments remains unchanged, what will be the new monthly payment, and what will be the total savings in interest? *Ans.* $269.24, $194.00

7.81 The Smiths buy a refrigerator and stove for $1400 in total. They finance the purchase at $j_{12} = 15\%$ to be repaid over 36 months (3 years). If they wish to pay off the loan early, they will incur a penalty equal to three times one month's interest on the effective outstanding balance. After 12 payments, they notice that interest rates at their local Credit Union are $j_{12} = 11\%$. Should they refinance? *Ans.* No

7.82 Mrs. Dent buys $5000 worth of home furnishings from the ABC Furniture Mart. She pays $500 down and agrees to pay the balance in monthly installments over 5 years at $j_{12} = 18\%$. The contract stipulates that in case of early repayment there is a penalty equal to three months' payments. After 2 years (24 payments), Mrs. Dent realizes that she can borrow the money from the bank at $j_{12} = 12\%$. Should she refinance? *Ans.* No

7.83 A couple buy a house and take out a $50 000 mortgage to be repaid with monthly payments over 20 years at $j_{12} = 9.6\%$. After $3\frac{1}{2}$ years they sell their house and pay off the mortgage. They find that in addition to repaying the loan balance, they must pay a penalty equal to three times one month's interest on the outstanding balance. What total amount must they repay? *Ans.* $47 672.25

7.84 Vera is repaying a $5000 loan at $j_{12} = 16\frac{1}{2}\%$ with monthly installments over 3 years. Owing to temporary unemployment, she misses the 13th through 18th payment, inclusive. Find the value of the revised monthly payment needed, starting in the 19th month, if the loan is still to be repaid at $j_{12} = 16\frac{1}{2}\%$ by the end of the original 3 years. *Ans.* $246.38

7.85 A loan effective January 1, 1994 is being amortized by equal monthly installments over 5 years using interest at a nominal annual rate of 12% compounded monthly. The first such installment was due February 1, 1994, and the last such installment was to be due January 1, 1999. Immediately after the 24th installment was made on January 1, 1996, a new level monthly installment is determined (using the same rate of interest) in order to shorten the total amortization period to $3\frac{1}{2}$ years, so the final installment will fall due on July 1, 1997. Find the ratio of the new monthly installment to the original monthly installment. *Ans.* 1.836017314

7.86 A loan of $50 000 was being repaid by monthly level installments over 20 years at $j_{12} = 9\%$ interest. Now, when 10 years of the repayment period are still to run, it is proposed to increase the interest rate to $j_{12} = 10\frac{1}{2}\%$. What should the new level payment be so as to liquidate the loan on its original due date? *Ans.* $479.18

7.87 The Mosers buy a camper trailer and take out a $15 000 loan. The loan is amortized over 10 years with monthly payments at $j_2 = 18\%$.

(a) Find the monthly payment needed to amortize this loan.

(b) Find the amount of interest paid by the first 36 payments.

(c) After 3 years (36 payments) they could refinance their loan at $j_2 = 16\%$ provided they pay a penalty equal to 3 months' interest on the outstanding balance. Should they refinance? Show the difference in their monthly payments.

Ans. (a) $264.13; (b) $7302.74; (c) Yes, $2.83

7.88 A couple buys a home and signs a mortgage contract for $120 000 to be paid with monthly payments over a 25-year period at $j_2 = 10\frac{1}{2}\%$. After 5 years, they renegotiate the interest rate and refinance the loan at $j_2 = 7\%$. Find (a) the monthly payment for the initial 5-year period; (b) the new monthly payment after 5 years; (c) the accumulated value of the savings for the second 5-year period at $j_{12} = 3\%$ valued at the end of the second 5-year period; (d) the outstanding balance at the end of 10 years.
Ans. (a) $1114; (b) $113 271.22; (c) $15 682.65; (d) $97 554.61

7.89 Give the repayment schedule for Problem 7.43, using the sum-of-digits method.

7.90 Give the repayment schedule for Problem 7.44, using the sum-of-digits method.

7.91 Redo Problem 7.46, using the sum-of-digits method.

7.92 Redo Problem 7.47, using the sum-of-digits method.

7.93 To pay off the purchase of a car, a woman gets a $6000, 3-year bank loan at $j_{12} = 18\%$ requiring monthly payments. Find the outstanding balance on the loan after the 24th payment, using the sum-of-digits method. *Ans.* $2390.98

7.94 Redo Problem 7.80, using the sum-of-digits method to determine the principal that is refinanced at the end of 1 year. *Ans.* $271.25, $145.81

7.95 A loan of $20 000 is to be retired with monthly payments over 10 years at $j_{12} = 15\%$. Using the sum-of-digits method, (a) construct the first two lines of the repayment schedule; (b) find the interest and principal portions of the 10th payment; (c) find the outstanding balance at the end of 3 years and compare it with the outstanding balance at the same time under the amortization method; (d) decide whether the loan should be refinanced at the end of 3 years at $j_{12} = 12\%$, with the term of the loan unchanged.
Ans. (b) $286.22, $36.45; (c) $17 898.79, $16 721.46; (d) yes

7.96 Consider a $10 000 loan being repaid with monthly payments over 15 years at $j_{12} = 15\%$. Find the outstanding balance (a) at the end of 2 years and (b) at the end of 5 years using both the sum-of-digits method and the amortization method.
Ans. (a) $10 412.52, $9584.29; (b) $10 024.06, $8674.92

7.97 Michelle has a $5000 loan that is being repaid by monthly payments over 4 years at $j_{12} = 15\%$. The lender uses the sum-of-digits method to determine outstanding balances. After 1 year of payments the lender's interest rate on new loans has dropped to $j_{12} = 12\%$. Will Michelle save money by refinancing the loan if the term of the loan remains the same?
Ans. Yes, $4.36 a month

7.98 Matthew can borrow \$15 000 at $j_4 = 15\%$ and repay the loan with monthly payments over 10 years. If he wants to pay the loan off early, the outstanding balance will be determined using the sum-of-digits method. He can also borrow \$15 000 with monthly payments over 10 years at $j_4 = 16\%$ and pay the loan off at any time without penalty. The outstanding balance will be determined using the amortization method. Matthew has an endowment insurance policy coming due in 4 years that could be used to pay off the outstanding balance at that time in full. Which loan should he take if he earns $j_{12} = 6\%$ on his savings? *Ans.* Borrow at $j_4 = 16\%$

7.99 A loan of \$18 000 is to be repaid with monthly payments over 10 years at $j_2 = 17\frac{1}{2}\%$. Using the sum-of-digits method (*a*) construct the first two and the last two lines of the repayment schedule; (*b*) find the interest and the principal portion of the 10th payment; (*c*) find the outstanding balance at the end of 2 years and compare it with the outstanding balance at the same time calculated by the amortization method; (*d*) advise whether the loan should be refinanced at the end of 2 years at current rate $j_2 = 16\%$ with the term of the loan unchanged.
Ans. (*b*) \$296.54, \$15.09; (*c*) \$17 477.61, \$16 351.36; (*d*) Do not refinance

7.100 A \$20 000 home renovation loan is to be amortized over 10 years by monthly payments, with each regular payment rounded up to the next dollar, and the last payment reduced accordingly. Interest on the loan is at $j_4 = 10\%$. After 4 years the loan is fully paid off with an extra payment. Find the amount of this final payment if the sum-of-digits method is used to calculate the outstanding principal. *Ans.* \$14 700.94

SINKING FUNDS

7.101 A couple is saving a down payment for a house. They want to have \$15 000 at the end of 4 years in an account that pays interest at $j_1 = 6\%$. (*a*) How much must be deposited in the fund at the end of each year? (*b*) Construct a complete sinking-fund schedule. *Ans.* (*a*) \$3428.87

7.102 A condominium high rise consists of 128 two-bedroom units of uniform size. The Owners' Association establishes a fund to save \$130 000 in 5 years to install a gymnasium. Assuming that the association can earn $j_{12} = 7\frac{1}{2}\%$ on its money, (*a*) what monthly sinking-fund assessment will be required per unit? (*b*) Show the first three and last two lines of the sinking-fund schedule. *Ans.* (*a*) \$14

7.103 A sinking fund earning interest at $j_\infty = 13\%$ now contains \$2000. (*a*) What quarterly deposits for the next 3 years will cause the fund to grow to \$10 000? (*b*) How much is in the fund 2 years from now? *Ans.* (*a*) \$487.98; (*b*) \$6980.14

7.104 A couple wants to save \$20 000 to buy some land. They can save \$150 a month in an account paying interest at $j_4 = 10\%$. How many deposits will be required, and what will be the size of the final smaller deposit? *Ans.* 91, \$0 [wrong]

7.105 A cottagers' association decides to set up a sinking fund to save enough money to have their cottage road widened. They need \$30 000 at the end of 5 years in a fund earning $9\frac{1}{2}\%$ per annum. What annual deposit is required per cottage, if there are 40 cottages on the road? *Ans.* \$124.08

7.106 A homeowners' association decided to set up a sinking fund to accumulate \$50 000 by the end of 3 years to improve recreational facilities. (*a*) What monthly deposits are required if the fund earns 5% compounded daily? (*b*) Show the first three and the last two lines of the sinking fund schedule. *Ans.* (*a*) \$1290.02

7.107 Consider an amount that is to be accumulated with equal deposits R at the end of each interest period for 5 periods at rate i per period. Hence, the amount to be accumulated is $Rs_{\overline{5}|i}$. Do a complete schedule for this sinking fund. Verify that the sum of the interest column plus the sum of the deposit-column equals the sum of the increase-in-the-fund column, and both sums equal the final amount in the fund.

7.108 In its manufacturing process, a company uses a machine that costs $75 000 and is scrapped at the end of 15 years with a value of $5000. The company sets up a sinking fund to finance the replacement of the machine, assuming no change in price, with level payments at the end of each year. Money can be invested at an annual effective interest rate of 4%. Find the value of the sinking fund at the end of the 10th year. *Ans.* $41 971.91

7.109 A sinking fund is being accumulated at $j_{12} = 6\%$ by deposits of $200 per month. If the fund contains $5394.69 just after the kth deposit, what did it contain just after the $(k-1)$st deposit? *Ans.* $5168.85

7.110 A man is repaying a $16 000 loan by the sinking-fund method. His total monthly expense is $350. Out of this $350, interest is paid monthly at $j_{12} = 12\%$ and the remainder of the money is deposited in a sinking fund earning interest at $j_{12} = 10\frac{1}{2}\%$. Find the duration of the loan and the final smaller payment. *Ans.* 64 months, $0

7.111 A borrower of $5000 agrees to pay interest semiannually at $j_2 = 10\%$ on the loan and to build up a sinking fund, which will repay the loan at the end of 5 years. If the sinking fund accumulates at $j_2 = 7\%$, (a) find his total semiannual expense. (b) How much is in the sinking fund at the end of 4 years? *Ans.* (a) $676.21; (b) $3857.92

7.112 A city borrows $250 000, paying interest annually on this sum at $j_1 = 9\frac{1}{2}\%$. An annual deposit must be made into a sinking fund earning interest at $j_1 = 6\%$ in order to pay off the entire principal at the end of 15 years. What is the total annual expense of the debt? *Ans.* $34 490.69

7.113 A company issues $500 000 worth of bonds, paying interest at $j_2 = 12\%$. A sinking fund with semiannual deposits accumulating at $j_2 = 9\%$ is established to redeem the bonds at the end of 20 years. Find (a) the semiannual expense of the debt, (b) the book value of the company's indebtedness at the end of the 15th year. *Ans.* (a) $34 671.57; (b) $215 001.20

7.114 On a debt of $10 000, interest is paid semiannually at $j_2 = 10\%$ and semiannual deposits are made into a sinking fund to retire the debt at the end of 5 years. If the sinking fund earns interest at $j_{12} = 6\%$, what is the semiannual expense of the debt? *Ans.* $1370.79

7.115 A 10-year loan of $10 000 at $j_1 = 11\%$ is to be repaid by the sinking-fund method, with interest and sinking-fund payments made at the end of each year. The rate of interest earned in the sinking fund is $j_1 = 5\%$. Immediately after the fifth year's payment, the lender requests that the outstanding principal be repaid in one lump sum. Calculate the amount of extra cash the borrower has to raise in order to extinguish the debt. *Ans.* $5606.85

7.116 Interest at $j_2 = 12\%$ on a loan of $3000 must be paid semiannually as it falls due. A sinking fund accumulating at $j_4 = 8\%$ is established to enable the debtor to repay the loan at the end of 4 years. (a) Find the semiannual sinking-fund deposit and construct the last two lines of the sinking-fund schedule, based on semiannual deposits. (b) Find the semiannual expense of the loan. (c) What is the outstanding principal (book value of the loan) at the end of 2 years? *Ans.* (a) $325.12; (b) $505.12; (c) $1618.57

7.117 Mr. White borrows $15 000 for 10 years. He makes total payments, annually, of $2000. The lender receives $j_1 = 10\%$ on his investment each year for the first 5 years and $j_1 = 8\%$ for the second 5 years. The balance of each payment is invested in a sinking fund earning $j_1 = 7\%$. (a) Find the amount by which the sinking fund is short of repaying the loan at the end of 10 years. (b) By how much would the sinking-fund deposit (in each of the first 5 years only) need to be increased so that the sinking fund at the end of 10 years will be just sufficient to repay the loan? *Ans.* (a) $6366.56; (b) $789.34

7.118 A $100 000 loan is to be repaid in 15 years, with a sinking fund accumulated to repay principal plus interest. The loan charges $j_2 = 12\%$, while the sinking fund earns $j_2 = 9\%$. What semiannual sinking-fund deposit is required? *Ans.* $9414.47

7.119 A loan of $20 000 bears interest on the amount outstanding at $j_1 = 10\%$. A deposit is to be made in a sinking fund earning interest at $j_1 = 4\%$, which will accumulate enough to pay one-half of the principal at the end of 10 years. In addition, the debtor will make level payments to the creditor, which will pay interest at $j_1 = 10\%$ on the outstanding balance first and the remainder will repay the principal. What is the total annual payment, including that made to the creditor and that deposited in the sinking fund, if the loan is to be completely retired at the end of 10 years? *Ans.* $4460.37

7.120 John borrows $10 000 for 10 years and uses a sinking fund to repay the principal. The sinking-fund deposits earn an annual effective interest rate of 5%. The total required payment for both the interest and the sinking-fund deposit made at the end of each year is $1445.05. Calculate the annual effective interest rate charged on the loan. *Ans.* 6.5%

7.121 A company borrows $10 000 for 5 years. Interest of $600 is paid semiannually. To repay the principal of the loan at the end of 5 years, equal semiannual deposits are made into sinking fund that credits interest at a nominal rate of 8% compounded quarterly. The first payment is due in 6 months. Calculate the annual effective rate of interest that the company is paying to service and retire the debt. *Ans.* 14.74%

7.122 On August 1, 1988, Mrs. Chan borrows $20 000 for 10 years. Interest at 11% per annum convertible semiannually must be paid as it falls due. The principal is replaced by means of level deposits on February 1 and August 1 in years 1989 to 1998 (inclusive) into a sinking fund earning $j_1 = 7\%$ in 1989 through December 31, 1993, and $j_1 = 6\%$ January 1, 1994 through 1998. (*a*) Find the semiannual expense of the loan. (*b*) How much is in the sinking fund just after the August 1, 1997, deposit? (*c*) Show the sinking-fund schedule entries at February 1, 1998, and August 1, 1998.
Ans. (*a*) $846.42; (*b*) $17 458.03

COMPARISON OF AMORTIZATION AND SINKING-FUND METHODS

7.123 A company borrows $50 000 to be repaid in equal installments at the end of each year for 10 years. Find the total annual cost under the following conditions:

(*a*) The debt is amortized at $j_1 = 9\%$

(*b*) Interest at 9% is paid on the debt and a sinking fund is set up at $j_1 = 9\%$

(*c*) Interest at 9% is paid on the debt and a sinking fund is set up at $j_1 = 6\%$

Ans. (*a*) $7791; (*b*) $7791; (*c*) $8293.40

7.124 A company can borrow $180 000 to be repaid over 15 years. They can amortize the debt at $j_1 = 10\%$ or they can pay interest on the loan at $j_1 = 9\%$ and set up a sinking fund at $j_1 = 7\%$. Which plan is cheaper, and by how much per annum? *Ans.* Sinking fund, $302.25

7.125 A firm borrows $60 000 to be repaid over 5 years. One source will lend them the money at $j_2 = 10\%$ if it is amortized by semiannual payments. A second source will lend them the money at $j_2 = 9.5\%$ if only the interest is paid semiannually and the principal is returned in a lump sum at the end of 5 years. Which source is cheaper, and how much will be saved each half-year if the required sinking fund earns interest at $j_2 = 8\%$? *Ans.* First source, $77.19

7.126 A city can borrow $500 000 for 20 years by issuing bonds on which interest will be paid semiannually at $j_2 = 9\frac{1}{8}\%$. The principal will be paid off by a sinking fund consisting of semiannual deposits invested at $j_2 = 8\%$. Find the nominal rate j_2 at which the loan could be amortized at the same semiannual cost. *Ans.* 9.42%

7.127 A firm can borrow $200 000 at $j_1 = 9\%$ and amortize the debt for 10 years. From a second source, the money can be borrowed at $j_1 = 8\frac{1}{2}\%$ if the interest is paid annually and the principal is repaid in a lump sum at the end of 10 years. What yearly rate j_1 must the sinking fund earn for the annual expense to be the same under the two options? *Ans.* 7.45%

7.128 A company wants to borrow \$500 000. One source of funds will agree to lend the money at $j_4 = 8\%$ if interest is paid quarterly and the principal is paid in a lump sum at the end of 15 years. The firm can set up a sinking fund at $j_4 = 6\%$ and will make quarterly deposits. (*a*) What is the total quarterly cost of the loan? (*b*) At what rate j_4 would it be less expensive to amortize the debt over 15 years?
Ans. (*a*) \$15 196.71; (*b*) $j_4 < 8.92\%$

7.129 You are able to repay a \$2000 loan by either (*a*) amortization at $j_{12} = 15\%$ with 12 months payments; or (*b*) at $j_{12} = 13\%$ using a sinking fund earning $j_{12} = 8\%$, and paid off in 1 year. Which method is cheaper? *Ans.* (*a*) is cheaper by \$1.79 a month

7.130 A company needs to borrow \$200 000 for 6 years. One source will lend them the money at $j_2 = 10\%$ if it is amortized by monthly payments. A second source will lend the money at $j_4 = 9\%$ if only the interest is paid monthly and the principal is returned in a lump sum at the end of 6 years. The company can earn interest at $j_{365} = 6\%$ on the sinking fund. Which source should be used for the loan and how much will be saved monthly? *Ans.* First source, \$117.61

7.131 Tanya can borrow \$10 000 by paying the interest on the loan as it falls due at $j_2 = 12\%$ and by setting up a sinking fund with semiannual deposits that accumulate at $j_{12} = 9\%$ over 10 years to repay the debt. At what rate j_4 would an amortization scheme have the same semiannual cost? *Ans.* 13.03\%

7.132 A loan of \$10 000 at 16\% per annum is to be repaid over 10 years: \$2000 by the amortization method and \$8000 be the sinking-fund method, where the sinking fund can be accumulated with annual deposits at $j_4 = 10\%$. What extra annual payment does the above arrangement require as compared to repayment of the whole loan by the amortization method? *Ans.* \$117.65

7.133 A company wants to borrow a large amount of money for 15 years. One source would lend the money at $j_2 = 9\%$, provided it is amortized over 15 years by monthly payments. The company could also raise the money by issuing bonds paying interest semiannually at $j_2 = 8\frac{1}{2}\%$ and redeemable at par in 15 years. In this case, the company would set up a sinking fund to accumulate the money needed for the redemption of the bonds at the end of 15 years. What rate j_{12} on the sinking fund would make the monthly expense the same under the two options? *Ans.* 7.25\%

7.134 A \$10 000 loan is being repaid by the sinking-fund method. Total annual outlay (each year) is \$1400 for as long as necessary, plus a smaller final payment made 1 year after the last regular payment. If the lender receives $j_1 = 8\%$ and the sinking fund accumulates at $j_1 = 6\%$, find the time and amount of the last irregular final payment. *Ans.* \$1278.04 at the end of 12 years

7.135 A \$5000 loan can be repaid quarterly for 5 years using amortization and an interest rate of $j_{12} = 10\%$ or by a sinking fund to repay both principal and accumulated interest. If paid by a sinking fund, the interest on the loan will be $j_{12} = 9\%$. What annual effective rate must the sinking fund earn to make the quarterly cost the same for both methods? *Ans.* 8.35\%

Chapter 8

Bonds

8.1 INTRODUCTION AND TERMINOLOGY

A *bond* is a written contract between the issuer (borrower) and the investor (lender), which specifies:

(i) The *face value*, or *denomination*, of the bond, which is stated on the front of the bond. This is usually a round figure such as $100, $500, $1000, $10 000.

(ii) The *redemption date*, or *maturity date*, which is the date on which the loan will be repaid.

(iii) The *bond rate*, or *coupon rate*, which is the rate at which the bond pays interest on its face value at regular time intervals until the maturity date. This rate is usually compounded semiannually.

(iv) The *redemption value*, which is the amount of money promised to be paid on the redemption date. In most cases, it is the same as the face value, and we say that the bond is *redeemed at par*.

Callable bonds contain a clause allowing the issuer to pay off the loan (redeem the bond) at a date earlier than the full redemption date. These are discussed in detail in Section 8.3.

Bonds may be bought and sold at any time. The buyer of a bond will want to realize a certain return on his investment, as specified by a desired *yield rate*.

In this chapter, we will use the following notation:

F $\equiv$ face value or par value of the bond
C $\equiv$ redemption value of the bond
r $\equiv$ bond rate or coupon rate per interest period
i $\equiv$ yield rate per interest period, often called the yield to maturity (YTM)
n $\equiv$ number of interest periods until the redemption date
P $\equiv$ purchase price of the bond to yield rate i
Fr $\equiv$ bond interest payment or coupon

8.2 PURCHASE PRICE TO YIELD A GIVEN INVESTMENT RATE

The investor who wishes to realize a rate of return i (until the bond is redeemed or matures) should pay a price equal to the discounted value of the n coupons Fr plus the discounted value of the redemption amount C:

$$P = Fr\ a_{\overline{n}|i} + C(1+i)^{-n} \tag{8.1}$$

In Problem 8.4, we show that (8.1) is equivalent to

$$P = C + (Fr - Ci)a_{\overline{n}|i} \qquad (8.2)$$

known as an *alternate purchase-price formula*; computationally, (8.2) is a little simpler than (8.1).

SOLVED PROBLEMS

8.1 A $1000 bond that pays interest at $j_2 = 12\%$ is redeemable at par at the end of 10 years. Find the purchase price to yield 10% compounded semiannually.

The bond pays $Fr = 1000(0.06) = \$60$ semiannually and $1000 at the end of 10 years, as shown in Fig. 8-1.

$$P = 60a_{\overline{20}|.05} + 1000(1.05)^{-20} = 747.73 + 376.89 = \$1124.62$$

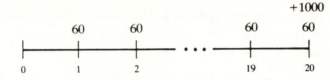

Fig. 8-1

8.2 Redo Problem 8.1 for a yield rate of $j_2 = 15\%$.

$$P = 60a_{\overline{20}|.075} + 1000(1.075)^{-20} = 611.67 + 235.41 = \$847.08$$

8.3 A $5000 bond maturing at 103 on October 1, 2002, has semiannual coupons at $10\frac{1}{2}\%$. Find the purchase price on April 1, 1995, to yield $9\frac{1}{2}\%$ compounded semiannually.

The bond pays 15 semiannual coupons of $Fr = 5000(0.0525) = \$262.50$. The bond matures on October 1, 2002, for $C = 5000(1.03) = \$5150$. See Fig. 8-2.

$$P = 262.50a_{\overline{15}|.0475} + 5150(1.0475)^{-15} = 2771.29 + 2567.42 = \$5338.71$$

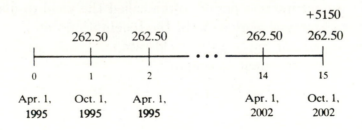

Fig. 8-2

8.4 Derive (8.2).

From (5.2),

$$a_{\overline{n}|i} = \frac{1 - (1+i)^{-n}}{i} \qquad \text{or} \qquad (1+i)^{-n} = 1 - ia_{\overline{n}|i}$$

Replacing $(1+i)^{-n}$ in (8.1), we get:

$$P = Fr\, a_{\overline{n}|i} + C(1 - ia_{\overline{n}|i}) = C + (Fr - Ci)a_{\overline{n}|i}$$

8.5 Use (8.2) in (*a*) Problem 8.1, (*b*) Problem 8.2, (*c*) Problem 8.3.

(*a*) With $C = 1000$, $Fr = 60$, $i = 0.05$, and $n = 20$,

$$P = 1000 + (60 - 50)a_{\overline{20}|.05} = 1000 + 124.62 = \$1124.62$$

(*b*) With $C = 1000$, $Fr = 60$, $i = 0.075$, and $n = 20$,

$$P = 1000 + (60 - 75)a_{\overline{20}|.075} = 1000 - 152.92 = \$847.08$$

(*c*) With $C = 5150$, $Fr = 262.50$, $i = 0.0475$, and $n = 15$,

$$P = 5150 + (262.50 - 244.625)a_{\overline{15}|.0475} = 5150 + 188.71 = \$5338.71$$

8.6 A corporation issues a 15-year, \$10 000 par value bond with semiannual coupons at 10% per annum. Find the price to yield 9% per annum compounded monthly.

First, find a rate of interest, i, per half-year such that

$$
\begin{aligned}
(1 + i)^2 &= \left(1 + \frac{0.09}{12}\right)^{12} \\
1 + i &= (1.0075)^6 \\
i &= (1.0075)^6 - 1 = 0.045852235
\end{aligned}
$$

Then, by (8.2),

$$
\begin{aligned}
P &= 10\ 000 + (500 - 458.52235)a_{\overline{30}|.045852235} \\
&= 10\ 000 + 668.90 = \$10\ 668.90
\end{aligned}
$$

8.7 Redo Problem 8.6 for a yield rate of $j_1 = 9\%$.

Find a rate of interest, i, per half-year such that

$$(1 + i)^2 = 1.09 \qquad \text{or} \qquad i = 0.044030651$$

Then
$$
\begin{aligned}
P &= 10\ 000 + (500 - 440.30651)a_{\overline{30}|.044030651} \\
&= 10\ 000 + 983.53 = \$10\ 983.53
\end{aligned}
$$

8.8 Derive *Makeham's purchase-price formula* for a bond redeemable at par,

$$P = F(1 + i)^{-n} + \frac{r}{i}[F - F(1 + i)^{-n}]$$

From (8.1), $P = Fr\, a_{\overline{n}|i} + F(1 + i)^{-n}$. Substituting for $a_{\overline{n}|i}$,

$$P = Fr\frac{1 - (1 + i)^{-n}}{i} + F(1 + i)^{-n} = F(1 + i)^{-n} + \frac{r}{i}[F - F(1 + i)^{-n}]$$

8.3   CALLABLE BONDS

Because *callable bonds* allow the issuer to pay off the loan (redeem the bond) prior to the maturity date, they present a problem with respect to the calculation of the purchase price, since the term of the bond is not certain. The investor will pay the price that will guarantee him his desired yield regardless of the call date. In determining the price, he must assume that the issuer of the bond will exercise his call option to the disadvantage of the investor.

For a bond *callable at par* $(C = F)$:

If the yield rate is greater than the coupon rate, the investor must calculate using the **latest** possible call date. (See Problem 8.9.)

If the yield rate is less than the coupon rate, the investor must calculate using the **earliest** possible call date. (See Problem 8.10.)

For any callable bond, even if the bond is not callable at par $(C \neq F)$, the investor can determine all possible purchase prices corresponding to his desired yield, and then pay the lowest of them. (See Problems 8.11 and 8.12.)

SOLVED PROBLEMS

8.9 The ABC Corporation issues a 20-year, $1000 bond with coupons at $j_2 = 12\%$. The bond can be called, at par, after 15 years. Find the purchase price to yield 13% compounded semiannually.

We must calculate the two purchase prices that correspond to the two possible redemption dates. If the bond is called after 15 years:

$$P_c = 1000 + (60 - 65)a_{\overline{30}|.065} = 1000 - 65.29 = \$934.71$$

If the bond matures after 20 years:

$$P_m = 1000 + (60 - 65)a_{\overline{40}|.065} = 1000 - 70.73 = \$929.27$$

The purchase price to guarantee a return of $j_2 = 13\%$ is the lower of these two answers, or $929.27. Paying that price, the investor realizes at least 13%, no matter what happens.

8.10 Redo Problem 8.9 for the yield rate $j_2 = 11\%$.

Again we calculate the two possible purchase prices. If the bond is called after 15 years:

$$P_c = 1000 + (60 - 55)a_{\overline{30}|.055} = 1000 + 72.67 = \$1072.67$$

If the bond matures after 20 years:

$$P_m = 1000 + (60 - 55)a_{\overline{40}|.055} = 1000 + 80.23 = \$1080.23$$

The price the investor must pay to guarantee a yield of $j_2 = 11\%$, regardless of what happens, is $1072.67.

8.11 For the bond of Problem 8.9, assume that the call feature specifies that if the bond is called after 15 years, the redemption value will be $1050. Otherwise, it matures at par in 20 years. Find the price to guarantee a yield of $j_2 = 11\%$.

If the bond is called after 15 years:

$$P_c = 1050 + (60 - 57.75)a_{\overline{30}|.055} = 1050 + 32.70 = \$1082.70$$

If the bond matures after 20 years, $P_m = \$1080.23$ (Problem 8.10). Hence, the price to guarantee a return of $j_2 = 11\%$ is $P_m = \$1080.23$.

8.12 A \$5000 callable bond pays interest at $j_2 = 9\frac{1}{2}\%$ and matures at par in 20 years. It may be called at the end of years 10 to 15 (inclusive) for \$5200. Find the price to yield at least $j_2 = 8\frac{1}{2}\%$ until the redemption.

If the bond is called after 10 years:

$$P_c = 5200 + (237.50 - 221)a_{\overline{20}|.0425} = \$5419.36$$

If the bond is called after 15 years:

$$P_c = 5200 + (237.50 - 221)a_{\overline{30}|.0425} = \$5476.85$$

Price P_c for call dates at the end of years 10 to 15 will gradually increase from \$5419.36 to \$5476.85.

If the bond matures after 20 years:

$$P_m = 5000 + (237.50 - 212.50)a_{\overline{40}|.0425} = \$5476.93$$

Hence, the price to guarantee a return of at least $j_2 = 8\frac{1}{2}\%$ until redemption is \$5419.36.

8.4 PREMIUM AND DISCOUNT

A bond is said to be purchased *at a premium* if its purchase price, P, exceeds its redemption value, C; the premium is $P - C$. A bond is said to be purchased *at a discount* if the purchase price, P, is less than its redemption value, C; the discount is $C - P$.

From (8.2), we can see that

$$\begin{aligned} \text{premium} &\equiv P - C = (Fr - Ci)a_{\overline{n}|i} \\ \text{discount} &\equiv C - P = (Ci - Fr)a_{\overline{n}|i} \end{aligned}$$

The *book value* of a bond at a given time is the sum recorded as being invested in the bond at that time. The book value of a bond on a date of purchase which coincides with an interest payment date (more precisely, the date of purchase precedes an interest payment date by one interest period) is just the purchase price of the bond. The book value on the redemption date is the redemption value of the bond.

When a bond is purchased at a premium ($P > C$), the book value of the bond will be written down (decreased) at each bond interest date, so that at the time of redemption the book value will equal the redemption value. This process is called *amortization of the premium* or *writing down*. When a bond is purchased at a discount ($C > P$), the book value of the bond will be written up (increased) at each bond interest date, so that at redemption the book value will equal the redemption value. This process is called *accumulation of the discount* or *writing up*.

A bond *amortization* (or *accumulation*) *schedule* shows the division of each bond coupon into its interest-yielded and principal-adjustment portions, together with the book value after each coupon is paid.

The payments made during the term of a bond can be regarded as loan payments made by the borrower (bond issuer) to the lender (the bondholder) to repay a loan amount equal to the purchase price of a bond. The bond purchase price is calculated as the discounted value of those payments (coupons plus redemption value) at a certain yield rate (the interest rate on the loan). Thus the bond transaction can be regarded as the amortization of a loan and an amortization schedule for the bond can be constructed like the general loan amortization schedule in Section 7.1. (See Problem 8.16.)

SOLVED PROBLEMS

8.13 A $1000 bond, redeemable at par on December 1, 1998, pays semiannual coupons at $j_2 = 9\%$. The bond is bought on June 1, 1996. Find the purchase price and construct a bond schedule, if the desired yield is 8% compounded semiannually.

From (8.2), the purchase price on June 1, 1996, is

$$P = 1000 + (45 - 40)a_{\overline{5}|.04} = \$1022.26$$

Thus the bond is purchased at a premium of $22.26. Table 8-1 gives the bond amortization schedule.

Table 8-1

Date	Bond Interest Payment	Interest on Book Value at Yield Rate	Principal Adjustment	Book Value
June 1, 1996	0	0	0	1022.26
Dec. 1, 1996	45.00	40.89	4.11	1018.15
June 1, 1997	45.00	40.73	4.27	1013.88
Dec. 1, 1997	45.00	40.56	4.44	1009.44
June 1, 1998	45.00	40.38	4.62	1004.82
Dec. 1, 1998	45.00	40.19	4.81	1000.01
TOTALS	225.00	202.74	22.25	

8.14 Redo Problem 8.13 if the desired yield is $j_2 = 10\%$.

Now the purchase price on June 1, 1996, is

$$P = 1000 + (45 - 50)a_{\overline{5}|.05} = \$978.35$$

and so the bond is purchased at a discount of $21.65. See Table 8-2.

Table 8-2

Date	Bond Interest Payment	Interest on Book Value at Yield Rate	Principal Adjustment	Book Value
June 1, 1996	0	0	0	978.35
Dec. 1, 1996	45.00	48.92	-3.92	982.27
June 1, 1997	45.00	49.11	-4.11	986.38
Dec. 1, 1997	45.00	49.32	-4.32	990.70
June 1, 1998	45.00	49.54	-4.54	995.24
Dec. 1, 1998	45.00	49.76	-4.76	1000.00
TOTALS	225.00	246.65	-21.65	

8.15 Verify the following properties of the bond schedules of Problems 8.13 and 8.14: (*a*) All book values can be reproduced using the purchase-price formula. (*b*) The total of the principal-adjustment column equals the original premium or discount. (*c*) Successive principal adjustments are in the ratio $1 + i$, where i is the desired yield rate.

(*a*) Consider the book value on June 1, 1997, for the bond purchased to yield $j_2 = 8\%$ (Table 8-1). Using (8.1),

$$P = 45a_{\overline{3}|.04} + 1000(1.04)^{-3} = \$1013.88$$

or using (8.2),

$$P = 1000 + (45 - 40)a_{\overline{3}|.04} = \$1013.88$$

Consider the book value on December 1, 1997, for the bond purchased to yield $j_2 = 10\%$ (Table 8-2). From (8.1),

$$P = 45a_{\overline{2}|.05} + 1000(1.05)^{-2} = \$990.70$$

or from (8.2),

$$P = 1000 + (45 - 50)a_{\overline{2}|.05} = \$990.70$$

(*b*) In Problem 8.13, the premium is \$22.26. The total of the principal-adjustment column of Table 8-1 is \$22.25. The 1¢ error is due to roundoff. In Problem 8.14, the discount is \$21.65 and the total of the principal-adjustment column of Table 8-2 is \$21.65.

(*c*) In Table 8-1,

$$\frac{4.27}{4.11} \approx \frac{4.44}{4.27} \approx \frac{4.62}{4.44} \approx \frac{4.81}{4.62} \approx 1.04$$

In Table 8-2,

$$\frac{4.11}{3.92} \approx \frac{4.32}{4.11} \approx \frac{4.54}{4.32} \approx \frac{4.76}{4.54} \approx 1.05$$

If we carried full decimal accuracy, these relationships would be exact.

8.16 Construct the amortization schedule for the loan of (*a*) Problem 8.13; (*b*) Problem 8.14.

(*a*) Table 8-3 is the amortization schedule for the loan of Problem 8.13.

Table 8-3

Date	Payment	Interest at $i = 0.04$	Principal Repaid	Outstanding Principal
June 1, 1996	0	0	0	1022.26
Dec. 1, 1996	45.00	40.89	4.11	1018.15
June 1, 1997	45.00	40.73	4.27	1013.88
Dec. 1, 1997	45.00	40.56	4.44	1009.44
June 1, 1998	45.00	40.38	4.62	1004.82
Dec. 1, 1998	1045.00	40.19	1004.81	0.01*
TOTALS	1225.00	202.74	1022.25	

*The 1¢ error is due to roundoff.

(*b*) Table 8-4 is the amortization schedule for the loan of Problem 8.14.

Table 8-4

Date	Payment	Interest at $i = 0.05$	Principal Repaid	Outstanding Principal
June 1, 1996	0	0	0	978.35
Dec. 1, 1996	45.00	48.92	−3.92	982.27
June 1, 1997	45.00	49.11	−4.11	986.38
Dec. 1, 1997	45.00	49.32	−4.32	990.70
June 1, 1998	45.00	49.54	−4.54	995.24
Dec. 1, 1998	1045.00	49.76	995.24	0
TOTALS	1225.00	246.65	978.35	

8.17 A 20-year bond with annual coupons is bought at a premium to yield $j_1 = 9.5\%$. If the amount of amortization of the premium in the 3rd bond interest payment is $50, determine the amount of amortization of the premium in the 14th payment.

The entries in the principal-adjustment column are in the ratio $1 + i = 1.095$. Thus, the amount of amortization of the premium in the 14th payment is

$$50(1.095)^{11} = \$135.68$$

8.18 A $1000 bond, redeemable at 105 on October 1, 1997, pays semiannual coupons at $10\frac{1}{2}\%$. The bond is bought on April 1, 1995, to yield $j_{365} = 14\%$. Find the purchase price and construct a bond schedule.

Find i per half-year such that

$$(1 + i)^2 = \left(1 + \frac{0.14}{365}\right)^{365} \quad \text{or} \quad i = 0.072493786$$

From (8.1), the purchase price on April 1, 1995, is

$$P = 52.50a_{\overline{5}|.072493786} + 1050(1.072493786)^{-5} = \$953.80$$

Table 8-5 is the bond schedule.

Table 8-5

Date	Bond Interest Payment	Interest on Book Value at Yield Rate	Principal Adjustment	Book Value
Apr. 1, 1995	0	0	0	953.80
Oct. 1, 1995	52.50	69.14	-16.64	970.44
Apr. 1, 1996	52.50	70.35	-17.85	988.29
Oct. 1, 1996	52.50	71.64	-19.14	1007.43
Apr. 1, 1997	52.50	73.03	-20.53	1027.96
Oct. 1, 1997	52.50	74.52	-22.02	1049.98
TOTALS	262.50	358.68	-96.18	

8.5   PRICE OF A BOND BETWEEN BOND INTEREST DATES

Suppose that a bond is purchased between bond interest dates to yield the buyer interest at rate i. Letting

$P_0 \equiv$ price of a bond on the preceding bond interest date (just after a coupon has been paid)

$k \equiv$ fractional part of an interest period that has elapsed ($0 < k < 1$)

$P \equiv$ purchase price of the bond on the actual purchase date, called the *flat price*

the purchase price of the bond, using the *theoretical method*, that is, compound interest for the fractional part of an interest period, is given by

$$P = P_0(1 + i)^k \tag{8.3}$$

However, in reality, the *practical method*, that is, simple interest for the fractional part of an interest period, is used, and the purchase price is given by

$$P = P_0(1 + ki) \tag{8.4}$$

The practical method yields slightly larger values (see Problem 8.19) and it will be used in this Outline, unless stated otherwise.

The purchase price P may be considered as composed of two parts: the *market price*, Q, which is always equal to the book value of the bond, plus the *accrued bond interest*, I, as of the date of purchase. Defining

$P_1 \equiv$ price of the bond on the next bond interest date (just after a coupon has been paid)

we have

$$P_1 = (1 + i)P_0 - Fr \tag{8.5}$$

We can obtain the market price by linear interpolation between P_0 and P_1

$$Q = P_0 + k(P_1 - P_0) \tag{8.6}$$

The accrued bond interest is given by

$$I = kFr \tag{8.7}$$

When P_0 and P_1 are eliminated among (8.4), (8.5), and (8.6), the result is

$$P = Q + I \tag{8.8}$$

as stated above. (See Problem 8.20.)

If the actual purchase price P were quoted, there would be a big discontinuity in price at each coupon date, when the accrued bond interest would abruptly change from Fr to zero (see Fig. 8-3). Therefore, market price Q is quoted; or, rather market price of a \$100 bond, called the *market quotation*, q, is given. It is rounded off to the nearest eighth, but published in its decimal equivalent form.

SOLVED PROBLEMS

8.19 A \$2000 bond, redeemable at par on October 1, 1998, pays bond interest at $j_2 = 10\%$. Find the purchase price on June 16, 1996, to yield $j_2 = 9\%$ using (a) the theoretical method; (b) the practical method.

The preceding bond interest date is April 1, 1996. The exact time elapsed from April 1, 1996, to June 16, 1996, is 76 days. The exact time elapsed from April 1, 1996 to October 1, 1996, is 183 days. Thus, $k = \frac{76}{183}$.
We have $Fr = 2000(.05) = \$100$, $Ci = 2000(.045) = \$90$ and from (8.2)

$$P_0 = 2000 + (100 - 90)a_{\overline{5}|.045} = \$2043.90$$

(a) From (8.3) $P = P_0(1+i)^k = 2043.90(1.045)^{\frac{76}{183}} = \$2081.61$

(b) From (8.4) $P = P_0(1+ki) = 2043.90[1 + (\frac{76}{183})(.045)] = \$2082.10$

8.20 Show that (8.8) is equivalent to (8.4).

Substituting (8.5), (8.6), and (8.7) into (8.8)

$$\begin{aligned}
P &= Q + I \\
&= P_0 + k(P_1 - P_0) + kFr \\
&= P_0 + kP_1 - kP_0 + kFr \\
&= P_0 + k[(1+i)P_0 - Fr] - kP_0 + kFr \\
&= P_0 + kP_0 + kiP_0 - kFr - kP_0 + kFr \\
&= P_0 + kiP_0 \\
&= P_0(1+ki)
\end{aligned}$$

8.21 A \$1000 bond, redeemable at par on October 1, 1998, is paying bond interest at rate $j_2 = 9\%$. Find the purchase price on August 7, 1996, to yield 10% per annum compounded semiannually and determine the market price, accrued bond interest, and market quotation on August 7, 1996.

$$\begin{aligned}
P_0(\text{on Apr. 1}) &= 1000 + (45 - 50)a_{\overline{5}|.05} = \$978.35 \\
P(\text{on Aug. 7}) &= (978.35)\left[1 + \frac{128}{183}(0.05)\right] = \$1012.57 \\
P_1(\text{on Oct. 1}) &= (1.05)(978.35) - 45 = \$982.27 \\
Q(\text{on Aug. 7}) &= 978.35 + \frac{128}{183}(982.27 - 978.35) = \$981.09
\end{aligned}$$

Scaling down the bond to \$100 gives us the market quotation

$$q = \frac{981.09}{10} = 98.11 \approx 98\frac{1}{8}$$

Finally
$$I(\text{on Aug. 7}) = \frac{128}{183} \times 45 = \$31.48$$

As a check: $Q + I = 981.09 + 31.48 = \$1012.57 = P$

The relationship between the purchase price, P, the market price, Q, and the accrued bond interest, I, is shown in Fig. 8-3.

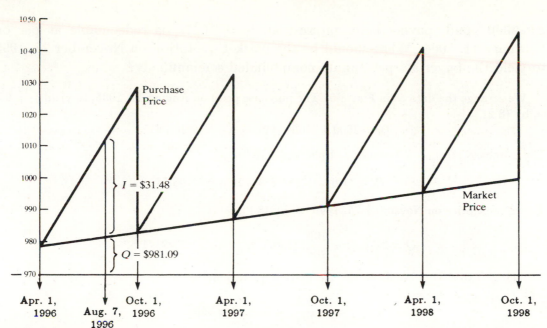

Fig. 8-3

8.22 A $1000 bond, redeemable at $1100 on November 9, 2006, has 11% coupons payable semiannually. Find the purchase price on April 29, 1996 if the desired yield is $j_\infty = 8\%$.

Find a rate of interest i per half-year such that

$$(1+i)^2 = e^{0.08}$$
$$i = e^{0.04} - 1 = 0.040810774$$

Then P_0 (Nov. 9, 1995) = $1100 + (55 - 1100i)a_{\overline{22}|i}$ = $1244.95. Now applying (8.4) and remembering that 1996 is a leap year,

$$P(\text{Apr. 29, 1996}) = 1244.95 \left(1 + \frac{172}{182}i\right) = \$1292.97$$

8.23 A $10\,000 bond with semiannual coupons at $9\frac{1}{2}\%$ is redeemable at par on August 25, 2005. This bond is sold on September 10, 1996, at a market quotation of $98\frac{7}{8}$. What did the buyer pay?

We arrange the data as in Fig. 8-4. The market price on September 10, 1996 is

$$Q = 100 \times 98\frac{7}{8} = \$9887.50$$

The accrued bond interest from August 25, 1996, to September 10, 1996, is

$$I = \frac{16}{184} \times 475 = \$41.30$$

Then, by (8.8), the total purchase price is

$$P = Q + I = 9887.50 + 41.30 = \$9928.80$$

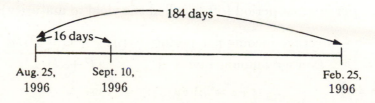

Fig. 8-4

8.24 A \$2000 bond, paying bond interest at $j_2 = 10\frac{1}{2}\%$, is redeemable at par on February 11, 2004. What should be the market quotation on November 25, 1995 to yield the buyer 9% per annum compounded semiannually?

We arrange the data as in Fig. 8-5. The purchase price on August 11, 1995, to yield $j_2 = 9\%$ is, by (8.2),

$$P_0 = 2000 + (105 - 90)a_{\overline{17}|.045} = \$2175.61$$

The purchase price on February 11, 1996, is, by (8.5),

$$P_1 = (1.045)(2175.61) - 105 = \$2168.51$$

The market price on November 25, 1995, is, by (8.6),

$$Q = 2175.61 + \frac{106}{184}(2168.51 - 2175.61) = \$2171.52$$

Reducing Q to a \$100 bond, we get the market quotation

$$q = \frac{2171.52}{20} = 108.58 \approx 108\frac{5}{8}$$

Alternate Solution

$$P(\text{Nov. 25, 1995}) = P_0(1 + ki) = 2175.61\left[1 + \frac{106}{184}(0.045)\right] = \$2232.01$$

The accrued bond interest from August 11, 1995 to November 25, 1995, is

$$I = \frac{106}{184} \times 105 = \$60.49$$

Then the market price on November 25, 1995, is

$$Q = P - I = 2232.01 - 60.49 = \$2171.52$$

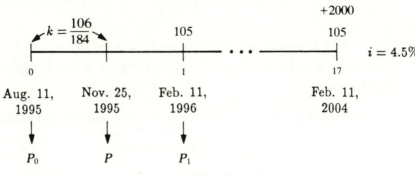

Fig. 8-5

8.6   FINDING THE YIELD RATE

Method of Averages (Bond Salesman's Method)

The yield rate per interest period (often called the yield to maturity) is approximated as

$$i \approx \frac{\text{average income per period}}{\text{average amount invested}} = \frac{(nFr + C - P)/n}{(P + C)/2}$$

(See Problem 8.24.)

If a more accurate answer is desired, the Method of Averages should be followed by the Method of Interpolation.

Method of Interpolation

This method requires determining market prices of a bond for two interest rates, such that one price is smaller and the other is greater than the given quoted price. Linear interpolation is then used to find i or j_m. If convenient, the interpolation can be on purchase price rather than market price.

The Method of Averages can be used to determine a starting point for the Method of Interpolation. (See Problem 8.26.)

SOLVED PROBLEMS

8.25 A $2000 bond, paying semiannual coupons at $9\frac{1}{2}\%$ and redeemable at par on July 20, 2009, is quoted at $96\frac{1}{2}$ on July 20, 1995. Find an approximate value of the yield rate to maturity, j_2, using the Method of Averages.

The purchase price is $P = 20 \times 96\frac{1}{2} = \$1930$, since the bond is sold on a bond interest date. If held to maturity (28 periods), the buyer will have realized $28 \times 95 = \$2660$ in coupons, plus $2000 - 1930 = \$70$ in capital gains; or average income per period of

$$\frac{2660 + 70}{28} = \$97.50$$

The average amount invested is

$$\frac{1930 + 2000}{2} = \$1965$$

Thus, the approximate value of the yield rate per half-year is

$$i \approx \frac{97.50}{1965} = 0.0496 = 4.96\%$$

or $j_2 \approx 9.92\%$.

8.26 Taking the answer to Problem 8.25 as a starting point, find a more accurate answer using the Method of Interpolation.

We compute the market prices (which, here, are equal to the purchase prices) to yield $j_2 = 9\%$ and $j_2 = 10\%$.

$$Q(\text{to yield } j_2 = 9\%) = 2000 + (95 - 90)a_{\overline{28}|.045} = \$2078.71$$
$$Q(\text{to yield } j_2 = 10\%) = 2000 + (95 - 100)a_{\overline{28}|.05} = \$1925.51$$

Arranging the data in an interpolation table, we have:

		Q	j_2		
		2078.71	9%		
	148.71				
153.20		1930.00	j_2	x	1%
		1925.51	10%		

$$\frac{x}{1\%} = \frac{148.71}{153.20}$$

$$x = 0.97\%$$

$$j_2 \approx 9.97\%$$

Note $Q(\text{to yield } j_2 = 9.97\%) = 2000(95 - 99.70)a_{\overline{28}|.04985} = \$1929.86$, or $q = 96.493$.

8.27 A \$1000 bond with semiannual coupons at 12% matures at par on June 1, 2006. On February 3, 1996, this bond is quoted at $94\frac{7}{8}$. Find the approximate yield rate to maturity, using the Method of Averages.

Since we are looking only for an approximate answer, we may assume that the bond was quoted on the nearest coupon date, December 1, 1995. Then the investor will receive 21 coupons of \$60 each, plus capital gains of $1000 - 948.75 = \$51.25$.

$$\text{average income per period} = \frac{21(60) + 51.25}{21} = \$62.44$$

$$\text{average amount invested} = \frac{948.75 + 1000}{2} = \$974.38$$

$$i \approx \frac{62.44}{974.38} = 0.0641 = 6.41\%$$

$$j_2 \approx 12.82\%$$

8.28 Find a more accurate estimate of the yield rate to maturity in Problem 8.27, by using the Method of Interpolation.

Select the two yield rates $j_2 = 12\%$ and $j_2 = 13\%$, and compute the corresponding market prices on February 3, 1996, using the time diagram given in Fig. 8-6.

At yield rate $j_2 = 12\%$, $Q = \$1000$ (since the yield rate coincides with the bond rate).

At $j_2 = 13\%$,
$$P_0 = 1000 + (60 - 65)a_{\overline{21}|.065} = \$943.58$$
$$P_1 = (1.065)(943.58) - 60 = \$944.91$$
$$Q = 943.58 + \frac{64}{182}(944.91 - 943.58) = \$944.05$$

Arranging the data in an interpolation table, we have:

$$
55.95 \left\{ 51.25 \left\{
\begin{array}{c|c}
Q \text{ (on Feb. 3, 1996)} & j_2 \\
\hline
1000.00 & 12\% \\
\\
948.75 & j_2 \\
\\
944.05 & 13\%
\end{array}
\right. \left. \right\} x \right. \right\} 1\%
$$

$$\frac{x}{1\%} = \frac{51.25}{55.95}$$

$$x = 0.92\%$$

$$j_2 \approx 12.92\%$$

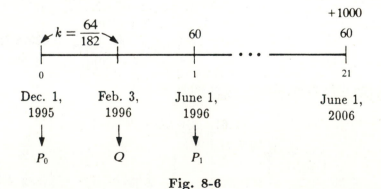

Fig. 8-6

8.29 A \$1000 bond, redeemable at par in 20 years, has semiannual coupons at 11%. It is callable at \$1050 at the end of 15 years. If it is quoted at 96, find the yield rate, j_2, using the Method of Interpolation and assuming (a) it is called, (b) it is not called.

(a) We want a yield rate $j_2 = 2i$ such that

$$960 = 1050 + (55 - 1050i)a_{\overline{30}|i}$$

By trial and error, we find

$$\text{at } j_2 = 11\%; \quad P = 1050 + (55 - 57.75)a_{\overline{30}|.055} = \$1010.03$$
$$\text{at } j_2 = 12\%; \quad P = 1050 + (55 - 63)a_{\overline{30}|.06} = \$939.88$$

Arranging the data in an interpolation table, we have:

	P	j_2
	1010.03	11%
50.03	960.00	j_2
	939.88	12%

$$\frac{x}{1\%} = \frac{50.03}{70.15}$$
$$x = 0.71\%$$
$$j_2 \approx 11.71\%$$

70.15 $\Big\{$ 50.03 $\Big\{$ ... $\Big\}$ x $\Big\}$ 1%

(b) We want a yield rate $j_2 = 2i$ such that

$$960 = 1000 + (55 - 1000i)a_{\overline{40}|i}$$

We know that at $j_2 = 11\%$ (which equals the bond rate), $P = \$1000$; and, at $j_2 = 12\%$,

$$P = 1000 + (55 - 60)a_{\overline{40}|.06} = \$924.77$$

Arranging the data in an interpolation table, we have:

	P	j_2
	1000.00	11%
40.00	960.00	j_2
	924.77	12%

$$\frac{x}{1\%} = \frac{40.00}{75.23}$$
$$x = 0.53\%$$
$$j_2 \approx 11.53\%$$

75.23 $\Big\{$ 40.00 $\Big\{$ ... $\Big\}$ x $\Big\}$ 1%

8.30 Mr. X buys a \$5000 bond that pays semiannual coupons at 10% and is redeemable at par in 10 years. The price he pays will provide a yield of $j_2 = 13\%$ if he holds the bond to maturity. After 5 years, Mr. X sells this bond to Mrs. Y, whose desired yield is $j_2 = 11\%$. (a) What price did Mr. X pay? (b) What price did Mrs. Y pay? (c) What yield, j_2, did Mr. X realize?

(a) $$P_X = 5000 + (250 - 325)a_{\overline{20}|.065} = \$4173.61$$

(b) $$P_Y = 5000 + (250 - 275)a_{\overline{10}|.055} = \$4811.56$$

(c) We want to find $j_2 = 2i$ such that

$$4173.61 = 250a_{\overline{10}|i} + 4811.56(1 + i)^{-10}$$

By trial and error, we find

$$\text{at } j_2 = 14\%; \quad 250a_{\overline{10}|.07} + 4811.56(1.07)^{-10} = \$4201.85$$
$$\text{at } j_2 = 15\%; \quad 250a_{\overline{10}|.075} + 4811.56(1.075)^{-10} = \$4050.56$$

Arranging the data in an interpolation table, we have

$$151.29 \left\{ 28.24 \left\{ \begin{array}{c|c} P & j_2 \\ \hline 4201.85 & 14\% \\ \\ 4173.61 & j_2 \\ \\ 4050.56 & 15\% \end{array} \right\} x \right\} 1\% \qquad \frac{x}{1\%} = \frac{28.24}{151.29}$$

$$x = 0.19\%$$

$$j_2 \approx 14.19\%$$

8.7   OTHER TYPES OF BONDS

Serial Bonds

To borrow money, companies sometimes issue a series of bonds with staggered redemption dates instead of with a common redemption date. Such bonds are called *serial bonds*. Serial bonds can be thought of simply as several bonds covered under one bond indenture. If the redemption date of each individual bond is known, then the valuation of any one bond can be performed by methods already described. The value of the entire issue of the bonds is just the sum of the values of the individual bonds. (See Problem 8.31.)

Strip Bonds

Investors have several reasons for choosing one investment over another. They may look for cash flows whose timing matches their needs. For some investors, the coupons on a bond match their cash flow needs (for example, a corporation responsible for paying monthly retirement benefits to workers). For other investors, the redemption value of a bond matches their cash flow needs (for example, an insurance company which needs to satisfy the maturity value of an endowment insurance contract).

For these reasons, some investors separate the coupons from the redemption value of the bond. An investor may buy the bond contract, "strip" the coupons from the bond and sell the remainder of the asset, that is, the redemption value. This redemption-value-only bond is called a "strip" bond. (See Problem 8.34.) The original buyer may also sell each coupon as a separate asset. (See Problem 8.35.)

Annuity Bonds

An *annuity bond*, with face value F, is a contract promising the payment of an annuity whose present value is F at the bond rate. When the face value and the bond rate are given, the periodic payment R of the bond is computed by the methods developed in Chapters 5 and 6. At any date, the price of the annuity bond is obtained as the present value of the future payments of the bond at the investor's interest rate. (See Problems 8.36 and 8.37.)

SOLVED PROBLEMS

8.31 The directors of a company authorized the issuance of $30 000 000 of serial bonds on September 1, 1994. Interest is to be paid annually on September 1 at rate 9%. The indenture provides that (i) $5 000 000 of the issue is to be redeemed on September 1, 1999; (ii) $10 000 000 of the issue is to be redeemed on September 1, 2004; (iii) $15 000 000 of the issue is to be redeemed on September 1, 2009. Find the purchase price of the issue on September 1, 1994, that would yield $j_1 = 8\%$.

We consider the serial issue to be composed of 3 subissues, as shown in Fig. 8-7. The purchase price of the entire serial issue, P, on September 1, 1994, is the sum of the purchase prices of the 3 subissues, $P^{(1)}$, $P^{(2)}$, $P^{(3)}$.

$$P^{(1)} = 5\ 000\ 000 + (450\ 000 - 400\ 000)a_{\overline{5}|.08} \qquad = \qquad 5\ 199\ 635.50$$
$$P^{(2)} = 10\ 000\ 000 + (900\ 000 - 800\ 000)a_{\overline{10}|.08} \qquad = \qquad 10\ 671\ 008.14$$
$$P^{(3)} = 15\ 000\ 000 + (1\ 350\ 000 - 1\ 200\ 000)a_{\overline{15}|.08} \qquad = \qquad \underline{16\ 283\ 921.70}$$
$$P \quad = \quad \$32\ 154\ 565.34$$

8.32 What would be the value of the serial bonds of Problem 8.31 on September 1, 2001?

On the given date, the first subissue will have already been redeemed. Therefore:

$$P^{(2)} = 10\ 000\ 000 + (900\ 000 - 800\ 000)a_{\overline{3}|.08} \qquad = \qquad 10\ 257\ 709.70$$
$$P^{(3)} = 15\ 000\ 000 + (1\ 350\ 000 - 1\ 200\ 000)a_{\overline{8}|.08} \qquad = \qquad \underline{15\ 861\ 995.83}$$
$$P \quad = \quad \$26\ 119\ 705.53$$

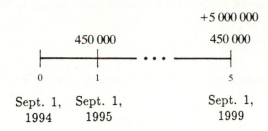

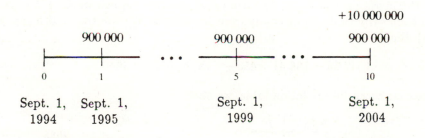

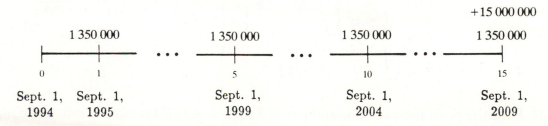

Fig. 8-7

8.33 The extension of Makeham's formula (Problem 8.8) to an entire serial issue, redeemable at par, is

$$P = \sum_k F^{(k)}(1+i)^{-t_k} + \frac{r}{i}\left[\sum_k F^{(k)} - \sum_k F^{(k)}(1+i)^{-t_k}\right]$$

where

$$P \equiv \text{purchase price of the entire issue}$$
$$F^{(k)}(1+i)^{-t_k} \equiv \text{present value of the } k\text{th redemption}$$
$$F^{(k)} \equiv \text{face value of the } k\text{th subissue}$$

Rework Problem 8.31 by means of this formula.

We have $r = 0.09$; $i = 0.08$; $F^{(1)} = 5\,000\,000$, $t_1 = 5$; $F^{(2)} = 10\,000\,000$, $t_2 = 10$; $F^{(3)} = 15\,000\,000$, $t_3 = 15$. Then, $\sum_k F^{(k)} = 30\,000\,000$,

$$\sum_k F^{(k)}(1+i)^{-t_k} = 5\,000\,000(1.08)^{-5} + 10\,000\,000(1.08)^{-10} + 15\,000\,000(1.08)^{-15} = \$12\,763\,476$$

and
$$P = 12\,763\,476 + \frac{0.09}{0.08}(30\,000\,000 - 12\,763\,476) = \$32\,154\,565$$

8.34 A \$1000 corporate bond, paying bond interest at $j_2 = 9\%$, is redeemable at par in 20 years. Investor A buys the bond to yield $j_2 = 8.5\%$. Investor A strips the coupons from the bond and sells the remaining "strip" bond to Investor B, who wishes a yield rate of $j_{12} = 9\%$. Find (a) the price Investor B paid for the strip bond; (b) the yield rate j_2 realized by Investor A.

(a) Investor B will pay the discounted value of \$1000 payable in 20 years at $j_{12} = 9\%$, or $i = 0.0075$ per month over 240 months.

$$P_B = 1000(1.0075)^{-240} = \$166.41$$

(b) First we calculate the price Investor A paid for the bond. With $F = C = 1000$, $r = 0.045$, $i = 0.0425$, and $n = 40$,

$$P_A = 1000 + (45 - 42.50)a_{\overline{40}|.0425} = \$1047.69$$

In return for the investment of \$1047.69, Investor A gets 40 coupons worth \$45 each over 20 years plus \$166.41 payable immediately from Investor B. We want to find a rate i per half-year, such that

$$1047.69 = 166.41 + 45a_{\overline{40}|i} \quad \text{or} \quad a_{\overline{40}|i} = 19.5840$$

Arranging the data in an interpolation table, we have:

		$a_{\overline{40}|i}$	j_2			
		19.7928	8%		$\dfrac{x}{1\%} = \dfrac{.2088}{1.3919}$	
1.3912	.2088	19.5840	j_2	x	$x = .15\%$	1%
		18.4016	9%		$j_2 \approx 8.15\%$	

8.35 Find the price Investor C would pay in Problem 8.34 if Investor A sells the coupons to Investor C whose desired yield is $j_{12} = 7.5\%$.

Investor C will pay the discounted value of 40 semiannual coupons worth \$45 each, at $j_{12} = 7.5\%$, or $i = (1 + \frac{0.075}{12})^6 - 1$

$$P_C = 45a_{\overline{40}|i} = \$916.55$$

8.36 An annuity bond promises to repay $50 000 principal with interest at $j_2 = 12\%$ in equal instalments at the end of each half-year for 10 years. How much will an investor offer for this bond if he wants to realize $j_2 = 13\%$?

First we find the semiannual payment, R, of an ordinary simple annuity:

$$R = \frac{50\ 000}{a_{\overline{20}|.06}} = \$4359.23$$

The price, P, the investor will pay is equal to the present value, at $j_2 = 13\%$, of the remaining payments (here, of all payments):

$$P = 4359.23 a_{\overline{20}|.065} = \$48\ 032.21$$

8.37 For the annuity bond of Problem 8.36, find the purchase price with 5 years remaining on the bond, if the prospective investor's desired yield is $j_{365} = 10\%$.

Find rate i per half-year such that

$$(1+i)^2 = \left(1 + \frac{0.10}{365}\right)^{365} \quad \text{or} \quad i = 0.051263897$$

Then
$$P = 4359.23 a_{\overline{10}|i} = \$33\ 455.17$$

Supplementary Problems

**PURCHASE PRICE TO YIELD A
GIVEN INVESTMENT RATE**

8.38 Find the purchase price of each bond in Table 8-6, using (8.1).

Table 8-6

	Face Value	Redemption	Bond Interest Rate, j_2	Years to Redemption	Yield Rate
(a)	$5 000	at par	11%	20	$j_2 = 14\%$
(b)	1 000	at par	10%	15	$j_2 = 9\%$
(c)	2 000	at par	12%	10	$j_4 = 13\%$
(d)	500	at par	$10\frac{1}{2}\%$	20	$j_{12} = 12\%$
(e)	1 000	at 105	9%	18	$j_2 = 13\%$
(f)	5 000	at 110	12%	7	$j_\infty = 10\%$
(g)	10 000	at 103	$12\frac{1}{2}\%$	14	$j_{12} = 9\%$
(h)	2 000	at 110	10%	$2\frac{1}{2}$	$j_2 = 12\frac{1}{2}\%$

Ans. (a) $4000.12; (b) $1081.44; (c) $1889.81; (d) $443.58; (e) $729.37; (f) $5676.82; (g) $12 681.51; (h) $2043.10

8.39 Redo Problem 8.38, using (8.2).

8.40 The ACME Corporation needs to build a new plant. It issues $500 000 worth of 20-year bonds with semiannual coupons at 11%. These bonds are redeemable at 105. The entire issue is purchased by an insurance company, whose required yield rate is $j_2 = 14\%$. Find the purchase price. *Ans.* $401 681.69

8.41 Redo Problem 8.40, assuming now that the bonds are redeemable at par, using Makeham's purchase-price formula (Problem 8.8). *Ans.* $400 012.18

8.42 (*a*) Mr. Green buys $2000 bond paying bond interest at $j_2 = 10\frac{1}{2}\%$ and redeemable at par in 20 years to yield $j_4 = 12\%$. Find the purchase price. (*b*) After exactly 5 years, he sells the bond to another investor, whose desired yield rate is $j_1 = 9\frac{1}{2}\%$. Find the sale price.
Ans. (*a*) $1750.06; (*b*) $2194.72

8.43 The XYZ Corporation needs to raise some funds to pay for new equipment. They issue $1 000 000 worth of 20-year bonds with semiannual coupons at $j_2 = 12\%$. These bonds are redeemable at 105. At the time of issue, interest rates in the market place are $j_{12} = 10.5\%$. How much money did they raise? *Ans.* $1 109 694.04

8.44 A corporation issues $600 000 worth of 12-year bonds with semiannual coupons at $j_2 = 10\%$. The bonds are priced to yield $j_2 = 9\%$. Determine the issue price per $100 unit. *Ans.* $107.25

8.45 A $1000 bond paying coupons at $j_2 = 12\%$ is redeemable at par in 20 years. Find the price to yield an investor (*a*) $j_2 = 14\%$; (*b*) $j_2 = 12\%$; (*c*) $j_2 = 10\%$.
Ans. (*a*) $866.68; (*b*) $1000; (*c*) $1171.59

8.46 An X bond quoted as redeemable at 105 in 10 years is purchased to yield $j_2 = 10\%$. If this same bond was redeemable at par, the actual purchase price would be $113.07 less. Determine the value of X. *Ans.* $6000

8.47 A corporation issues a special 20-year bond issue that has no coupons. Rather, interest will accumulate on the bond at rate $j_2 = 11\%$ for the life of the bond. At the time of maturity, the total value of the loan will be paid off, including all accumulated interest. Find the price of a $1000 bond of this issue to yield $j_2 = 10\%$. *Ans.* $1209.28

8.48 A $1000 bond bearing coupons at $j_2 = 13\%$ and redeemable at par is bought to yield $j_2 = 12\%$. If the present value of the redemption value at this yield is $140, what is the purchase price? *Ans.* $1071.67

8.49 Two $1000 bonds redeemable at par at the end of n years are bought to yield $j_2 = 10\%$. One bond costs $1153.72 and has semiannual coupons at $j_2 = 12\%$. The other bond has semiannual coupons at $j_2 = 8\%$. Find the price of the second bond. *Ans.* $846.28

8.50 A $1000 bond with semiannual coupons at $j_2 = 12\%$ and redeemable at the end of n years at $1050 sells at $930 to yield $j_2 = 15\%$. Find the price of a $1000 bond with semiannual coupons at $j_2 = 10\%$ redeemable at the end of $2n$ years at $1040 to yield $j_2 = 15\%$. *Ans.* $767.62

8.51 A bond with a par value of $100 000 has coupons at the rate $j_2 = 13\%$. It will be redeemed at par when it matures a certain number of years hence. It is purchased for a price of $94 702.99. At this price, the purchaser who holds the bond to maturity will realize a yield rate $j_2 = 14\%$. Find the number of years to maturity. *Ans.* 10

8.52 A $1000 bond with annual coupons at 8.5% and maturing in 20 years at par is purchased to yield an annual effective rate of interest of 9% if held to maturity. The book value of the bond at any time is the discounted value of all remaining payments, using the 9% rate. Ten years later, just after a coupon payment, the bond is sold to yield the new purchaser a 10% annual effective rate of interest if held to maturity. Find the excess of the book value over the second sale price. *Ans.* $60.08

8.53 A corporation has an issue of bonds with annual coupons at $j_1 = 10\%$ maturing in 5 years at par which are quoted at a price that yields $j_1 = 12\%$. (*a*) What is the price of a $1000 bond? (*b*) It is proposed to replace this issue of bonds with an issue of bonds with annual coupons at $j_1 = 11\%$. How long must the new issue run so that the bond holders will still yield $j_1 = 12\%$? Express your answer to the nearest year. *Ans.* (*a*) $927.90; (*b*) 18

8.54 A $1000 bond bearing semiannual coupons at $j_2 = 10\%$ is redeemable at par. What is the minimum number of whole years that the bond should run so that a person paying $1100 for it would earn at least $j_2 = 8\%$? *Ans.* 7

8.55 If the coupon rate on a bond is 1.5 times the yield rate when it sells for a premium of $10 per $100, find the price per $100 for a bond with the same number of coupons to run and the same yield with coupons equal to $\frac{3}{4}$ of the yield rate. *Ans. $95*

CALLABLE BONDS

8.56 A $5000 bond with semiannual coupons at $j_2 = 11\%$ is redeemable at par in 20 years; it is callable at par in 15 years. Find the price to guarantee a yield of (a) $j_2 = 13\%$; (b) $j_2 = 9\%$. *Ans.* (a) $4292.72; (b) $5814.44

8.57 A $2000 bond paying interest at $j_2 = 10\%$ is redeemable at par in 20 years. It is callable at par in 15 years. Find the price to guarantee a yield rate of (a) $j_2 = 8\%$; (b) $j_2 = 12\%$. *Ans.* (a) $2345.84; (b) $1699.07

8.58 A $1000 bond paying bond interest at $j_2 = 13\%$ is redeemable at par in 20 years or callable at 105 in 15 years. Find the price to guarantee a yield of (a) $j_2 = 15\%$, (b) $j_2 = 11\%$. *Ans.* (a) $874.06; (b) $1160.46

8.59 A $2000 bond with semiannual coupons at 12% is redeemable at par in 20 years. It is callable after 10 years at 110, and after 15 years at 105. Find the price to guarantee a yield of $j_\infty = 11\%$. *Ans. $2108.81*

8.60 A $2000 bond with semiannual coupons at $j_2 = 13\%$ is redeemable at par in 20 years. It is callable at a 5% premium in 15 years. Find the price to guarantee a yield rate of (a) $j_4 = 16\%$; (b) $j_1 = 11\%$. *Ans.* (a) $1610.79; (b) $2358.61

8.61 A $1000 bond with coupons at $j_2 = 11\%$ is redeemable at par in 20 years. It also has the following call options:

Call Date	Redemption
15 Years	105
16 years	104
17 years	103
18 years	102
19 years	101

Find the price to guarantee a yield rate of (a) $j_1 = 10\%$; (b) $j_{12} = 12\%$. *Ans.* (a) $1107.63; (b) $903.75

PREMIUM AND DISCOUNT

8.62 For each bond in Table 8-7, find the purchase price and make out a complete bond schedule showing the amortization of the premium or accumulation of the discount.

Table 8-7

	Face Value	Redemption	Bond Interest Rate, j_2	Years to Redemption	Yield Rate
(a)	$500	at par	11%	3	$j_2 = 9\%$
(b)	10 000	at par	$10\frac{1}{2}\%$	$2\frac{1}{2}$	$j_2 = 12\%$
(c)	1 000	at par	12%	3	$j_4 = 10\%$
(d)	2 000	110	9 %	3	$j_{12} = 12\%$
(e)	1 000	105	12%	$2\frac{1}{2}$	$j_{365} = 14\%$
(f)	5 000	103	11%	$2\frac{1}{2}$	$j_2 = 10\%$

Ans. (a) $525.79; (b) $9684.07; (c) $1047.49; (d) $1978.09; (e) $984.35; (f) $5225.77

8.63 A \$1000-par-value bond, with semiannual coupons at 12% payable March 1 and September 1, has a book value of \$1075 on March 1, 1995, the underlying yield rate having been $j_2 = 10\%$. Find the amount for amortization of the premium on September 1, 1995, and the new book value at that time. *Ans.* \$6.25, \$1068.75

8.64 A 20-year bond with semiannual coupons is bought at a premium to yield $j_2 = 12\%$. If the amount of amortization of the premium at the time of the 4th coupon is \$4, determine the amount of amortization of the premium at the time of the 15th coupon. *Ans.* \$7.59

8.65 A \$1000 bond, redeemable at par, with annual coupons at 13%, is purchased for \$1184.34. If the write-down in the book value is \$11.57 at the end of the first year, what is the write-down at the end of the fifth year? *Ans.* \$16.94

8.66 A bond with \$65 semiannual coupons is purchased to yield $j_2 = 12\%$. If the first write-down is \$0.87, find the purchase price of the bond. *Ans.* \$1068.83

8.67 A \$1000 bond, redeemable at par, with annual coupons at 10% is purchased for \$1060. If the write-down in the book value is \$7 at the end of the first year, what is the write-down at the end of the 4th year? *Ans.* \$9.01

8.68 A bond with \$160 annual coupons is purchased at a discount to yield $j_1 = 15\%$. The write-up for the first year is \$44. What was the purchase price? *Ans.* \$1360

8.69 A \$1000 bond redeemable at \$1050 on December 1, 1998, pays interest at $j_2 = 13\%$. The bond is bought on June 1, 1996. Find the price and construct a bond schedule if the desired yield is (a) $j_{12} = 12\%$; (b) $j_1 = 11\%$. *Ans.* (a) \$1051.69; (b) \$1087.54

8.70 A \$1000, 20-year par value bond with semiannual coupons is bought at a discount to yield $j_2 = 10\%$. If the amount of the write-up of the discount in the last entry in the schedule is \$5, find the purchase price of the bond. *Ans.* \$909.91

8.71 A \$1000 bond pays coupons at $j_2 = 14\%$ on January 1 and July 1 and will be redeemed at par on July 1, 2000. If the bond was bought on January 1, 1992, to yield 12% per annum compounded semiannually, find the interest due on the book value on January 1, 1996. *Ans.* \$64.42

8.72 A \$1000 bond providing annual coupons at $j_1 = 9\%$ is redeemable at par on November 1, 2000. The write-down in the first year was \$5.63. The write-down in the eleventh year was \$19.08. Determine the book value of the bond on November 1, 1996. *Ans.* \$881.53

8.73 A \$2000 bond with annual coupons matures at par in 5 years. The first interest coupon is \$400, with subsequent coupons reduced by 25% of the previous year's coupon, each year. (a) Find the price to yield $j_1 = 10\%$. (b) Draw up the bond schedule. *Ans.* (a) \$2216.30

8.74 A 10-year bond matures for \$2000 and has annual coupons. The first coupon is \$100, and each increases by 10%. The bond is priced to yield $j_1 = 9\%$. Find the price and draw up the bond schedule. *Ans.* \$1801.07

8.75 A \$1000 bond with semiannual coupons at $j_2 = 10\%$ is redeemable for \$1100. If the amount for the 16th write-up is \$5, calculate the purchase price to yield $j_2 = 12\%$. *Ans.* \$868.11

8.76 You are told that a \$1000 bond with semiannual coupons at $j_2 = 8\%$ redeemable at par will be sold at \$700 to an investor requiring 12% per annum compounded semiannually. (a) Find the price of this bond to the same investor if the above coupon rate were changed to $j_2 = 11\%$. (b) Is the 11% bond purchased at a premium or a discount? (c) For the 11% bond show the write-up (or write-down) entries at the first two coupon dates.
Ans. (a) \$925; (b) Discount; (c) \$0.50, \$0.53

8.77 A \$10 000, 15-year bond is priced to yield $j_4 = 12\%$. It has quarterly coupons of \$200 each the first year, \$215 each the second year, \$230 each the third year, ..., \$410 each the fifteenth year. (a) Show that the price is \$9267.05. (b) Find the book value after 14 years. (c) Draw up a partial bond schedule showing the first and last year's entries only.
Ans. (b) \$10 408.88

PRICE OF A BOND BETWEEN BOND INTEREST DATES

8.78 Find the purchase price of the bonds of Table 8-8.

Table 8-8

	Face Value	Bond Interest Rate, j_2	Yield Rate	Redemption	Date of Purchase
(a)	$500	$9\frac{1}{2}\%$	$j_2 = 11\%$	Jan. 1, 2013, at par	May 8, 1995
(b)	1000	11%	$j_2 = 10\%$	Jan. 1, 2008, at par	Oct. 3, 1994
(c)	2000	10%	$j_2 = 14\%$	Nov. 1, 2004, at par	July 20, 1995
(d)	1000	12%	$j_4 = 10\%$	Feb. 1, 2013, at par	Oct. 27, 1995
(e)	5000	$10\frac{1}{2}\%$	$j_{12} = 12\%$	July 1, 2004, at 105	July 30, 1994
(f)	2000	13%	$j_2 = 10\%$	Oct. 1, 2009, at 110	Apr. 17, 1995

Ans. (a) $438.62; (b) $1100.63; (c) $1472.05; (d) $1179.89; (e) $4609.03; (f) $2502.82

8.79 Find the purchase price of each $1000 bond in Table 8-9 if bought at the given market quotation.

Table 8-9

	Redemption Value	Bond Interest Rate, j_2	Market Quotation	Redemption Date	Date of Purchase
(a)	par	11%	$101\frac{7}{8}$	Sept. 1, 2002	June 8, 1995
(b)	par	$10\frac{1}{2}\%$	$96\frac{1}{2}$	Feb. 1, 1998	Oct. 2, 1995
(c)	$1050	9%	$92\frac{1}{8}$	Oct. 1, 1999	Nov. 29, 1996
(d)	1100	13%	$104\frac{1}{4}$	Apr. 1, 1998	Jan. 12, 1995

Ans. (a) $1048.34; (b) $982.69; (c) $935.76; (d) $1079.08

8.80 What would be the market quotations on the $1000 bonds of Table 8-10?

Table 8-10

	Redemption Value	Bond Interest Rate, j_2	Yield Rate	Redemption Date	Date of Purchase
(a)	par	9%	$j_2 = 12\%$	Nov. 1, 2012	Feb. 8, 1995
(b)	par	13%	$j_2 = 10\%$	Mar. 1, 2014	Aug. 19, 1997
(c)	$1050	$10\frac{1}{2}\%$	$j_4 = 12\%$	June 1, 2008	Oct. 30, 1995
(d)	1100	11%	$j_2 = 9\%$	Oct. 1, 2006	Nov. 2, 1996

Ans. (a) $78\frac{1}{8}$; (b) 124; (c) $90\frac{1}{2}$; (d) $117\frac{1}{8}$

8.81 A $1000 bond, redeemable at par on October 1, 1997, is paying semiannual coupons at $j_2 = 10\%$. Find the purchase price on August 7, 1995 to yield $j_2 = 13\%$. Draw a diagram similar to Fig. 8-3 for this bond. *Ans.* $980.30

8.82 A $1000 bond, redeemable at $1100 on November 7, 2004, has coupons at $j_2 = 11\%$. Find the purchase price on April 18, 1995, if the desired yield is (a) $j_{12} = 13\%$; (b) $j_1 = 9\%$. *Ans.* (a) $953.16; (b) $1232.88

8.83 An investment company is being audited and must locate the complete records of a specific bond transaction. The bond was a $500 bond with bond interest at rate $j_2 = 12\%$ redeemable at par on January 1, 2010. It was purchased for $550.89 sometime between July 1, 1995, and January 1, 1996. Find the exact date if the bond was purchased to yield $j_2 = 11\%$. *Ans.* October 3, 1995

8.84 A National Auto Company Limited $1000 bond is due at par on December 1, 2006. Interest is payable at $j_2 = 12\%$ on June 1 and December 1. The bond may be called at 104 on December 1, 2000. Find the flat (purchase) price and the market price for this bond on August 8, 1995 if the yield is to be $j_2 = 10.5\%$, (a) assuming the bond is called at December 1, 2000; (b) assuming the bond matures at par on December 1, 2006. *Ans.* (a) $1083.12; (b) $1097.96

8.85 A $5000 bond with semiannual coupons at $j_2 = 9\%$ is redeemable at par on November 1, 2013. (a) Find the price on November 1, 1993 to yield $j_4 = 10\%$. (b) Find the book value of the bond on May 1, 1996, (just after the coupon is cashed). (c) What should the market quotation of this bond be on August 17, 1996 if the buyer wants a yield of $j_1 = 7\%$? *Ans.* (a) $4521.50; (b) $4543.09; (c) $121\frac{1}{8}$

8.86 (a) A $1000 bond paying interest at $j_2 = 10\%$ is redeemable at par on September 1, 2013. Find the price on its issue date of September 1, 1993 to yield $j_2 = 12\%$. (b) Find the book value of the bond on September 1, 1995, (just after the coupon is cashed). (c) Find the sale price of this bond on September 1, 1995, if the buyer wants a yield of (i) 9% compounded semiannually; (ii) 15% compounded semiannually. (d) What should the market quotation of this bond be on October 8, 1995 to yield a buyer $j_2 = 11\%$? *Ans.* (a) $849.54; (b) $853.79; (c) (i) $1088.33; (ii) $691.34; (d) $92\frac{1}{4}$

8.87 The ABC Corporation $5000 bond that pays interest at $j_2 = 11\%$ matures at par on October 1, 2007. (a) What did the buyer pay for the bond if it was sold on July 28, 1995, at a market quotation of $89\frac{3}{8}$? (b) What should be the market quotation of this bond on July 28, 1997, to yield a buyer $j_{12} = 9\%$? (c) What should be the market quotation of this bond on December 13, 1997, to yield a buyer $j_1 = 12\%$? *Ans.* (a) $4646.07; (b) $113\frac{1}{4}$; (c) $96\frac{1}{4}$

FINDING THE YIELD RATE

8.88 Find the yield rate of each bond in Table 8-11 by the Method of Averages.

Table 8-11

	Face Value	Redemption Value	Bond Interest Rate, j_2	Years to Redemption	Purchase Price
(a)	$2000	at par	$10\frac{1}{2}$%	12	$1920
(b)	1000	at par	9%	10	910
(c)	5000	at 105	13%	$8\frac{1}{2}$	5860
(d)	500	at 110	11%	15	580

Ans. (a) 11.05%; (b) 10.37%; (c) 10.41%; (d) 9.38%

8.89 Redo Problem 8.88, using the Method of Interpolation.
Ans. (a) 11.12%; (b) 10.48%; (c) 10.30%; (d) 9.33%

8.90 A $1000 bond with semiannual coupons at 8% matures at par on September 1, 2002. If this bond was quoted at 76 on September 1, 1995, what was the yield rate, j_2, (a) using the Method of Averages? (b) Using the Method of Interpolation? *Ans.* (a) 12.99%; (b) 13.40%

8.91 A $1000 bond, redeemable at par at the end of 20 years and paying bond interest at $j_2 = 10$%, is callable in 15 years at $1050. Find the yield rate, j_2, if the bond is quoted now at $92\frac{3}{4}$ and (a) it is called, (b) it is not called. *Ans.* (a) 11.15%; (b) 10.90%

8.92 A $2000 bond with semiannual coupons at 11% matures at par on June 1, 2011. On August 17, 1995, this bond is quoted at $96\frac{1}{2}$. What is the yield rate (a) j_2? (b) j_∞?
Ans. (a) 11.50%; (b) 11.18%

8.93 The ABC Corporation has a $1000 bond that pays bond interest at $j_2 = 12\frac{1}{2}$%. The bond is redeemable at par on September 1, 2003. On September 1, 1987, an investor buys this bond on the open market at 96; on September 1, 1995, he sells the bond on the open market at 106. Find his yield rate (a) j_2, (b) j_1. *Ans.* (a) 13.78%; (b) 14.25%

8.94 Ms. Holman buys a $1000 bond that pays bond interest at $j_2 = 11$% and is redeemable at par in 15 years. The price she pays will give her a yield of $j_2 = 12$% if held to maturity. After 5 years, Ms. Holman sells this bond to Mr. Dawson, who desires a yield of $j_2 = 10$% on his investment. (a) What price did Ms. Holman pay? (b) What price did Mr. Dawson pay? (c) What yield j_2 did Ms. Holman realize? *Ans.* (a) $931.18; (b) $1062.31; (c) 13.86%

8.95 The XYZ Corporation issues a $1000 bond with semiannual coupons at $j_2 = 11$% redeemable at par in 20 years or callable at par in 15 years. Mr. LaBelle buys the bond to guarantee a yield rate $j_{12} = 12$%. (a) Find the purchase price. (b) After 15 years, the XYZ Corporation calls the bond in and pays Mr. LaBelle his $1000. Find his overall yield stated as a rate j_{12}.
Ans. (a) $903.75; (b) 12.13%

8.96 A corporation has a bond due December 4, 2008, paying bond interest at $j_2 = 10\frac{3}{8}$%. The price bid on December 14, 1995, is 104. What yield j_{12} does the investor desire? *Ans.* 9.63%

8.97 Mr. Hunter buys a $1000 bond with semiannual coupons at $j_2 = 12$%. The bond is redeemable at par in 20 years. The price he pays will guarantee him a yield of $j_4 = 16$% if held to maturity. After 5 years, Mr. Hunter sells this bond to Miss Schnarr, who desires a yield of $j_1 = 11$% on her investment. (a) What price did Mr. Hunter pay? (b) What price did Miss Schnarr pay? (c) What yield, j_4, did Mr. Hunter realize? *Ans.* (a) $746.78; (b) $1095.02; (c) 21.13%

8.98 The ABC Corporation issues a $1000 bond, with coupons payable at $j_2 = 11\%$ on January 1 and July 1, redeemable at par on July 1, 2005. (*a*) How much would an investor pay for this bond on September 1, 1995, to yield $j_2 = 13\%$? (*b*) Given the purchase price from part (*a*), if each coupon is deposited in a bank account paying interest at $j_4 = 10\%$ and the bond is held to maturity, what is the effective annual yield rate j_1 on this investment?
Ans. (*a*) $909.30; (*b*) 12.24%

8.99 A bond paying semiannual coupons at $j_2 = 14\%$ matures at par in n years and is quoted at 110 to yield rate j_2. A bond paying semiannual coupons at $j_2 = 14\frac{1}{2}\%$ matures at par in n years and is quoted at 112 on the same yield basis. (*a*) Find the unknown yield rate, j_2. (*b*) Determine n to the nearest half-year. *Ans.* (*a*) 11.5%; (*b*) $5\frac{1}{2}$ years

8.100 An issue of bonds, redeemable at par in n years, is to bear coupons at $j_2 = 9\%$. An investor offers to buy the entire issue at a premium of 15%. At the same time, she advises the issuer that if the coupon rate were raised from $j_2 = 9\%$ to $j_2 = 10\%$, she would offer to buy the entire issue at a premium of 25%. At what yield rate j_2 are these two offers equivalent? *Ans.* $7\frac{1}{2}\%$

8.101 In August, 1991, interest rates in the bond markets were such that an investor could expect a yield of $j_2 = 10\%$. On August 1, 1991, an investor bought a bond with semiannual coupons at $j_2 = 9\%$ which was to mature in 20 years at par. By August 1, 1992, interest rates had fallen to $j_2 = 8\%$. This same investor sold his bond on August 1, 1992, on the bond market. Find the yield j_2 on this investment. *Ans.* 28.50%

OTHER TYPES OF BONDS

8.102 A $10 000 serial bond is to be redeemed in instalments of $2000 at the end of each of the 21st through 25th year from the date of issue. The bonds pay interest at $j_2 = 11\%$. What is the price to yield 9% per annum compounded semiannually? *Ans.* $11 926.56

8.103 A $100 000 issue of serial bonds issued July 1, 1995, and paying interest at rate $j_2 = 9\%$ is to be redeemed in instalments of $25 000 each on the first day of July in 2000, 2002, 2004, and 2006. Find the price to yield 7% per annum compounded semiannually. *Ans.* $111 898.48

8.104 Redo Problems 8.102 and 8.103 using Makeham's formula (Problem 8.33).

8.105 To finance an expansion of production capacity, Minicorp Ltd. will issue, on March 15, 1996, $30 000 000 of serial bonds. Bond interest at 13% per annum is payable half yearly on March 15 and September 15, and the contract provides for redemption as follows:

> $10 000 000 of the issue to be redeemed March 15, 2001;
> $10 000 000 of the issue to be redeemed March 15, 2006;
> $10 000 000 of the issue to be redeemed March 15, 2011.

Calculate the purchase price of the issue to the public to yield $j_1 = 12\%$ on those bonds redeemable in 5 years and $j_1 = 14\%$ on the remaining bonds. *Ans.* $29 861 132.81

8.106 A $210 000 issue of serial bonds with annual coupons of 10% on the balance outstanding is to be redeemed in 20 annual instalments beginning at the end of 1 year. The amount redeemed at the end of 1 year is to be $1000; at the end of 2 years, $2000; at the end of 3 years, $3000; and so on, so that the last redemption amount is $20 000 at the end of 20 years. Find the price to be paid for this issue to yield $j_1 = 8\%$. [Note that Makeham's formula may prove to be useful.] *Ans.* $242 773.02

8.107 A $10 000 serial bond is to be redeemed in $1000 instalments of principal per year over the next 10 years. Coupons at the rate of 9% on the outstanding balance over the next 10 years are also to be paid annually. Calculate the purchase price to yield an investor a yield (*a*) $j_1 = 10\%$; (*b*) $j_2 = 8\%$. *Ans.* (*a*) $9614.46; (*b*) $10 343.62

8.108 A corporation issues a $1000, 20-year bond with bond interest at $j_2 = 10\%$. It is purchased by an investor A who wishes a yield of $j_2 = 9.5\%$. This investor A keeps the coupons but sells the strip bond to another investor whose desired yield is $j_{12} = 10.5\%$. Determine the overall rate j_2 to investor A. *Ans.* 8.99%

8.109 A corporation issued a $1000, 15-year bond with bond interest at $j_2 = 9.5\%$ and redeemable at 105. It is purchased by Investor A who wishes a yield of $j_2 = 10\%$. Investor A sells the coupons to Investor B whose desired yield is $j_{12} = 10.5\%$ and Investor A keeps the strip bond. Determine the overall yield rate j_2 to Investor A. *Ans.* 9.20%

8.110 An annuity bond offers to repay $10 000 principal and interest at $j_2 = 9\%$ by equal payments at the end of each half-year for 10 years. How much will an investor offer for this bond if he wants to realize (a) $j_2 = 8\%$; (b) $j_2 = 10\%$? *Ans.* (a) $10 447.70; (b) $9580.45

8.111 An annuity bond promises to repay $50 000 principal with interest at $j_2 = 7\%$ by 20 equal semiannual payments, first payment in 3 years. How much should an investor, who wants to yield $j_1 = 8\%$, pay for this bond (a) now; (b) in 2 years; (c) in 5 years?
Ans. (a) $47 167.03; (b) $55 015.62; (c) $46 707.38

Chapter 9

Capital Budgeting and Depreciation

9.1 NET PRESENT VALUE

Business enterprises and investors are often forced to choose between alternative projects or investments to decide which is best or which is more profitable.

As introduced in Section 5.3, one method of assessment is to evaluate the future cash flows presented by the project or investment by calculating the present value of these cash flows at a particular rate of interest. The interest rate used is known as the **cost of capital** and may be the cost of borrowing money or may be the desired rate of return for the investor. Finding the **net present value** can be represented as:

$$\textbf{Net Present Value (or NPV)} = F_0 + F_1(1+i)^{-1} + F_2(1+i)^{-2} + \cdots + F_n(1+i)^{-n}$$
$$(9.1)$$

where $F_t \equiv$ estimated cash flow at time t (may be positive, negative, or zero)

$\quad\quad i \equiv$ cost of capital per period

A positive net present value indicates a profitable investment, while a negative result indicates a loss (which would normally mean the project would not proceed). F_0 often represents one's investment in the project and, hence, will be negative.

The following guidelines are often used:

1. Because the future cash flows are often only estimates, the net present value is rounded to the nearest hundred or thousand dollars.

2. More than one future cash flow estimate may be evaluated. It is common to evaluate three scenarios: one optimistic, one pessimistic, and one best guess.

As seen in the following examples, the cost of capital can be of extreme importance in deciding between alternatives.

SOLVED PROBLEMS

9.1 A project is expected to provide the cash flows indicated below. Would you invest $100 000 in this project if the cost of capital is: (*a*) $j_1 = 7\%$, (*b*) $j_1 = 14\%$?

Year end	1	2	3	4
Cash flow	$40 000	$25 000	$35 000	$30 000

(*a*) We have $F_0 = -100\ 000$, $F_1 = 40\ 000$, $F_2 = 25\ 000$, $F_3 = 35\ 000$, $F_4 = 30\ 000$, and $j_1 = 0.07$. From (9.1)

$$\begin{aligned} \text{NPV at } 7\% \quad = \quad & -100\ 000 + 40\ 000(1.07)^{-1} + 25\ 000(1.07)^{-2} \\ & +35\ 000(1.07)^{-3} + 30\ 000(1.07)^{-4} = +\$10\ 676 \end{aligned}$$

This NPV indicates an expected profit of $10 676.

(b) With the same F_t's but with $j_1 = 0.14$,

$$\begin{aligned} \text{NPV at 14\%} \;=\; & -100\,000 + 40\,000(1.14)^{-1} + 25\,000(1.14)^{-2} \\ & + 35\,000(1.14)^{-3} + 30\,000(1.14)^{-4} = -\$4289 \end{aligned}$$

This NPV indicates an expected loss of $4289.

9.2 An investor is presented with alternative projects, A and B with the following end-of-year cash flows. Each project requires an investment of $200 000. Which project would be chosen if (a) $j_1 = 6\%$; (b) $j_1 = 8\%$?

Year	1	2	3	4
Project A	$80 000	$70 000	$60 000	$35 000
Project B	$30 000	$40 000	$40 000	$150 000

(a)
$$\begin{aligned} \text{NPV (A) at 6\%} \;=\; & -200\,000 + 80\,000(1.06)^{-1} \\ & + 70\,000(1.06)^{-2} + 60\,000(1.06)^{-3} + 35\,000(1.06)^{-4} = +\$15\,871 \end{aligned}$$

$$\begin{aligned} \text{NPV (B) at 6\%} \;=\; & -200\,000 + 30\,000(1.06)^{-1} \\ & + 40\,000(1.06)^{-2} + 40\,000(1.06)^{-3} + 150\,000(1.06)^{-4} = +\$16\,301 \end{aligned}$$

While both projects are profitable, project B is slightly preferred.

(b) Using the same equations as in (a) but at $j_1 = 8\%$,

$$\text{NPV (A)} = +\$7444$$

$$\text{NPV (B)} = +\$4079$$

Now Project A is preferred given its higher indicated profit.

9.3 A mutual fund can invest $1 000 000 in a series of $100 government bonds paying semiannual bond interest at $j_2 = 12\%$; redeemed at par in 4 years and priced at $95; or a zinc mine with 3 years of ore left, but which requires environmental clean up in year 4. The expected pay-off is:

Year	1	2	3	4
Cash flow	$400 000	$600 000	$500 000	−$200 000

Which investment should be preferred?

Determine the yield that would be realized on the bond investment using:

$$95 = 6a_{\overline{8}|i} + 100(1+i)^{-8}$$

We find:
$$P(\text{at } j_2 = 13\%) = 96.96$$
$$P(\text{at } j_2 = 14\%) = 94.03$$

By interpolation:

$$\frac{d}{1\%} = \frac{1.96}{2.93}$$

$$d = .67\%$$

$$j_2 \approx 13.67\%$$

Find i per annum equivalent to $j_2 = 13.67\%$,

$$1 + i = \left(1 + \frac{.1367}{2}\right)^2$$

$$i \approx 14.14\%$$

Using $i = 14.14\%$, determine the profitability of the zinc mine.

NPV at 14.14% $= -1\,000\,000 + 400\,000(1.1414)^{-1}$
$+600\,000(1.1414)^{-2} + 500\,000(1.1414)^{-3} - 200\,000(1.1414)^{-4} = +\$29\,405$

While the positive NPV indicates an expected profit, the investor will still have to factor the risk associated with each option (as analyzed in Chapter 10).

9.4 A company is considering whether or not to develop a mine site. It will take a capital investment of \$14 000 000 to start mining production. The mine is expected to produce \$3 400 000 of profits from the ore each year for 10 years. At that time (that is, at the end of 11 years) an expenditure of \$6 400 000 will be needed to bring the site to environmental standards. If the company wishes to earn $j_1 = 20\%$, what should they do?

Find the NPV at $j_1 = 20\%$:

$$\begin{aligned} \text{NPV (at 20\%)} &= -14\,000\,000 + 3\,400\,000a_{\overline{10}|.20} - 6\,400\,000(1.20)^{-11} \\ &= -14\,000\,000 + 14\,254\,405 - 861\,363 \\ &= -\$606\,958 \end{aligned}$$

The decision would be not to go ahead.

If the company's desired yield were $j_1 = 15\%$, then:

$$\begin{aligned} \text{NPV (at 15\%)} &= -14\,000\,000 + 3\,400\,000a_{\overline{10}|.15} - 6\,400\,000(1+i)^{-11} \\ &= \$1\,688\,177 \end{aligned}$$

and the project would be viable.

9.2 INTERNAL RATE OF RETURN

The net present value method of comparing alternative investments assumes the investor starts with a target yield rate and then chooses to invest, or not, by looking at the net present value of the project at the target interest rate.

Another method that can be used to choose between alternative investments is to determine the rate of return that the project will produce. This is called the **internal rate of return** (IRR) and is defined as the rate of interest that produces a zero net present value. That is, it is the rate of interest that is the solution to the following equation:

$$F_0 + F_1(1+i)^{-1} + F_2(1+i)^{-2} + F_3(1+i)^{-3} + \cdots + F_n(1+i)^{-n} = 0$$

This is often written as:

$$F_1(1+i)^{-1} + F_2(1+i)^{-2} + F_3(1+i)^{-3} + \cdots + F_n(1+i)^{-n} = -F_0 = A$$

where $-F_0 = A$ is the original investment (i.e., negative cash flow) and $F_1, F_2, ..., F_n$ are all positive.

A bond's "yield to maturity (YTM)" and a capital budgeting project's "internal rate of return (IRR)" are mathematically equivalent.

If one of the cash flows, $F_1, F_2, \ldots, F_n$, is also negative then the possibility of multiple IRRs exists. (See Problem 9.8.)

Having calculated the IRR, the potential investor will compare this IRR to existing interest rates in the market place or to the cost of capital to the investor. The investor should also factor in to the decision the relative risk of each alternative investment (this will be discussed more fully in Chapter 10).

In most IRR problems, the future cash flows are based on very rough estimates. Hence, accurate solutions to the IRR are not the norm, as will be seen in the solved problems that follow.

Some financial calculators will calculate IRRs directly. However, students should work through the problems presented in this section from first principles so as to gain an understanding of the issues surrounding the method.

SOLVED PROBLEMS

9.5 Find the IRR for a project requiring an investment of \$100 000 which is expected to produce cash flows of \$50 000 at the end of year 1, \$30 000 at the end of year 2, and \$50 000 at the end of year 3.

We have $F_0 = -100\,000$ or $A = 100\,000$, $F_1 = 50\,000$, $F_2 = 30\,000$, and $F_3 = 50\,000$, and

$$-100\,000 + 50\,000(1+i)^{-1} + 30\,000(1+i)^{-2} + 50\,000(1+i)^{-3} = 0$$

or 　　　　$$50\,000(1+i)^{-1} + 30\,000(1+i)^{-2} + 50\,000(1+i)^{-3} = 100\,000$$

Using the latter equation, and by trial and error, we find:

$$
\begin{aligned}
\text{NPV (at 14\%)} &= 100\,692 \\
\text{NPV (at 15\%)} &= 99\,038
\end{aligned}
$$

We would estimate an IRR of approximately $14\tfrac{1}{2}\%$ for this project, which is sufficient given the future cash flows are only estimates. This rate would then be compared to alternative investments and a decision to proceed or not would be made.

This solved problem indicated one issue surrounding this method, namely, how to proceed quickly, by trial and error, to the range of applicable rates of return.

As a first approximation, take the equation:

$$F_1(1+i)^{-1} + F_2(1+i)^{-2} + F_3(1+i)^{-3} + \cdots + F_n(1+i)^{-n} = F_0 = A$$

Then using a binomial expansion to one term, we get:

$$F_1(1-i) + F_2(1-2i) + F_3(1-3i) + \cdots + F_n(1-n_i) \simeq A$$

or 　　$$\sum_{t=1}^{n} F_t(1-ti) \simeq A$$

or 　　$$\sum_{t=1}^{n} F_t - \sum_{t=1}^{n} tF_t \cdot i \simeq A.$$

Solving for i, we get

$$i \simeq \frac{\sum_{t=1}^{n} F_t - A}{\sum_{t=1}^{n} t \cdot F_t} \qquad (9.2)$$

We can use this equation as a starting value for i.

9.6 An investment of \$10 000 returns \$3000 at the end of years 1 and 2 and \$3500 at the end of years 3 and 4. Calculate the IRR.

First find an approximation for i using (9.2) given:

$$3000(1+i)^{-1} + 3000(1+i)^{-2} + 3500(1+i)^{-3} + 3500(1+i)^{-4} = 10\ 000$$

$$i \simeq \frac{13\ 000 - 10\ 000}{33\ 500} = 8.95\%$$

Now   NPV (at 9%) = 10 459
NPV (at 10%) = 10 227
NPV (at 11%) = 10 002
NPV (at 12%) =   9 786

Hence the IRR is 11%.

9.7 An investment of \$5 million is expected to produce the following cash-flows at each year-end (in millions):

Year	1	2	3	4
Cash flow	\$1.25	2.00	2.50	.75

Find the internal rate of return, IRR.

From (9.2) $i \simeq \dfrac{6.50 - 5.00}{15.75} = 9.52\%$

NPV (at 10%) = \$5.15
NPV (at 11%) = \$5.04
NPV (at 12%) = \$4.93

Thus the IRR is approximately $11\frac{1}{2}\%$.

9.8 For an investment of \$7200 today and \$27 000 in 2 years time, an investor expects to receive \$24 200 in 1 year and \$10 000 in 3 years. Determine the IRR.

We have:

$$NPV = -7200 + 24\ 200(1+i)^{-1} - 27\ 000(1+i)^{-2} + 10\ 000(1+i)^{-3}$$

We then calculate net present values for this investment at 5% intervals from 0% to 30% inclusive and get:

i	NPV
0%	0
5%	−$4
10%	−$1
15%	+$3
20%	+$4
25%	0
30%	−$9

Thus we find exact IRRs of 0% and 25% and see that there must be another solution for an IRR between 10% and 15% (actually 11.1%). The reason for these multiple solutions is that the cumulative F_t's change sign more than once. That is, $F_0 < 0$; $F_0 + F_1 > 0$; $F_0 + F_1 + F_2 < 0$; and $F_0 + F_1 + F_2 + F_3 = 0$. For there to be a single IRR, there can be only one change of signs in the cumulative F_t's. This will be explored further in Problem 9.46.

9.3 CAPITALIZED COST AND CAPITAL BUDGETING

The *capitalized cost* of an asset may be defined as its original cost, C, plus the present value of an unlimited number of replacements, plus the present value of an unlimited number of maintenance costs in perpetuity. (See Problem 9.9.) In general, the capitalized cost K of an asset of original cost C, with salvage value S after n years and annual maintenance cost M—all at rate i per year—is:

$$K = C + \frac{C - S}{(1+i)^n - 1} + \frac{M}{i} = C + \frac{C - S}{i^*} + \frac{M}{i} \qquad (9.3)$$

where i^* is an equivalent interest rate convertible every n years.

The concept of capitalized cost can be used to decide which of several alternatives is economically the best. (See Problem 9.10.)

Different assets may produce revenue at different rates; if so, it is necessary to account for this difference. If U_1 and U_2 are the numbers of items produced per unit of time by machines 1 and 2, having capitalized costs K_1 and K_2, respectively, then the two machines are economically equivalent if

$$\frac{K_1}{U_1} = \frac{K_2}{U_2} \qquad (9.4)$$

(See Problem 9.11.)

SOLVED PROBLEMS

9.9 A certain machine costs $25 000 and lasts 6 years, after which time it has a scrap value of $5000. Annual maintenance costs are $800. If money is worth 8% per annum, find the capitalized cost of the machine.

The annual maintenance costs form a perpetuity whose present value is, by (6.2),

$$\frac{800}{0.08} = \$10\ 000$$

The cost to the company of a replacement is $25\ 000 - 5000 = \$20\ 000$ every 6 years. This forms a perpetuity at a rate of interest i' per 6 years,

$$i' = (1.08)^6 - 1 = 0.586874323$$

with present value

$$\frac{20\ 000}{0.586874323} = \$34\ 078.85$$

Hence, the capitalized cost is $25\ 000 + 34\ 078.85 + 10\ 000 = \$69\ 078.85$.

9.10 Machine A costs \$36 000, will last 15 years, and will have salvage value \$4800 at that time. Its cost of maintenance is \$3000 a year. Machine Z costs \$40 000, will last 20 years, and will have salvage value \$4000 at that time. Its annual maintenance cost is \$2400. If money is worth $j_1 = 11\%$, which machine should be purchased?

The capitalized cost of machine A is by (9.3)

$$K_A = 36\ 000 + \left(\frac{36\ 000 - 4800}{(1.11)^{15} - 1}\right) + \frac{3000}{0.11} = \$71\ 516.69$$

and the capitalized cost of machine Z is by (9.3)

$$K_Z = 40\ 000 + \left(\frac{40\ 000 - 4000}{(1.11)^{20} - 1}\right) + \frac{2400}{0.11} = \$66\ 915.66$$

Therefore, you should buy machine Z.

9.11 If, in Problem 9.10, machine A produces 11 000 units in a year and machine Z produces 10 000 units, which machine should be purchased?

From (9.4)

$$\text{per-unit capitalized cost for machine A} = \frac{71\ 516.69}{11\ 000} = \$6.50$$

$$\text{per-unit capitalized cost for machine Z} = \frac{66\ 915.66}{10\ 000} = \$6.69$$

Therefore, you should buy machine A.

9.12 A machine costing \$40 000 has scrap value \$5000 after 10 years. It produces 2000 units of output per year, and the annual maintenance cost is \$1500. How much can be spent on increasing its productivity to 3000 units per year, if its period of service and maintenance costs remain unchanged? Assume $j_1 = 12\%$.

By (9.4),

$$\frac{K_1}{U_1} = \frac{K_2}{U_2}$$

$$\frac{40\ 000 + \left(\dfrac{40\ 000 - 5000}{(1.12)^{10} - 1}\right) + \dfrac{1500}{0.12}}{2000} = \frac{(40\ 000 + x) + \left(\dfrac{40\ 000 + x - 5000}{(1.12)^{10} - 1}\right) + \dfrac{1500}{0.12}}{3000}$$

$$\frac{69\ 120.38}{2000} = \frac{69\ 120.38 + 1.474868035x}{3000}$$

$$1.474868035x = 34\ 560.19$$

$$x = \$23\ 432.73$$

9.4 DEPRECIATION

Most assets (land being a notable exception) provide services or revenues for a finite period of time. The accountant will allocate the cost of the asset to the accounting periods in which it generates revenues—a periodic expense item called *depreciation*.

The difference between the original cost of an asset and its accumulated depreciation to date is its *book value*. This book value is not necessarily the same as the market value, representing, as it does, the portion of the original cost of the asset which has yet to be charged off as an expense. At the end of the asset's useful lifetime, its book value will equal its estimated *scrap* or *salvage value*.

The *depreciation base* of an asset is its original cost less its estimated scrap value; that is, the total amount that will be written off over its estimated lifetime as depreciation expense.

In this chapter, we will use the following notation:

C $\equiv$ original cost of the asset

S $\equiv$ estimated scrap, salvage, or residual value of the asset at the end of its estimated useful lifetime (S can be negative)

n $\equiv$ estimated useful lifetime of the asset, in years

R_k $\equiv$ depreciation expense for year k ($1 \leq k \leq n$)

B_k $\equiv$ book value of the asset at the end of k years ($0 \leq k \leq n$), where $B_0 = C$ and $B_n = S$

D_k $\equiv$ accumulated depreciation expense at the end of k years ($0 \leq k \leq n$), where $D_0 = 0$ and $D_n = C - S$

In each method of accounting for depreciation, we set up a *depreciation schedule* showing the yearly depreciation expense, the book value at the end of each year, and the accumulated depreciation to date. Note that, in all cases, the accumulated depreciation, D_k, plus the book value of the asset, B_k, equals the original cost of the asset, C:

$$D_k + B_k = C$$

THE STRAIGHT-LINE METHOD

The simplest and most popular method of accounting for depreciation is the *straight-line method*, whereby the depreciation base is evenly allocated over the lifetime of the asset, making

$$R_k = \frac{C - S}{n} \equiv R \quad \text{(independent of } k\text{)} \tag{9.5}$$

$$D_k = kR = k\left(\frac{C - S}{n}\right) \tag{9.6}$$

$$B_k = C - kR = C - k\left(\frac{C - S}{n}\right) \tag{9.7}$$

Other depreciation methods, which charge more of the depreciation expense to the early years, or, conversely, to the later years, may be preferred to the straight-line method because they may match more suitably the perceived service consumption of the asset.

SOLVED PROBLEMS

9.13 A machine costing \$40 000 is estimated to have a useful lifetime of 5 years and scrap value of \$5000. Prepare a depreciation schedule using the straight-line method.

We have $C = 40\ 000$, $S = 5000$, and $n = 5$; by (9.5),

$$R = \frac{40\ 000 - 5000}{5} = \$7000$$

Hence, the accumulated depreciation increases by \$7000 each year and the book value of the asset decreases by \$7000 each year, as shown in the depreciation schedule, Table 9-1.

Table 9-1

End of Year	Yearly Depreciation	Accumulated Depreciation	Book Value
0	0	0	40 000
1	7000	7 000	33 000
2	7000	14 000	26 000
3	7000	21 000	19 000
4	7000	28 000	12 000
5	7000	35 000	5 000

9.14 A machine costing \$300 000 has an estimated lifetime of 15 years and zero scrap value at that time. At the end of 6 years, the machine becomes obsolete because of the development of a better machine. What are the total accumulated depreciation and the book value of the asset at that time, under the straight-line method?

With $C = 300\ 000$, $S = 0$, and $n = 15$, we have from (9.5)

$$R = \frac{300\ 000}{15} = \$20\ 000$$

Then, from (9.6),

$$D_6 = 6(20\ 000) = \$120\ 000$$

and from (9.7)

$$B_6 = 300\ 000 - 120\ 000 = \$180\ 000$$

THE CONSTANT-PERCENTAGE METHOD

Under the *constant-percentage method*, the depreciation expense accounted for in any year is a fixed percentage of the book value at the beginning of that year. When this method is used, it is customary to evaluate the *rate of depreciation*, d, rather than to estimate the useful lifetime, n, and scrap value, S. The yearly depreciation for the kth year is given by

$$R_k = B_{k-1}d \qquad (9.8)$$

It is clear (cf. Problem 2.7) that the successive book values over the life of the asset are the terms of a geometric progression with common ratio $1-d$. Hence

$$B_k = C(1-d)^k \qquad (9.9)$$

and

$$S = C(1-d)^n \qquad (9.10)$$

Also

$$D_k = C - B_k = C - C(1-d)^k$$

Formula (9.10) may be used to solve for the rate of depreciation, d, given n and S. (See Problem 9.16.)

The constant-percentage method can be used only if S (the salvage value) is positive. A modified version of the constant-percentage method is illustrated in Problem 9.19.

SOLVED PROBLEMS

9.15 A car costing \$24 000 depreciates 25% of its value each year. Make a depreciation schedule for the first 3 years; find the book value at the end of 5 years and the depreciation expense for the 6th year.

Table 9-2

End of Year	Yearly Depreciation	Accumulated Depreciation	Book Value
0	0	0	24 000
1	6000	6 000	18 000
2	4500	10 500	13 500
3	3375	13 875	10 125

See Table 9-2. We calculate B_5 using (9.9):

$$B_5 = 24\,000(1 - 0.25)^5 = 24\,000(0.75)^5 = \$5695.31$$

The depreciation expense in year 6 can now be found from (9.8):

$$R_6 = B_5 d = (5695.31)(0.25) = \$1423.83$$

9.16 Determine the rate of depreciation and construct the depreciation schedule for the machine of Problem 9.13, using the constant-percentage method.

We have $C = 40\,000$, $S = 5000$, $n = 5$; and from (9.10),

$$\begin{aligned} 5000 &= 40\,000(1-d)^5 \\ (1-d)^5 &= 0.125 \\ 1-d &= (0.125)^{1/5} = 0.659753955 \\ d &= 0.340246045 \end{aligned}$$

Table 9-3

End of Year	Yearly Depreciation	Accumulated Depreciation	Book Value
0	0	0	40 000.00
1	13 609.84	13 609.84	26 390.16
2	8 979.15	22 588.99	17 411.01
3	5 924.03	28 513.02	11 486.98
4	3 908.40	32 421.42	7 578.58
5	2 578.58	35 000.00	5 000.00

9.17 Equipment costing \$60 000 depreciates 10% of its value each year. How long will it take before the equipment is worth less than \$30 000?

By (9.9), we have

$$60\ 000(0.90)^k < 30\ 000$$
$$(0.90)^k < 0.5$$
$$k \log 0.90 < \log 0.5$$

where the last step follows from the fact that the logarithm is an increasing function. Then, since $\log 0.90$ is negative,

$$k > \frac{\log 0.5}{\log 0.90} = \frac{-0.301039996}{-0.045757491} \approx 6.58 \text{ years}$$

9.18 For the equipment of Problem 9.17, find (*a*) the book value at the end of 6 years, and (*b*) the depreciation expense in year 7.

(*a*) $B_6 = 60\ 000(1 - 0.10)^6 = \$31\ 886.46$

(*b*) $R_7 = (31\ 886.46)(0.10) = \$3188.65$

9.19 One method of accelerated depreciation allowed by the Internal Revenue Service for new assets having a life of at least 3 years is the *double-declining-balance method*, a combination of the straight-line and constant-percentage methods. Redo Problems 9.13 and 9.16 by this method.

In Problems 9.13 and 9.16, we have $C = 40\ 000$, $S = 5000$, $n = 5$. The I.R.S. Code allows depreciation of the asset at twice the rate allowable under the straight-line method. Under the straight-line method, we would depreciate $C - S = \$35\ 000$ over 5 years, or a constant 20% a year. Under the double-declining-balance method, we use a constant-percentage approach, with $d = 40\%$ (double the straight-line 20%). This leads to the schedule of Table 9-4.

Table 9-4

End of Year	Yearly Depreciation	Accumulated Depreciation	Book Value
0	0	0	40 000.00
1	16 000.00	16 000.00	24 000.00
2	9 600.00	25 600.00	14 400.00
3	5 760.00	31 360.00	8 640.00
4	3 456.00	34 816.00	5 184.00
5	184.00	35 000.00	5 000.00

Note that in year 5 the asset may not be depreciated at 40%, since that would bring the book value below the estimated salvage value, $5000. Instead, the depreciation for year 5 is

$$5184.00 - 5000.00 = \$184.00$$

Like the constant-percentage method, the double-declining-balance method can never reduce the book value to zero. Thus, if an asset has zero scrap value, the taxpayer who is using the double-declining-balance method is permitted to switch to the straight-line method at any convenient stage.

THE SUM-OF-DIGITS METHOD

The *sum-of-digits method* is another accelerated method that assigns more depreciation to the earlier service years of the asset. The yearly depreciation expenses are specified fractions of the depreciation base, $C - S$; the denominator of all the fractions is the sum of the digits from 1 to n (n being the estimated lifetime of the asset), which we know from (2.2) to be

$$s = \frac{n(n + 1)}{2}$$

The numerators of the fractions are the year numbers in reverse order. Thus,

$$R_1 = \frac{n}{s}(C - S) \quad R_2 = \frac{n - 1}{s}(C - S) \quad \cdots \quad R_n = \frac{1}{s}(C - S)$$

or, generally,

$$R_k = \frac{n - k + 1}{s}(C - S) \tag{9.11}$$

We have, of course, $\sum_k R_k = (s/s)(C - S) = C - S$.

SOLVED PROBLEMS

9.20 Construct the depreciation schedule for the machine of Problem 9.13, using the sum-of-digits method.

We have $C = 40\,000$, $S = 5000$, $n = 5$, $C - S = 35\,000$, and

$$s = \frac{5(5 + 1)}{2} = 15$$

Hence

$$R_1 = \frac{5}{15}(35\,000) = 11\,666.67$$

$$R_2 = \frac{4}{15}(35\,000) = 9\,333.33$$

$$\cdots\cdots\cdots\cdots\cdots\cdots\cdots\cdots\cdots$$

$$R_5 = \frac{1}{15}(35\,000) = 2\,333.33$$

leading to Table 9-5.

Table 9-5

End of Year	Yearly Depreciation	Accumulated Depreciation	Book Value
0	0	0	40 000.00
1	11 666.67	11 666.67	28 333.33
2	9 333.33	21 000.00	19 000.00
3	7 000.00	28 000.00	12 000.00
4	4 666.67	32 666.67	7 333.33
5	2 333.33	35 000.00	5 000.00

9.21 Redo Problem 9.14 by the sum-of-digits method.

We have $C = 300\ 000$, $S = 0$, $n = 15$, and

$$s = \frac{15(15+1)}{2} = 120$$

Then

$$D_6 = \frac{15 + 14 + \cdots + 10}{120}(300\ 000)$$

$$= \frac{75}{120}(300\ 000) = \$187\ 500$$

and $B_6 = 300\ 000 - 187\ 500 = \$112\ 500$.

THE PHYSICAL-SERVICE METHOD AND DEPLETION

If an asset is purchased to perform a service, the cost of that asset might reasonably be charged off in relation to the amount of service rendered in each period. Amount of service might be measured in hours of work, units of production, mileage, etc.

Certain kinds of assets, such as mines, oil wells, and timber tracts, diminish in value as the natural resources they originally held are used up. This gradual loss in value is called *depletion*. While any of the preceding methods of calculating depreciation expenses could also be used for calculating depletion expenses, it is usual to apply the physical-service approach. (See Problem 9.23.)

SOLVED PROBLEMS

9.22 The machine of Problem 9.13 produced 4000, 3800, 4500, 3750, and 1450 units in years 1 to 5, respectively. (*a*) Find the depreciation expense per unit of production. (*b*) Prepare a full depreciation schedule.

(*a*) We have $C - S = 40\ 000 - 5000 = \$35\ 000$, and the total number of units produced is

$$4000 + 3800 + 4500 + 3750 + 1450 = 17\ 500$$

Hence, the depreciation expense per unit of production is

$$\frac{35\ 000}{17\ 500} = \$2.00$$

(*b*) See Table 9-6.

Table 9-6

End of Year	Production Units	Yearly Depreciation	Accumulated Depreciation	Book Value
0	0	0	0	40 000
1	4000	8000	8 000	32 000
2	3800	7600	15 600	24 400
3	4500	9000	24 600	15 400
4	3750	7500	32 100	7 900
5	1450	2900	35 000	5 000

9.23 A gravel pit is purchased for $300 000. After excavation as follows:

Year	Truckloads of Gravel
1	8000
2	9100
3	7300
4	5600

the pit is exhausted, leaving a property worth $30 000. Produce a depletion schedule.

Here, $C - S = 300\ 000 - 30\ 000 = \$270\ 000$, and the total number of truckloads excavated is 30 000. Then the depletion expense per truckload is

$$\frac{270\ 000}{30\ 000} = \$9.00$$

Table 9-7 gives the depletion schedule.

Table 9-7

End of Year	Truckloads of Gravel	Yearly Depreciation	Accumulated Depreciation	Book Value
0	0	0	0	300 000
1	8000	72 000	72 000	228 000
2	9100	81 900	153 900	146 100
3	7300	65 700	219 600	80 400
4	5600	50 400	270 000	30 000

THE SINKING-FUND METHOD

To understand this method, one may assume that a sinking fund is set up at rate j to accumulate an amount equal to the depreciation base, $C - S$, at the end of n years. By (5.1), the constant annual sinking-fund deposit equals $(C - S)/s_{\overline{n}|i}$.

The depreciation expense for the kth year, R_k, is equal to this level sinking-fund deposit plus the interest earned on the fund in year k. Thus, the interest being compound, the depreciation expense increases with time.

The depreciation schedule here is the same as an ordinary sinking-fund schedule, except for an appended "Book Value" column. In particular, the accumulated depreciation, D_k, at the end of k years is equal to the accumulated value of the sinking fund at that time; that is, from (5.1),

$$D_k = \left(\frac{C - S}{s_{\overline{n}|i}}\right) s_{\overline{k}|i} \tag{9.12}$$

SOLVED PROBLEMS

9.24 Construct the depreciation schedule for the machine of Problem 9.13, using the sinking-fund method with $j_1 = 8\%$.

With $C = 40\,000$, $S = 5000$, $n = 5$, $i = 0.08$, we calculate the required annual sinking-fund deposit:

$$\frac{C - S}{s_{\overline{n}|i}} = \frac{40\,000 - 5000}{s_{\overline{5}|.08}} = \$5965.98$$

This leads to the depreciation schedule of Table 9-8.

Table 9-8

End of Year	Sinking-Fund Deposit	Interest on Fund	Yearly Depreciation	Accumulated Depreciation	Book Value
0	0	0	0	0	40 000.00
1	5965.98	0	5965.98	5 965.98	34 034.02
2	5965.98	477.28	6443.26	12 409.24	27 590.76
3	5965.98	992.74	6958.72	19 367.96	20 632.04
4	5965.98	1549.44	7515.42	26 883.38	13 116.62
5	5965.98	2150.67	8116.65	35 000.03	4 999.97

The discrepancy of 3¢ in our final answer is due to rounding off all entries to the nearest cent. We can check the accumulated depreciation—for example, D_4—by using (9.12)

$$D_4 = 5965.98 s_{\overline{4}|.08} = \$26\,883.37$$

Again we have a 1¢ discrepancy due to roundoff.

9.25 Redo Problem 9.14, using the sinking-fund method with $j_1 = 11\%$.

We have $C = 300\,000$, $S = 0$, $n = 15$, $i = 0.11$; from (9.12),

$$D_6 = \left(\frac{300\,000}{s_{\overline{15}|.11}}\right) s_{\overline{6}|.11} = \$68\,996.75$$

and $$B_6 = 300\,000 - 68\,996.75 = \$231\,003.25$$

Supplementary Problems

NET PRESENT VALUE

9.26 Calculate the net present value for a $100 000 investment with the following alternative cash flows and thereby determine whether the investment should proceed.

Alternative	Cost of Capital per Annum	End-of-Year Cash Flow			
		Year 1	Year 2	Year 3	Year 4
(a)	10%	$40 000	$30 000	$40 000	$30 000
(b)	7%	$50 000	$60 000	$20 000	–
(c)	12%	$20 000	$40 000	$60 000	$80 000
(d)	9%	$80 000	$60 000	$20 000	–$20 000

Ans. (a) +$11 700.02, yes; (b) +$15 461.25, yes; (c) +$43 293.16, yes; (d) +$25 170.46, yes

9.27 Redo 9.26(a) at $j_1 = 16\%$. *Ans.* $506.38, yes

9.28 A company is considering a $200 000 investment. It is expected that project A will return $50 000 at the end of each year for the next 5 years. On the other hand, it is expected that project B will return nothing for the next 2 years but $100 000 at the end of years 3, 4, and 5. Which project should be chosen if the cost of capital is (a) $j_1 = 4\%$; (b) $j_1 = 7\%$? *Ans.* (a) B; (b) B

9.29 Which of the following projects should a company choose if each proposal costs $50 000 and the cost of capital is $j_1 = 10\%$?

	End-of-Year Cash Flow				
	Year 1	Year 2	Year 3	Year 4	Year 5
Project A	$20 000	$10 000	$5 000	$10 000	$20 000
Project B	$5 000	$20 000	$20 000	$20 000	$5 000

Ans. B

9.30 The Northeast Mining Company is considering the exploitation of a mining property. If the company goes ahead, the estimated cash flows are as follows:

	Cash Inflow	Cash Outflow
Now	0	$3 000 000
End of year 1	$1 000 000	$2 000 000
End of year 2	$1 000 000	0
End of year 3	$1 000 000	0
End of year 4	$1 000 000	0
End of year 5	$1 000 000	0
End of year 6	$1 000 000	0
End of year 7	$1 000 000	0
End of year 8	$1 000 000	0

The project would be financed out of working capital, on which Northeast expects to earn at least 16% per annum. Advise whether Northeast should proceed. *Ans.* NPV = −$380 547, no

9.31 A pension fund is able to earn $j_2 = 10\%$ on its investments in government bonds. Should it proceed with an investment that costs $100 000 and returns $17 000 at the end of each half-year for the next 4 years? *Ans.* NPV = +$9874.62, yes

9.32 A company is able to borrow money at $j_1 = 14\%$. A new machine costing $60 000 is now available and is estimated to produce the following savings over the next 6 years.

End of year	1	2	3	4	5	6
Saving	$20 000	$18 000	$16 000	$14 000	$12 000	$8000

Should the company borrow money to buy this machine? *Ans.* NPV = $360.06, yes

9.33 An investment of $100 000 produces the following year-end cash flows.

Year	1	2	3	4	5
Cash flow	$30 000	$30 000	$20 000	$30 000	$20 000

Calculate the net present value for this project if the cost of capital is $j_1 = 2\%$, 4%, 6%, 8%, 10%, or 12%. Graph your six answers showing the NPV against the cost of capital.
Ans. $22 923.25; $16 445.44; $10 502.14; $5037.15; $1.24; −$4648.79

9.34 A company is able to borrow money at $j_1 = 12\frac{1}{2}\%$. It is considering the purchase of a machine costing $100 000, which will save the company $5000 at the end of every quarter for 7 years and may be sold for $14 000 at the end of the term. Should it borrow the money to buy the machine? *Ans.* NPV = +$92.71, yes

Internal Rate of Return

9.35 Having calculated net present values in Problem 9.26, calculate the internal rate of return (to the nearest $\frac{1}{2}\%$) in each case using the same data. *Ans.* (a) $15\frac{1}{2}\%$; (b) $16\frac{1}{2}\%$; (c) 27%; (d) 29%

9.36 An investment of $50 000 provides an estimated cash flow of $15 000 for the next 2 years and $17 500 for the following 2 years. Calculate this investment's internal rate of return. *Ans.* 11%

9.37 Find the internal rate of return for a project that costs $100 000 and is estimated to return the following year-end cash flows:

Year	1	2	3	4	5
Cash flow	$10 000	$20 000	$30 000	$40 000	$50 000

Ans. 12%

9.38 An insurance company must choose between the two investments, which each cost $10 000 and produce the following cash flows at the end of each year.

Year	1	2	3	4	5
Project A	$5000	$4000	$3000	$2000	$1000
Project B	$1000	$3000	$4000	$5000	$6000

Calculate (a) the internal rate of return for each project; (b) the net present value for each project at $j_1 = 15\%$.
Ans. (a) $IRR(A) \approx 20\%$, $IRR(B) \approx 20\%$; (b) $NPV(A) = +\$10\,985.62$, $NPV(B) = +\$11\,609.89$

9.39 An investment of $100 000 generates income of $75 000 at the end of the first and second years but requires an outlay of $25 000 at the end of the third year. (a) Calculate this project's internal rate of return. (b) As the final cash flow is a negative one, a sinking fund is set up at the end of the second year so that this payment will be met. If the sinking fund earns $j_1 = 5\%$ (and not the IRR), what is the project's rate of return under this condition?
Ans. (a) 20%; (b) 19%

9.40 An investment project requires $55 000 now, $69 000 in 1 year's time, and $1.147m in 3 years' time. It is estimated that it will return $772 000 in 2 years' time and half a million dollars 2 years after that. (*a*) Calculate the net present values for this project at 6% intervals from 0% to 60% inclusive. (*b*) Find three positive internal rates of return for this project.
Ans. (*b*) 5.7.%, 9.2%, 58.0%

CAPITALIZED COST AND CAPITAL BUDGETING

9.41 Benjamin can purchase a steel barn having an estimated lifetime of 20 years. It will cost $200 000 and will require $1000 a year to maintain. Alternatively, he can purchase a wooden barn having an estimated lifetime of 14 years. It will cost $160 000 and require $2000 a year to maintain. If Benjamin requires a return of $j_1 = 15\%$ on his investment, what should he do?
Ans. Purchase the wooden barn.

9.42 A machine costs $8000 and has an expected lifetime of 10 years, after which it will be worth $1000. Maintenance costs $1500 per annum. Find the capitalized cost of the machine if $j_1 = 12\%$. *Ans.* $23 824.08

9.43 A machine costing $4000 has an estimated useful lifetime of 10 years and estimated salvage value $500. It produces 1600 items per year and has an annual maintenance cost of $800. How much can be spent on increasing its productivity to 2000 units per year, if its lifetime, salvage value, and maintenance costs remain the same? Assume that the owners want a rate of return of $j_1 = 18\%$. *Ans.* $1874.93

9.44 The XYZ Company uses special lights that cost $22.50 each and have a useful lifetime of 2 years. They are approached by a firm selling lights that cost $33 but last 3 years. Should they switch, if money is worth (*a*) $j_1 = 11\%$? (*b*) $j_1 = 4\%$? *Ans.* (*a*) No; (*b*) yes

9.45 Widget-producing machines can be purchased from Manufacturer A or Manufacturer Z. Machine A costs $18 000, will last 15 years, and will have a salvage value of $2400 at that time. The cost of maintenance is $1500 a year. Machine Z costs $30 000, will last 20 years, and will have a salvage value of $2000 at that time. The annual maintenance cost is $1200. If money is worth $j_1 = 15\%$, which machine should be purchased? *Ans.* Machine A

9.46 NYTel uses poles costing $200 each. These poles last 15 years and have no salvage value. How much per pole would NYTel be justified in spending on a preservative to lengthen the life of the pole to 20 years? Assume that the annual maintenance costs are equal and that money is worth $j_1 = 8\%$. *Ans.* $29.41

9.47 A town is putting a new roof on its arena. One type of roof costs $600 000 and has an expected lifetime of 20 years with no salvage value. Another type costs $700 000 but will last 30 years with no salvage value. Maintenance costs are equal. If money is worth $j_1 = 7\%$, how much can be saved each year by purchasing the most economical roof? *Ans.* $225.27

9.48 A new car can be purchased for $16 000. If it is kept for 4 years, its trade-in value will be $5200. What should the trade-in value be at the end of 3 years so that trading would be economically equivalent to keeping it 4 years, if money is worth $j_1 = 6\%$? *Ans.* $8140.37

9.49 Machine 1 sells for $100 000, has an annual maintenance expense of $2000 and an estimated lifetime of 25 years with a salvage value of $5000. Machine 2 has an annual maintenance expense of $4000 and an estimated lifetime of 20 years with no salvage value. Machine 2 produces output twice as fast as Machine 1. If money is worth $j_1 = 8\%$, find the price of Machine 2 so that it would be economically equivalent to Machine 1. *Ans.* $182 607.41

9.50 A machine costing $40 000 has a scrap value of $5000 after 10 years. It produces 2000 units of output per year. The annual maintenance cost is $1500. How much can be spent on increasing its productivity to 3000 units per year if its period of service, scrap value, and maintenance cost remain unchanged? Assume money is worth $j_1 = 7\%$. *Ans.* $23 996.81

9.51 An oil furnace should be serviced annually to maintain efficient combustion. (a) Find the capitalized cost of this servicing one year before the next servicing, assuming interest is $j_1 = 6\%$ and the cost of servicing is $100. (b) If the first servicing is due right now, costing $100, and the cost of servicing increases by 2% per year, what is the capitalized cost at $j_1 = 6\%$? Ans. (a) $1666.67; (b) $2650

9.52 Egbert can purchase a car with an estimated lifetime of 7 years. It will cost $20 000 and will require $1000 a year to maintain. Alternatively, he can purchase a car with an estimated lifetime of 10 years. It will cost $25 000 and will require $750 a year to maintain. Using $j_1 = 8\%$, how much can be saved by purchasing the most economical car? Ans. $4571.34

9.53 A company is thinking of buying a machine that will replace five skilled workmen. These workmen are paid $3600 each at the end of each month. The machine costs $2400 a month to maintain, will last 20 years, and will have no scrap value at that time. If money is worth $j_1 = 8\%$, what is the largest amount (in thousands of dollars) that can be spent for the machine while remaining profitable? Ans. $1 905 412.62

9.54 Show that the capitalized cost K of an asset with zero salvage value is

$$K = \frac{1}{i}\left[\frac{C}{a_{\overline{n}|i}} + M\right]$$

9.55 An asset, which costs C dollars when new, has a service life of n years. At the rate of interest i per year, show that the maximum amount to be spent to extend the life of this asset by an additional m years, given $S = 0$, is

$$\frac{C\,a_{\overline{m}|i}}{s_{\overline{n}|i}} \text{ dollars}$$

DEPRECIATION

9.56 Equipment costing $60 000 is estimated to have a useful lifetime of 5 years and a scrap value of $8000. Prepare a depreciation schedule using (a) the straight-line method; (b) the constant-percentage method (you must find d); (c) the sum-of-digits method; (d) the sinking-fund method, using $j_1 = 6\%$; (e) double-declining balance method. Ans. (b) $d = 0.331674938$

9.57 A machine costing $26 000 is installed. Its expected useful lifetime is 6 years. At the end of that time it will have no scrap value. In fact, it is estimated that it will cost the company $1000 to remove the old machine (i.e., $S = -\$1000$). Prepare a depreciation schedule using (a) the straight-line method; (b) the sum-of digits method; (c) the sinking-fund method, using $j_1 = 9\%$.

9.58 A machine costing $50 000 has an estimated useful lifetime of 20 years and a scrap value of $2000 at that time. Find the accumulated depreciation and the book value of this asset at the end of 8 years using (a) the straight-line method; (b) the constant-percentage method (you must find d); (c) the sum-of-digits method; (d) the sinking-fund method, using $j_1 = 8\%$.
Ans. (a) $19 200, $30 800; (b) $36 202.70, $13 797.30; (c) $30 171.43; $19 828.57; (d) $11 156.87, $38 843.13

9.59 A machine costing $20 000 has an estimated lifetime of 15 years and zero scrap value at that time. At the end of 6 years, the machine becomes obsolete because of the development of a better machine. What is the accumulated depreciation and the book value of the asset at that time using (a) the straight-line method; (b) the sum-of-digits method; (c) the sinking-fund method, using $j_{12} = 6\%$?
Ans. (a) $8000, $12 000; (b) $12 500; $7500; (c) $5942.43, $14 057.57

9.60 A machine costing $45 000 depreciates 20% of its remaining value each year. Make out a depreciation schedule for the first 5 years. Ans. $B_5 = \$14 745.60$

9.61 A machine that costs $40 000 will depreciate to $3000 in 12 years. Find its book value at the end of 7 years and the depreciation expense for the eighth year using the constant-percentage method. *Ans.* $B_7 = \$8827.66$; $R_8 = \$1713.88$

9.62 Equipment worth $30 000 depreciates 10% of its remaining value each year. How long in years will it take before the equipment is worth less than $15 000? *Ans.* 7

9.63 A machine costing $28 000 depreciates to $4000 over 15 years. How long does it take for the machine to depreciate to less than half of its original value under the constant-percentage method? *Ans.* 5.343 years

9.64 A machine costing $50 000 will last 5 years, at which time it will be worth $5000. In that time it is expected to produce 90 000 units. If its production follows that pattern outlined below, do a depreciation schedule.

Year	Units Produced
1	23 000
2	16 000
3	21 000
4	17 000
5	13 000

9.65 An airplane costing $10 000 000 is purchased and will be used for 3 years by a company. At the end of that time it is estimated that it can be sold for $6 000 000. The accountant sets up a depreciation schedule based on miles flown and estimates that the plane will fly 450 000 miles in the next 3 years. Given the following, do a depreciation schedule.

Year	Miles Flown
1	130 000
2	210 000
3	110 000

9.66 A mine is purchased for $300 000 and has an expected reserve of 400 000 units of ore. After mining is complete, the land will be worth $20 000. If 100 000 units of ore are extracted in the first year, find the depletion deduction in 1 year. *Ans.* $70 000

9.67 Texaco purchases some land containing oil wells for $20 000 000. The total reserves of crude oil are estimated at 4 000 000 barrels. After the oil is gone, the land can be sold for $140 000. If Texaco removes 400 000 barrels in year 1 and 560 000 barrels in year 2, determine the respective depletion deductions. *Ans.* $1 986 000, $2 780 400

9.68 An $18 000 car will have a scrap value of $800 nine years from the date of purchase. (*a*) Use the constant-percentage depreciation method to find the book value of the car at the end of 3 years. (*b*) An identical car can be leased for 3 years at a cost of $390 paid at the beginning of each month. If money is worth $j_1 = 9\%$, determine whether it is more economical to lease or buy the car over a 3-year period. Assume the car can be sold at the end of 3 years for its book value found in (*a*) if you buy the car. *Ans.* (*a*) $6375.95; (*b*) lease

9.69 An automobile with an initial cost of $24 000 has an estimated depreciated value for tax purposes of $4033 at the end of 5 years. Our tax system uses the constant-percentage method of depreciation. (*a*) Find the annual rate of depreciation allowed. (*b*) Show the depreciation schedule. *Ans.* (*a*) 30%

9.70 Under the straight-line method, the book value at the end of k years is

$$B_k = C - k\left(\frac{C-S}{n}\right)$$

Because of the straight-line nature of this depreciation method, it is also possible to find the book value at the end of year k by linear interpolation between C, the original cost of the asset and S, the salvage value of the asset at the end of n years. Thus,

$$B_k = \left(1 - \frac{k}{n}\right) C + \frac{k}{n} S$$

Prove this formula is equivalent to

$$B_k = C - k \left(\frac{C - S}{n}\right)$$

and try this formula on Problems 9.58(a) and 9.59(a).

9.71 Show that

$$d = 1 - \left(\frac{S}{C}\right)^{\frac{1}{n}}$$

where n is the estimated lifetime of the asset.

9.72 Given s_n is the sum-of-digits to n, prove

$$B_k = S + \frac{s_{n-k}}{s_n}(C - S)$$

9.73 Show that, under the sinking-fund method, (a) the book value B_k of the asset at the end of k years is

$$B_k = C - \left(\frac{C - S}{s_{\overline{n}|i}}\right) s_{\overline{k}|i}$$

(b) the depreciation expense R_k in year k is

$$R_k = \left(\frac{C - S}{s_{\overline{n}|i}}\right) (1 + i)^{k-1}$$

9.74 Show that if $i = 0$, the sinking-fund method is equivalent to the straight-line method.

9.75 An \$18 000 car will have a scrap value of \$500 nine years from the date of purchase.

(a) Using the constant-percentage depreciation method, find the book value of the car at the end of 4 years.

(b) If money can be invested at $j_{12} = 6\%$, what is the monthly sinking-fund deposit necessary to replace the car if it is sold for its book value in (a) after 4 years and the price of new cars has increased at an annual rate of $j_1 = 5\%$.

(c) An identical car can be leased for 4 years at a cost of \$405 paid at the beginning of each month. The contract calls for the driver to pay for fuel, repairs, licenses, and maintenance (but **not** for insurance). If a person buys the car, there will be a cost of \$800 at the time of purchase for 1 year's insurance, with insurance costs increasing by 4% at the beginning of each of the next 3 years. Still using $j_{12} = 6\%$, determine whether it is more economical to buy or lease the car over a 4-year period. Assume that it can be sold for the book value found in (a) if you buy the car.

Ans. (a) \$3660.85; (b) \$366.77; (c) lease

9.76 An office machine is being depreciated over 10 years to an eventual scrap value of \$500, using the constant-percentage method. If the book value after four years is \$4573.05, what was the original value of the machine? *Ans.* \$20 000

Chapter 10

Contingent Payments

10.1 INTRODUCTION

It was seen in Chapter 8 that an investor's return from the purchase of a bond may be uncertain, for example, because of the possibility of default on the part of the issuer of the bond. Bonds backed by the federal government are often referred to as "risk-free" bonds as no payment default is assumed. At the other extreme, bonds with the lowest rating, and, therefore, the highest probability of some default, are called "junk" bonds and are extremely risky investments.

To analyze such *contingent payments* satisfactorily (not just on a "worst-case" basis) requires a combination of probability theory and compound interest.

10.2 PROBABILITY

To a given event E (belonging to a collection or "universe" of related events) we assign a number, $P(E)$, called its *probability*. This number, which is between 0 and 1, measures the likelihood that, out of the whole collection, it is precisely the event E which occurs. If E never occurs, $P(E) = 0$; if it inevitably occurs, $P(E) = 1$.

Events are said to be *mutually exclusive* if the occurrence of one of the events precludes the occurrence of any of the other events. For example, in flipping a coin, getting a head rules out getting a tail. Given n mutually exclusive events, $E_1, E_2, ..., E_n$, the probability of occurrence of any one of the events (that is, E_1 or E_2 or ... or E_n) is the sum of the respective probabilities of the individual events. Symbolically,

$$P(E_1 \text{ or } E_2 \text{ or } ... \text{ or } E_n) = P(E_1) + P(E_2) + \cdots + P(E_n) \tag{10.1}$$

Events are said to be *independent* if the occurrence of one of the events has no effect on the occurrence of any of the other events. Successive flips of a fair coin are an example; the outcome of one toss has no effect on the outcome of any other toss. The probability of the occurrence of all n independent events $E_1, E_2, ..., E_n$ is the product of the respective probabilities of the individual events. Symbolically,

$$P(E_1 \text{ and } E_2 \text{ and } ... \text{ and } E_n) = P(E_1) \times P(E_2) \times \cdots \times P(E_n) \tag{10.2}$$

The question remains as to how the numbers $P(E)$ are assigned. Some probabilities can be determined *a priori* (before the event), such as the probability of tossing a head with a fair coin or a "6" with a fair die. The rule is:

If an event E can happen in h ways, and fail to happen in f ways, *all of these ways being equally likely*, then the probability $p = P(E)$ of the occurrence of the event is given by

$$p = \frac{h}{h+f}$$

and the probability that the event will fail to occur is

$$q = 1 - p = \frac{f}{h+f}$$

However, in the case of such events as an individual's dying or a borrower's defaulting, we cannot assign a probability *a priori*, but must instead make a prediction on the basis of past experience. Specifically, we observe a (large) number, N, of trials of the event E in question, and let

$$p_N = \frac{\text{number of occurrences of } E}{N} \tag{10.3}$$

Then, according to the *Law of Large Numbers* (the "law of averages"), as N becomes very large, the fraction p_N approaches p, the true underlying probability of the event E.

SOLVED PROBLEMS

10.1 Find the probability that in throwing one die I get either a five or a six.

There are two sides of the die that qualify, and four sides that lead to failure; thus,

$$p = \frac{h}{h+f} = \frac{2}{2+4} = \frac{2}{6} = \frac{1}{3}$$

Equivalently, the probability of throwing a five is 1/6 and of throwing a six is 1/6, so that by (10.1) the total probability is $1/6 + 1/6 = 1/3$.

10.2 Find the probability that in drawing two cards from a standard deck of 52 cards, I get an ace and a king.

It is helpful here to draw what is referred to as a *probability tree* (Fig. 10-1). We can draw an ace first, following by a king, with probability

$$E_1 = \frac{4}{52} \times \frac{4}{51} = \frac{4}{663}$$

or we can draw a king first, followed by an ace, with probability

$$E_2 = \frac{4}{52} \times \frac{4}{51} = \frac{4}{663}$$

The total probability is

$$P(E_1 \text{ or } E_2) = P(E_1) + P(E_2) = \frac{4}{663} + \frac{4}{663} = \frac{8}{663}$$

Note that, in using Fig. 10-1, we first computed the probability of each path not terminating in a failure, by multiplying together the probabilities of the branches composing the path. Then we added together the probabilities of these paths.

Stage I

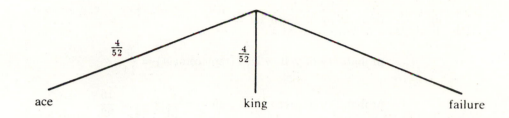

Stage II

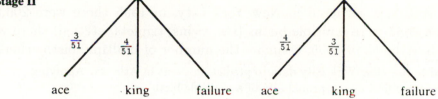

Fig. 10-1

Alternate Solution

There are $h + f = 52 \cdot 51$ equally likely ways of drawing two cards from the deck, and there are $h = 4 \cdot 4 \cdot 2$ ways of drawing an ace and a king (each ace can pair with each king, in either order). Hence,

$$p = \frac{4 \cdot 4 \cdot 2}{52 \cdot 51} = \frac{8}{663}$$

10.3 A box contains six green, three blue, and five yellow marbles. Determine the probabilities of (*a*) drawing a green marble (on one draw); (*b*) drawing two yellow marbles (no replacement); (*c*) drawing a blue marble twice (with replacement); (*d*) drawing a green marble followed by a yellow marble (with replacement); (*e*) drawing a blue marble and a green marble (no replacement).

(*a*)
$$p = \frac{6}{6 + 8} = \frac{3}{7}$$

(*b*)
$$\text{probability of 1st yellow marble} = \frac{5}{14}$$

$$\text{probability of 2nd yellow marble (no replacement)} = \frac{4}{13}$$

$$\text{probablity of two yellow marbles} = \frac{5}{14} \times \frac{4}{13} = \frac{10}{91}$$

(c)   probability of 1st blue ball $= \dfrac{3}{14}$

                 probability of 2nd blue ball (replacement) $= \dfrac{3}{14}$

                 probability of drawing twice $= \dfrac{3}{14} \times \dfrac{3}{14} = \dfrac{9}{196}$

(d)   probability of green 1st $= \dfrac{6}{14}$

                 probability of yellow 2nd (replacement) $= \dfrac{5}{14}$

                 probability of green then yellow $= \dfrac{6}{14} \times \dfrac{5}{14} = \dfrac{15}{98}$

(e)   As in the alternate solution to Problem 10.2:

$$p = \frac{3 \cdot 6 \cdot 2}{14 \cdot 13} = \frac{18}{91}$$

10.4 According to statistics gathered in New York City, in 1995 there were 200,000 births, of which 2087 were multiple births (i.e., twins, triplets, etc.). If there were 24 000 births in Atlanta in 1995, estimate the number of multiple births there.

We are forced to use New York City data to predict an event in Atlanta. Applying (10.3) to the New York data, we find as the probability of a multiple birth

$$\frac{2087}{200\,000} = 0.010435$$

Therefore, our best estimate for the number of multiple births in Atlanta is

$$(24\,000)(0.010435) = 250$$

10.5 Life insurance data tell us that the probability of surviving 10 years (i.e., living *at least* 10 years more) for persons aged 30, 40, and 50 is 0.9, 0.8, and 0.75, respectively. Determine the probability that (*a*) a 40-year-old dies in the next 10 years, (*b*) a 30-year-old survives to age 60, (*c*) a 30-year old dies between ages 50 and 60.

(*a*) $p \;=\; 1 - 0.8 = 0.2$

(*b*) $p \;=\;$ P (having lived to 30, lives to 40)
                 $\times P$ (having lived to 40, lives to 50)
                 $\times P$ having lived to 50, lives to 60)
                 $=\; 0.9 \times 0.8 \times 0.75 = 0.54$

(*c*) $p \;=\;$ P (lives to 50) $\times P$ (then dies within 10 years)
                 $=\; (0.9 \times 0.8) \times (1 - 0.75) = 0.18$

10.3   MATHEMATICAL EXPECTATION

Let X denote the numerical outcome of an experiment, whose possible outcomes are the numerical values $x_1, x_2, \ldots$. Then X will take on the values $x_1, \; x_2, \ldots$ with given probabilities $f(x_1), f(x_2), \ldots$. The *mathematical expectation* or *expected value* or *mean value* of X is defined by:

$$E(X) = x_1 f(x_1) + x_2 f(x_2) + x_3 f(x_3) + \cdots \qquad (10.4)$$

In the special case of N equally likely outcomes, (10.4) becomes

$$E(X) = \frac{x_1 + x_2 + \cdots + x_N}{N}$$

i.e., the arithmetic mean of the x's.

SOLVED PROBLEMS

10.6 A box contains four \$10 bills, six \$5 bills, and three \$1 bills. You are allowed to pull one bill from the box and keep it. What is the expected value of your winnings?

$$E(X) = \$10 \left(\frac{4}{13} \right) + \$5 \left(\frac{6}{13} \right) + \$1 \left(\frac{3}{13} \right) = \frac{\$73}{13} = \$5.62$$

Note that the expected value of this game is not one of the possible outcomes.

10.7 Mr. A pays \$10 to enter a betting game. If he can get 3 tails in a row by tossing a fair coin, he wins \$50; otherwise, he loses his \$10. What is his expected gain?

$$P(\text{three tails in a row}) = \left(\frac{1}{2} \right) \left(\frac{1}{2} \right) \left(\frac{1}{2} \right) = \frac{1}{8}$$

$$E(X) = \$50 \left(\frac{1}{8} \right) + (-\$10) \left(\frac{7}{8} \right) = -\$2.50$$

That is, his expectation is a loss of \$2.50.

10.8 A and B play a dice game. A wins if, when the dice are thrown, the two faces showing are identical; otherwise, B wins. If A wagers \$15, how much should B wager, to make the game *fair*?

Let B's bet be \$$x$.

$$P(A \text{ wins}) = \frac{6}{36} = \frac{1}{6} = P(B \text{ loses})$$

$$P(A \text{ loses}) = \frac{30}{36} = \frac{5}{6} = P(B \text{ wins})$$

Since one player gains what the other loses, either player's expected winnings must be zero if the game is to be fair. Thus,

$$\text{for A:} \quad 0 = x\frac{1}{6} + (-15)\frac{5}{6}$$
$$\text{for B:} \quad 0 = 15\left(\frac{5}{6} \right) + (-x)\frac{1}{6}$$

Either equations gives $x = 75$.

10.9 In a state lottery, 5 million \$10 tickets are issued each week. Cash prizes are awarded each week according to the odds listed in the table that follows. Also 250 000 tickets that reveal a bonus mark entitle the holder to 5 free tickets in the next draw. Find the expected value of any \$10 lottery ticket.

Prize	No. Awarded	Probability
\$1 000 000	5	$\frac{1}{1\,000\,000}$
250 000	20	$\frac{1}{250\,000}$
50 000	50	$\frac{1}{100\,000}$
10 000	100	$\frac{1}{50\,000}$
1000	2000	$\frac{1}{2500}$
100	25 000	$\frac{1}{200}$
Bonus	1 250 000	

Solution: Let X be the value of the state lottery ticket. Its expected value is $E(X)$ and the value of each bonus ticket is $E(X)$

$$
\begin{aligned}
E(X) = \; & \$1\,000\,000\left(\frac{1}{1\,000\,000}\right) + \$250\,000\left(\frac{1}{250\,000}\right) \\
& + 50\,000\left(\frac{1}{100\,000}\right) + \$10\,000\left(\frac{1}{50\,000}\right) \\
& + \$1000\left(\frac{1}{2500}\right) + \$100\left(\frac{1}{200}\right) + E(X)\left(\frac{1\,250\,000}{5\,000\,000}\right)
\end{aligned}
$$

Solving, $E(X) = \$4.80$.

10.4 CONTINGENT PAYMENTS WITH TIME VALUE

We can now mathematically analyze situations where investors discount the value of an investment to account for the probability of receipt of payment.

If p is the probability of receiving a sum S and $1-p$ is the corresponding probability of receiving nothing, then the amount X actually received has expected value

$$E(X) = Sp + 0(1-p) = pS$$

by (10.4).

The *discounted value* of the expectation pS to be received n periods from today, assuming interest at rate i per period, is:

$$pS(1+i)^{-n} \tag{10.5}$$

More generally, if sums of money $S_1, S_2, ..., S_n$ are received with probabilities $p_1, p_2, ..., p_n$ at times $t_1, t_2, ..., t_n$, then the *discounted value* of the expectation is

$$p_1 S_1(1+i)^{-t_1} + p_2 S_2(1+i)^{-t_2} + \cdots + p_n S_n(1+i)^{-t_n} \tag{10.6}$$

SOLVED PROBLEMS

10.10 Mr. Jones wants to borrow some money. He can repay the loan with a single payment of \$8000 in 1 year's time. The lending institution determines that there is a 5% chance that Mr. Jones will not repay the loan. The normal lending rate at that time is $j_1 = 10\%$. (a) How much will they lend Mr. Jones? (b) If he repays the loan in full, what rate of interest was realized?

(a) The expected value of the repayment is $pS = (0.95)(8000) = \$7600$. Then,

$$\text{discounted value} = 7600(1.10)^{-1} = \$6909.09$$

(b) The experienced rate of interest, i, is such that

$$
\begin{aligned}
6909.09 &= 8000(1+i)^{-1} \\
1+i &= 1.1579 \\
i &= 15.79\%
\end{aligned}
$$

This may seem like a very high rate of interest; however, if 5 borrowers out of 100 do not repay their loans, then the lending institution makes exactly 10%. This assumes total default; that is, either you pay in full or you pay nothing at all.

10.11 Mr. Smith wants to borrow some money and repay in a lump sum at the end of one year. The bank's past experience with repayments is indicated in Table 10-1.

Table 10-1

Proportion of Total Debt Repaid	Probability
0%	0.05
50%	0.05
75%	0.10
90%	0.10
100%	0.70

If the "risk-free" rate of interest the bank uses is $j_1 = 12\%$, what rate of interest will they charge Mr. Smith?

The expected proportion repaid by Mr. Smith is

$$E(X) = (0\%)(0.05) + (50\%)(0.05) + (75\%)(0.10) + (90\%)(0.10) + (100\%)(0.70) = 89\%$$

He will therefore be charged i such that

$$
\begin{aligned}
(1+i)^{-1} &= (0.89)(1.12)^{-1} \\
1+i &= \frac{1.12}{0.89} = 1.2584
\end{aligned}
$$

or

$$j_1 = 25.84\%$$

10.12 Robert's Aunt Mary dies and leaves him an inheritance of $50\,000 in a bank account earning interest at 8% per annum. He gains title to the money if he lives to age 21; he is now aged 8. If he dies before age 21 (probability 0.05), the money goes to his sister Sandra, aged 5, if she survives to age 21 (probability 0.97). Determine Sandra's expectation.

Sandra receives a sum $S = 50\,000(1.08)^{21-5}$ with probability $p = (0.05)(0.97)$. Her expectation is

$$pS = 50\,000(1.08)^{16}(0.05)(0.97) = \$8307.90$$

10.13 Find a fair price, ignoring expenses, for a term insurance policy that will pay $100 000 at the end of 1 year if the policyholder dies during the year. The policy is issued to a female, aged 30, whose probability of death is 0.00135; assume $j_2 = 10\%$.

Using (10.5),
$$\text{price} = (0.00135)(100\ 000)(1.05)^{-2} = \$122.45$$

10.14 Edna D., who is in very poor health, is to receive $5000 at the end of each year as long as she is alive. Her survival probabilities are:

n	Probability of Surviving n Years
1	0.75
2	0.40
3	0

If $j_1 = 7\%$, what is the expected value of these payments?

By (10.6),
$$\text{expected value} = (0.75)(5000)(1.07)^{-1} + (0.40)(5000)(1.07)^{-2} + 0 = \$5251.55$$

Alternate Solution

If she dies in year 1 (probability $1 - 0.75 = 0.25$), she gets 0. If she dies in year 2 (probability $0.75 - 0.40 = 0.35$), she gets one payment worth $5000(1.07)^{-1}$ now. If she dies in year 3 (probability 0.40), she gets two payments worth $5000a_{\overline{2}|.07}$ now. Hence,

$$\text{expected value} = 0(0.25) + 5000(1.07)^{-1}(0.35) + 5000a_{\overline{2}|.07}(0.40) = \$5251.55$$

10.15 Find the discounted expected value of a series of payments of $1000 made at the end of each year for the next 10 years, if $j_1 = 8\%$ and there is probability of 0.9 that the first payment will be made; probability $(0.9)^2 = 0.81$ that the second payment will be made; probability $(0.9)^3 = 0.729$ that the third payment will be made; and so on.

By (10.6),
$$\text{value} = (0.9)(1000)(1.08)^{-1} + (0.9)^2(1000)(1.08)^{-2} + \cdots + (0.9)^{10}(1000)(1.08)^{-10}$$

This could be summed as a geometric progression, but it is simpler to find an equivalent rate of interest, j_1, for which the series becomes an ordinary simple annuity at j_1. Thus,

$$1 + j_1 = \frac{1.08}{0.9} \qquad \text{or} \qquad j_1 = 0.20$$

and
$$\text{value} = 1000a_{\overline{10}|.20} = \$4192.47$$

10.16 A 20-year, $1000 coupon bond is offered for sale. The bond interest is $j_2 = 10\%$ and the bond matures for $1050. (*a*) Find the purchase price to yield $j_2 = 12\%$. (*b*) Find the purchase price to yield $j_2 = 12\%$ if the issuer, going into a six-month period, will default in that period with probability 0.02; beyond the point of default the investor receives nothing.

(a) By (8.1),

$$P = Fr\, a_{\overline{n}|i} + C(1+i)^{-n} = 50 a_{\overline{40}|.06} + 1050(1.06)^{-40}$$
$$= 752.31 + 102.08 = \$854.39$$

(b) The situation resembles that of Problem 10.14, except that now the probability that the kth payment is made is given by $(0.98)^k$. Thus, we find the new interest rate, i, per half-year that satisfies

$$1 + i = \frac{1.06}{0.98} \quad \text{or} \quad i = 8.16327\%$$

and
$$P = Fr\, a_{\overline{n}|i} + C(1+i)^{-n} = 50 a_{\overline{40}|i} + 1050(1+i)^{-40}$$
$$= 585.96 + 45.50 = \$631.46$$

Supplementary Problems

PROBABILITY

10.17 An urn contains 4 red balls, 2 white balls, and 1 yellow ball. Find the probabilities of the following events (for multiple drawings, assume no replacement). (a) You draw one ball and it is white. (b) You draw either a white or a yellow ball. (c) You draw two balls which are red. (d) You draw two balls; the first is yellow and the second is white. (e) You draw two balls of the same color.
Ans. (a) $\frac{2}{7}$; (b) $\frac{3}{7}$; (c) $\frac{2}{7}$; (d) $\frac{1}{21}$; (e) $\frac{1}{3}$

10.18 Given a standard deck of 52 cards, determine the probabilities of drawing (a) the queen of hearts; (b) a queen or jack; (c) the queen of hearts in two draws with no replacement; (d) the queen of hearts in two draws with replacement of the first draw; (e) the queen of hearts on the second draw with no replacement.
Ans. (a) $\frac{1}{52}$; (b) $\frac{2}{13}$; (c) $\frac{1}{26}$; (d) $1 - \left(\frac{51}{52}\right)^2$; (e) $\frac{1}{52}$

10.19 Given two fair dice, determine the probabilities of the following events; (a) rolling a total equal to 9; (b) rolling a total which exceeds 5; (c) rolling an even total; (d) in *two* rolls, getting an overall total equal to 5.
Ans. (a) $\frac{1}{9}$; (b) $\frac{13}{18}$; (c) $\frac{1}{2}$; (d) $\frac{1}{324}$

10.20 Given that the probability of having a male baby is equal to the probability of having a female baby, and assuming independent events, determine (a) the probability of having two girls in a family of three children, (b) the probability of all boys in a family of five children, (c) the probability of two girls and then a boy.
Ans. (a) $\frac{3}{8}$; (b) $\frac{1}{32}$; (c) $\frac{1}{8}$

10.21 A baseball team presently has a 0.600 record (they win 60% of their games). Assuming that games are independent events, determine the following probabilities: (a) they win their next two games, but then lose; (b) they win exactly two of their next three games; (c) they win at least one of their next four games.
Ans. (a) $\frac{18}{125}$; (b) $\frac{54}{125}$; (c) $\frac{609}{625}$

10.22 The probabilities that Adam, Bob, Colin, and Derek survive the next 10 years are 9/10, 5/8, 3/4, and 4/5, respectively. Assuming that deaths are independent events, determine the probability that (a) all four die in the period; (b) all four survive the period; (c) at least one survives the period; (d) at least one dies in the period.
Ans. (a) $\frac{3}{1600}$; (b) $\frac{27}{80}$; (c) $\frac{1597}{1600}$; (d) $\frac{53}{80}$

10.23 From the data

Age	Probability of Surviving 10 years
60	3/4
70	3/5
80	1/3

determine the probability that: (a) someone aged 60 lives to age 90, (b) someone aged 60 dies between ages 80 and 90.

Ans. (a) $\frac{3}{20}$; (b) $\frac{3}{10}$

10.24 A coin is weighted in such a way that the probability of a head is 1/3 and a tail is 2/3. Determine the probability of (a) 3 tails in a row; (b) no heads in two tosses; (c) a tail followed by two heads.

Ans. (a) $\frac{8}{27}$; (b) $\frac{1}{9}$; (c) $\frac{2}{27}$

10.25 A frequent flyer must change planes three times in one trip. If the probability of making any one connection successfully is 90%, what is the probability: (a) the trip is completed successfully; (b) the trip is not completed successfully.

Ans. (a) .729; (b) .271

MATHEMATICAL EXPECTATION

10.26 Susan is willing to bet Henry $1 that she can flip a coin 3 times and get three heads. How much should Henry wager to make the game fair? *Ans.* $7

10.27 A box contains five $10 bills, three $5 bills, and four $1 bills. You are to pull out two bills at random from the box (no replacement). If both bills are of the same denomination, you can keep them. What is your expectation?

Ans. $\$\frac{121}{33} = \$3.6\dot{6}$

10.28 A 31-year-old female wants to buy $120 000 worth of insurance on her life for 1 year. If the probability that she will die in 1 year is 0.0014, what is a fair price for this contract, ignoring any effect of compound interest and expenses? *Ans.* $168

10.29 In a certain lottery the prize is $20 000 and 100 tickets have been sold. What is the expectation of B, who holds 8 tickets? *Ans.* $1600

10.30 The manager of an outdoor circus wants to insure his next show against bad weather. He estimates that if it doesn't rain, 20 000 people will pay $16 each to attend; if it does rain, attendance will be only 10 000. The weatherman tells him there is a 10% chance of rain for the day of the event. What is a fair price for the insurance, ignoring expenses? *Ans.* $16 000

10.31 For a $5 ticket a lottery offers the following prizes:

1	prize of $100 000
20	prizes of $10 000
200	prizes of $1000
and 2000	prizes of 5 free tickets

100 000 tickets are issued each week. Find the expected value of any ticket. *Ans.* $3.56

10.32 Students with a special driver training course have a probability of 0.02 of having an accident in the next year. Students without driver training have a probability of 0.07 of having an accident in the next year. Statistics indicate that the cost of any one accident averages $8000. Ignoring expenses, how much of a discount should an insurance company offer an applicant with driver training versus without? (Assume no one has two accidents in the next year). *Ans.* $400

10.33 A TV quiz show gives five questions. A contestant gets $100 for each correct answer and $10 000 extra if all five answers are correct. What is the expectation for a contestant who guesses every answer?
Ans. $\dfrac{16\ 500}{32} = \$515.63$

CONTINGENT PAYMENTS WITH TIME VALUE

10.34 Mrs. Anderson is approached by an agent selling a special college scholarship plan. If Mrs. Anderson's new-born daughter survives to age 18 and enters college, she will be given $16 000 on her 18th birthday. The cost of the plan is $1600. If interest rates are 9% per annum and the probability that a new-born female survives to age 18 is 0.9840, what probability must Mrs. Anderson attach to her daughter's entering college to make the scheme worthwhile?
Ans. 47.94%

10.35 A 25-year-old male buys a one-year insurance policy which promises to pay $100 000 at the end of the year if he dies. If the probability of death at age 25 is 0.00177 and if $j_1 = 8\%$, what is a fair price for this contract, ignoring expenses? *Ans.* $163.89

10.36 How much would you lend a person today if she promises to repay $500 at the end of each year for 10 years, if $j_1 = 12\%$, and if there is a 10% chance of her defaulting in any one year (after which no monies would be paid)? *Ans.* $1815.82

10.37 The Friendly Finance Company charges $j_1 = 15\%$ on loans that are fully backed by collateral (which assumes that their value will be received). If they charge $j_1 = 20\%$ for unsecured loans, what is their expected default rate? *Ans.* 4.16%

10.38 Mrs. Zelenka dies and leaves an estate of $250 000 in trust at 8% per annum. There are two heirs to the estate, Tom, aged 7, and Susan, aged 3. The accumulated value of the trust will be divided equally between the survivors 18 years from now, when Susan turns 21. The probability that Tom will survive 18 years is 0.95, and for Susan it is 0.97. Assuming independence of events, find the expected value of inheritance for (*a*) Tom, (*b*) Susan.
Ans. (*a*) $488 763; (*b*) $508 743

10.39 Based on past experience, the ABC Savings and Loan has determined the following probability distribution of repayments:

Proportion of Total Debt Repaid	Probability
0%	0.05
50%	0.10
80%	0.10
90%	0.15
100%	0.60

If their "risk-free" rate of interest is $j_1 = 9\%$, what rate of interest should they charge on unsecured consumer loans? *Ans.* 26.01%

10.40 Mr. Jones, who is in poor health, wants to buy an annuity that will pay him $20 000 at the end of each year for as long as he lives. If $j_1 = 12\%$, find a fair price for this annuity, given the following probabilities of survival for Mr. Jones:

n	Probability of Surviving n Years
1	0.85
2	0.65
3	0.35
4	0

Ans. $30 524.55

10.41 A 20-year, $1000 bond is offered for sale March 1, 1996. The bond carries coupons at 12% per annum payable semiannually, and will mature at par. Find the purchase price (*a*) to yield $j_2 = 14\%$, (*b*) to yield $j_2 = 14\%$ if the probability of default in any 6-month period entered is 5%.
Ans. (*a*) $866.68; (*b*) $479.51

10.42 A $1000 bond with semiannual coupons at 10% matures for $1050 in 20 years. Find its purchase price to yield 9% per annum compounded semiannually, if the probability of default in any 6-month period entered is 1%. *Ans.* $917.25

10.43 For the bond in Problem 10.41, find the purchase price if all coupons are received with certainty but there is only a 95% chance that the redemption value will be paid at maturity. *Ans.* $1091.58

10.44 A $1000 par value "junk" bond is issued with semiannual coupons at 9% and matures in 20 years. (*a*) Find the purchase price to yield 10% per annum compounded semiannually, if the probability of default in any 6-month period is 2%. (*b*) Find the yield rate, j_2, if there are no payment defaults and the bond is held to maturity. *Ans.* (*a*) $653.42; (*b*) 14.29% ($i = 7.14\%$)

Chapter 11

Life Annuities and Life Insurance

11.1 INTRODUCTION

Life annuities and life insurance constitute one of the more important examples of contingent payments with a time factor (Section 10.4). The required probability distribution is specified in the form of a *mortality table*.

11.2 MORTALITY TABLES

A mortality table is a record of the observed number of deaths in a large population. If that base population is the population of a country, then we have what is called a *life table*, which is used by the census statisticians and demographers of the country in question. For example, the base population for the U.S. 1989-91 Life Tables was the tabulated population of the United States as of Census Day, April 1, 1990. Mortality rates were calculated by recording the age at death for all deaths in the U.S. in 1989, 1990, and 1991, and comparing the results ($\div 3$) to the base population.

If, on the other hand, the base population was a group of life annuity policyholders, then the derived mortality table would be used to develop the value of life annuities.

The mortality table (Appendix B) that we will use is the 1980 Commissioners Standard Ordinary Mortality Table, based on a population of North American life insurance policyholders. Strictly speaking, this table should be used for life insurance calculations only; but we will apply it in life annuity problems as well. Separate tables are given for males and for females.

Column 4, labeled $1000q_x$, is the first column derived, where q_x is the probability that an individual alive at age x will die before age $x + 1$. Thus, for a male aged 30, we have $1000q_{30} = 1.73$, which means that from life insurance data the probability of death at age 30 has been observed to be 0.00173.

Column 2 is denoted l_x, where l_0, the *radix*, has been arbitrarily set at 10 000 000. The l_x-column can be considered to give the number of people who attain age x out of the original 10 000 000 people aged zero. Column 3, the d_x-column, gives the number of deaths at age x (i.e., between ages x and $x + 1$). Columns 2 and 3 can be derived from column 4 and l_0 via the relations

$$d_x = l_x \cdot q_x \qquad \text{and} \qquad l_{x+1} = l_x - d_x$$

As a matter of convenience, q_{99} is set equal to unity, whence $l_{100} = 0$.

We will also define the following notations:

$p_x \equiv$ probability that an individual aged x will survive for at least one year

$$p_x = \frac{l_{x+1}}{l_x}$$

213

$_np_x \equiv$ probability that an individual aged x will survive for at least n years $(p_x = {}_1p_x)$

$$_np_x = \frac{l_{x+n}}{l_x} \tag{11.1}$$

$q_x \equiv$ probability that an individual aged x will die before age $x + 1$

$$q_x = 1 - p_x = \frac{l_x - l_{x+1}}{l_x} = \frac{d_x}{l_x}$$

$_nq_x \equiv$ probability that an individual aged x will die before age $x + n$ $(q_x = {}_1q_x)$

$$_nq_x = 1 - {}_np_x = \frac{l_x - l_{x+n}}{l_x} \tag{11.2}$$

SOLVED PROBLEMS

11.1 Find the probability that a male aged 30 will survive for at least one year.

$$p_{30} = \frac{l_{31}}{l_{30}} = \frac{9\ 563\ 425}{9\ 579\ 998} = 0.99827$$

or $p_{30} = 1 - q_{30} = 1 - 0.00173 = 0.99827$.

11.2 Find the probability that a female aged 30 will survive for at least 20 years.

$$_{20}p_{30} = \frac{l_{50}}{l_{30}} = \frac{9\ 219\ 130}{9\ 707\ 590} = 0.94968$$

11.3 Find the probability that a female aged 50 will die before age 60.

$$_{10}q_{50} = \frac{l_{50} - l_{60}}{l_{50}} = \frac{9\ 219\ 130 - 8\ 603\ 801}{9\ 219\ 130} = 0.06674$$

11.4 Find the probability that a female aged 30 dies between ages 50 and 60.

By the results of Problems 11.2 and 11.3, the answer is

$$_{20}p_{30} \cdot {}_{10}q_{50} = (0.94968)(0.06674) = 0.06338$$

Equivalently,
$$_{20}p_{30} \cdot {}_{10}q_{50} = \left(\frac{l_{50}}{l_{30}}\right)\left(\frac{l_{50} - l_{60}}{l_{50}}\right) = \frac{l_{50} - l_{60}}{l_{30}}$$
$$= \frac{9\ 219\ 130 - 8\ 603\ 801}{9\ 707\ 590} = 0.06339$$

(The small difference is due to roundoff error.)

11.3 PURE ENDOWMENTS

A *pure endowment* is the discounted value of a payment that will be made to an individual at some specified future date if that individual is then alive. By proportionality, it suffices to consider the *pure endowment factor*, $_nE_x$, which represents the discounted expected value of a payment of \$1 to be made to an individual now aged x

if and when he attains age $x + n$. The probability that an individual now aged x will survive to age $x + n$ is

$$_nP_x = \frac{l_{x+n}}{l_x}$$

and so

$$_nE_x = \$1 \cdot \frac{l_{x+n}}{l_x} \cdot (1+i)^{-n} = (1+i)^{-n}\,_nP_x \tag{11.3}$$

Another explanation of this formula is instructive. Assume that l_x individuals aged x each put the same amount of money in a fund. This fund is left to accumulate for n years at rate i, and is then divided up among the l_{x+n} survivors. If we call the unknown deposit $\$y$, the payment to each survivor is given by

$$\frac{yl_x(1+i)^n}{l_{x+n}} = \frac{y}{_nE_x}$$

That is, $y = {}_nE_x$ is the deposit which provides each survivor $1.

SOLVED PROBLEMS

11.5 Find the discounted expected value of $50\,000$ to be paid to Mr. Saujani, now aged 30, if he survives to age 65. Let $j_1 = 10\%$.

Here, $x = 30$, $n = 65 - 30 = 35$, $i = 0.10$; using (11.3),

$$50\,000\,_{35}E_{30} = 50\,000\frac{l_{65}}{l_{30}}(1.10)^{-35}$$

$$= 50\,000\left(\frac{7\,329\,740}{9\,579\,998}\right)(1.10)^{-35} = \$1361.29$$

11.6 Mary Smith inherits $10\,000$ on her 47th birthday. She uses the money to buy a pure endowment payable if and when she attains age 60. (*a*) Assuming she survives, how much will she receive if $j_4 = 8\%$? (*b*) Compare the endowment to investing the money in a savings account at $j_4 = 8\%$.

(*a*) Find i per annum such that

$$1 + i = (1.02)^4 \qquad \text{or} \qquad i = 0.08243216$$

Now find Y such that

$$10\,000 = Y \cdot {}_{13}E_{47} = Y\frac{l_{60}}{l_{47}}(1.08243216)^{-13}$$

$$Y = \frac{10\,000 l_{47}(1.08243216)^{13}}{l_{60}} = 10\,000\left(\frac{9\,340\,119}{8\,603\,801}\right)(1.08243216)^{13} = \$30\,399.82$$

(*b*) $$Y = 10\,000(1.08243216)^{13} = \$28\,003.28$$

11.4   LIFE ANNUITIES

A *life annuity* is an annuity that makes payments on a regular basis to an individual, the *annuitant*. Payments may be made either for as long as the annuitant is alive (a *whole life annuity*) or for a specified number of years (a *temporary life annuity*). Life annuities can be paid in any frequency—monthly, quarterly, annually—but, because of the form of our mortality table, Appendix B, we shall restrict our consideration to annual payments.

In Chapter 5, we defined an ordinary annuity as one in which payments are made at the ends of the payment intervals. An ordinary whole life annuity, then, has payments at the end of each year for as long as the annuitant survives. As such, it is equivalent to a set of level pure endowments payable at the end of 1, 2, 3,... years (the series terminating with the death of the annuitant). Hence, the expected discounted value (ignoring expenses) of an ordinary whole life annuity of $1 per annum (first payment at the end of the first year) issued to an individual now aged x is

$$a_x = {}_1E_x + {}_2E_x + {}_3E_x + \cdots = \sum_{t=1}^{\infty}(1+i)^{-t}\frac{l_{x+t}}{l_x} = \sum_{t=1}^{\infty}(1+i)^{-t}\,{}_tp_x \qquad (11.4)$$

Since, according to our mortality table, no individual reaches age 100, the summation in (11.4) actually terminates when $x + t = 99$, or $t = 99 - x$. Thus, for a prospective annuitant aged 30, there would be 69 terms in the sum.

The quantity Pa_x is the cost of an ordinary whole life annuity of $$P$ per annum to a person aged x; it is referred to as the *net single premium* ("net", because expenses are ignored). Obviously, most annuity calculations require a computer.

SOLVED PROBLEMS

11.7 Find the net single premium for an ordinary whole life annuity of $1000 per annum issued to a male aged 95, if $j_1 = 7\%$.

$$1000a_{95} = 1000\sum_{t=1}^{\infty}(1+i)^{-t}\frac{l_{x+t}}{l_{95}} = \$1217.30$$

Note: There are four terms to this calculation.

11.8 Find the net single premium for a whole life annuity due of $3000 per year issued to a male now aged 95, using $j_1 = 7\%$.

From Chapter 5, we know that an annuity due has payments at the beginning of each payment interval. Using Fig. 11-1, we see that the cost of a whole life annuity due of $1 per annum to a person aged x, denoted $\ddot{a}_x$, is just the first payment ($1) plus a_x:

$$\ddot{a}_x = 1 + a_x = \sum_{t=0}^{\infty}(1+i)^{-t}\frac{l_{x+t}}{l_x} \qquad (11.5)$$

From problem 11.7: $a_{95} = 1.21730$ which gives

$$\ddot{a}_{95} = 2.21730$$

and $3000\ddot{a}_{95} = \$6651.89$.

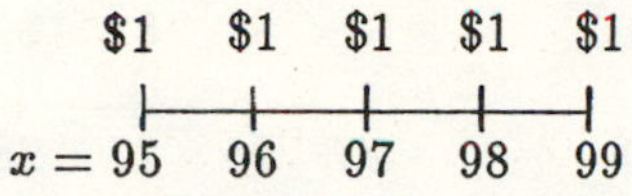

$$\begin{array}{ccccc} \$1 & \$1 & \$1 & \$1 & \$1 \\ \vdash & + & + & + & \dashv \\ x=95 & 96 & 97 & 98 & 99 \end{array}$$

Fig. 11-1

11.9 Find the net single premium for a whole life annuity of \$50 000 per annum at $j_1 = 8\%$ issued to a female aged 70, if the first payment is to be at age 95.

This is referred to as a *deferred annuity*; it can be evaluated by using Fig. 11-2. We have level annual payments starting at age 95 whose discounted expectation at age 70 is

$$_{25}E_{70} + \ _{26}E_{70} + \cdots = \sum_{t=0}^{\infty} (1+i)^{-(25+t)} \frac{l_{95+t}}{l_{70}}$$

The notation for this is either $_{24|}a_{70}$ or $_{25|}\ddot{a}_{70}$, depending on whether the first payment is considered to occur at the end of the 24th year (an ordinary deferred annuity) or at the beginning of the 25th year (a deferred annuity due). In general,

$$_{k|}a_x = \sum_{t=1}^{\infty} (1+i)^{-(k+t)} \frac{l_{x+k+t}}{l_x} \tag{11.6}$$

$$_{k|}\ddot{a}_x = \sum_{t=0}^{\infty} (1+i)^{-(k+t)} \frac{l_{x+k+t}}{l_x} \tag{11.7}$$

For the present data, the answer is:

$$50\ 000 \sum_{t=0}^{\infty} (1+i)^{-(25+t)} \frac{l_{95+t}}{l_{70}} = \$872.01$$

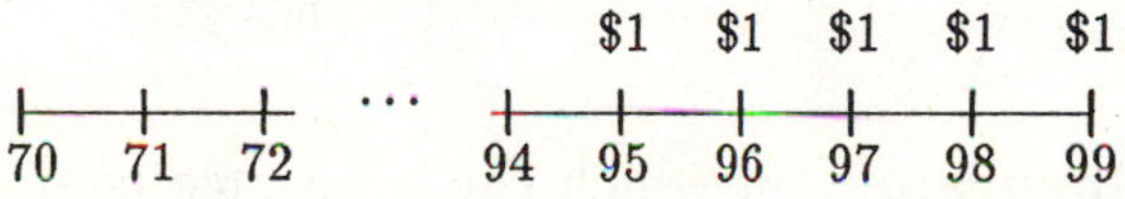

$$\begin{array}{ccccccccc} & & & & \$1 & \$1 & \$1 & \$1 & \$1 \\ \vdash & + & + & \cdots & + & + & + & + & \dashv \\ 70 & 71 & 72 & & 94 & 95 & 96 & 97 & 98 & 99 \end{array}$$

Fig. 11-2

11.10 Find the net single premium for a 5-year temporary life annuity of \$1000 per year for a female now aged 25, if $j_1 = 9\%$.

Regardless of how long the woman lives, she will receive at most 5 annual payments. If payments are made at the end of each year [an ordinary annuity, Fig. 11-3(a)], the expected discounted value at age 25 is, for $P = \$1$

$$_{1}E_{25} + \ _{2}E_{25} + \cdots + \ _{5}E_{25} = \sum_{t=1}^{5} (1+i)^{-t} \frac{l_{25+t}}{l_{25}}$$

The notation for this is $a_{25:\overline{5}|}$; in general,

$$a_{x:\overline{n}|} = a_x - \ _{n|}a_x = \sum_{t=1}^{n} (1+i)^{-t} \frac{l_{x+t}}{l_x} \tag{11.8}$$

Note that $a_{x:\overline{n}|}$ is the sum of the first n terms of the series (11.4) for a_x.

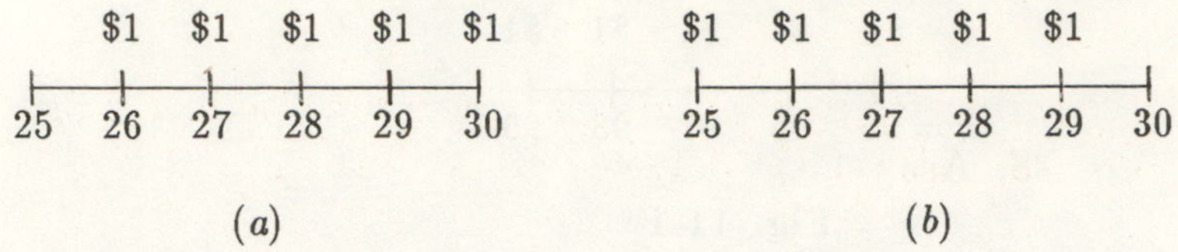

$1 $1 $1 $1 $1 $1 $1 $1 $1 $1

25 26 27 28 29 30   25 26 27 28 29 30

(a)   (b)

Fig. 11-3

If, instead, payments are made at the beginning of each year [an annuity due, Fig. 11-3(b)], the answer is

$$_0E_{25} + {}_1E_{25} + \cdots + {}_4E_{25} = \sum_{t=0}^{4}(1+i)^{-t}\frac{l_{25+t}}{l_{25}}$$

which is denoted $\ddot{a}_{25:\overline{5}|}$. In general,

$$\ddot{a}_{x:\overline{n}|} = \ddot{a}_x - {}_n|\ddot{a}_x = \sum_{t=0}^{n-1}(1+i)^{-t}\frac{l_{x+t}}{l_x} \tag{11.9}$$

For our particular data, and assuming an ordinary annuity, the net single premium is

$$1000a_{25:\overline{5}|} = 1000\sum_{t=1}^{5}(1+i)^{-t}\frac{l_{25+t}}{l_{25}} = 1000\frac{[(1.09)^{-1}l_{26} + (1.09)^{-2}l_{27} + \cdots + (1.09)^{-5}l_{30}]}{l_{25}} = \$3876.47$$

11.5 LIFE INSURANCE

The principal types of life insurance are:

Whole life insurance, in which the insurance company promises to pay the face value of the policy to the beneficiary upon the death of the insured, whenever that may occur.

n-Year term insurance, in which the insurance company promises to pay the face value of the policy to the beneficiary upon the death of the insured only if the insured dies within n years of the issuance of the policy.

n-Year endowment insurance, in which the insurance company promises to pay the face value of the policy to the beneficiary upon the death of the insured, if the insured dies within n years of the issuance of the policy, or, if the insured is alive after n years, to pay the face value of the policy to the insured at that time. Thus, an n-year endowment insurance policy combines the features of an n-year term insurance policy and an n-year pure endowment.

Insurance benefits are actually paid promptly upon proof of the insured's death. However, because of the form of our mortality tables, we shall assume that the benefits are paid at the end of the year of death. As in the case of life annuities, only net premiums will be considered here (i.e., expenses will be ignored).

SOLVED PROBLEMS

11.11 Find the net single premium for a one-year term insurance policy of face $1000 is issued to a male aged 28. Assume $j_1 = 10\%$.

The probablity of payment is

$$q_{28} = \frac{1.70}{1000}$$

and so the premium is

$$1000(1.10)^{-1}q_{28} = \$1.55$$

This is denoted $\$1000A^1_{28:\overline{1}|}$, where $A^1_{x:\overline{n}|}$ can be interpreted to mean that the insured, aged x, must die before n years go by for there to be a payment.

11.12 Find the net single premium for a 10-year term insurance policy of face $15\,000 issued to a female aged 31, assuming $j_1 = 4\%$.

We can think of this policy as a series of 10 *consecutive* 1-year term insurance policies (each for $15\,000). The notation for such a policy (of face $1) would be $A^1_{31:\overline{10}|}$, which signifies that an insured, aged 31, must die before the end of 10 years if there is to be a payment (of $1). Using the time diagram, Fig. 11-4, we have for the net single premium

$$A^1_{31:\overline{10}|} = (1+i)^{-1}q_{31} + (1+i)^{-2}\,{}_1p_{31}\cdot q_{32} + (1+i)^{-3}\,{}_2p_{31}\cdot q_{33} + \cdots + (1+i)^{-10}\,{}_9p_{31}\cdot q_{40}$$

$$= (1+i)^{-1}\frac{d_{31}}{l_{31}} + (1+i)^{-2}\frac{d_{32}}{l_{31}} + (1+i)^{-3}\frac{d_{33}}{l_{31}} + \cdots + (1+i)^{-10}\frac{d_{40}}{l_{31}}$$

where, in the second line, we have simplified the death probabilities by applying the definition of Section 11.2. The formula is still sufficiently cumbersome to require a computer; so, other problems with this length of policy period will not be presented. However, the answer to our particular problem is:

$$\begin{aligned} A^1_{31:\overline{10}|} &= \frac{1}{l_{31}}[(1+i)^{-1}d_{31} + \cdots + (1+i)^{-10}d_{40}] \\ &= \$212.00 \end{aligned}$$

More generally, the net single premium for an n-year term insurance policy of face $1 issued to an individual aged x is

$$A^1_{x:\overline{n}|} = \sum_{t=0}^{n-1}(1+i)^{-(t+1)}\frac{d_{x+t}}{l_x} \tag{11.10}$$

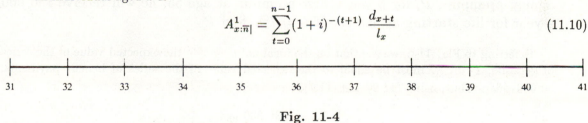

Fig. 11-4

11.13 Find the net single premium for a $50\,000 whole life insurance policy issued to a male aged 95, assuming $j_1 = 8\%$.

From the definitions of the different insurance coverages, we can see that a whole life policy is just a term policy that has no termination point. For a $1 face, the net single premium could then be symbolized $A^1_{x:\overline{\infty}|}$, but the conventional notation is A_x. Our answer is then:

$$50\,000\sum_{t=0}^{\infty}(1+i)^{-(t+1)}\frac{d_{95+t}}{l_{95}} = \$41\,859.26$$

In general, the net single premium for a $1 whole life insurance policy is

$$A_x = \sum_{t=0}^{\infty} (1+i)^{-(t+1)} \frac{d_{x+t}}{l_x} \tag{11.11}$$

11.14 Find the net single premium for a $5000, 5-year endowment insurance policy bought for a male on his fifteenth birthday, assuming $j_1 = 9\%$.

As pointed out in the verbal definition, this policy is a combination of a 5-year term insurance policy and a 5-year pure endowment. The net single premium for a $1, n-year endowment insurance policy is $A_{x:\overline{n}|}$; and we have

$$A_{x:\overline{n}|} = A^1_{x:\overline{n}|} + {}_nE_x = \sum_{t=0}^{n-1} (1+i)^{-(t+1)} \frac{d_{x+t}}{l_x} + (1+i)^{-n} {}_np_x \tag{11.12}$$

In particular, the solution to the present problem is:

$$5000 A_{15:\overline{5}|} = 5000 \left[\sum_{t=0}^{4} (1+i)^{-(t+1)} \frac{d_{15+t}}{l_{15}} + (1+i)^{-5} {}_5p_{15} \right] = \$3254.42$$

11.6 ANNUAL-PREMIUM POLICIES

In real life, people do not pay one large lump sum for their insurance policies; rather, they make periodic (yearly, monthly, etc.) payments to the insurance company, called *premiums*. We will assume annual premiums in all our examples, which means that premiums are paid to the insurance company at the *beginning* of each year for a specified period of time, but only if the policyholder is alive. Hence, the stream of premium payments must be valued using the symbols developed in Sections 11.4 and 11.5.

SOLVED PROBLEMS

11.15 Mr. Aiken, at age 55, buys a special chronic care insurance policy. By paying an annual premium, P, for 5 years, first premium at age 55, he will receive $30\,000 a year for life starting at age 95. If $j_1 = 7\%$, find P.

Referring to Fig. 11-5, we see that on the focal date, age 55, the expected value of the series of deposits, $P\,\ddot{a}_{55:\overline{5}|}$, must be equal to the expected value of the series of benefit payments, $30\,000 {}_{40|}\ddot{a}_{55}$. Thus, using (11.9) and (11.7),

$$P \cdot \ddot{a}_{55:\overline{5}|} = 30\,000\ {}_{40|}\ddot{a}_{55}$$

$$P \sum_{t=0}^{4} (1+i)^{-t} \frac{l_{55+t}}{l_{55}} = 30\,000 \sum_{t=0}^{\infty} (1+i)^{-(40+t)} \frac{l_{95+t}}{l_{55}}$$

$$P = \frac{30\,000 \displaystyle\sum_{t=0}^{\infty} (1+i)^{-(40+t)}\, l_{95+t}}{\displaystyle\sum_{t=0}^{4} (1+i)^{-t}\, l_{55+t}}$$

$$= \$17.62$$

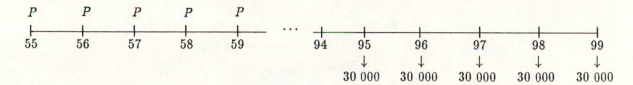

Fig. 11-5

11.16 Find the net annual premium, P, for a \$10 000 whole life policy issued to a female aged 95, if $j_1 = 8\%$.

Equating expectations at issue, we have, by (11.5) and (11.11),

$$P\ddot{a}_{95} = 100\ 000A_{95}$$

$$P = \frac{A_{95}}{\ddot{a}_{95}} = 100\ 000\frac{\displaystyle\sum_{t=0}^{\infty}(1+i)^{-(t+1)}d_{95+t}}{\displaystyle\sum_{t=0}^{\infty}(1+i)^{-t}l_{95+t}} = \$3740.14$$

In general, the symbol for the net annual premium for a whole life insurance policy of \$1 issued to a person aged x is

$$P_x = \frac{A_x}{\ddot{a}_x} = \frac{\displaystyle\sum_{t=0}^{\infty}(1+i)^{-(t+1)}d_{x+t}}{\displaystyle\sum_{t=0}^{\infty}(1+i)^{-t}\ l_{x+t}} \qquad (11.13)$$

11.17 Find the net annual premium, P, payable for a maximum of 20 years, for \$40 000 whole life policy issued to a male aged 25, if $j_1 = 4\%$. (This is sometimes called "20-payment life insurance.")

At age 25:

$$P\ddot{a}_{25:\overline{20|}} = 40\ 000A_{25}$$

$$P = 40\ 000\frac{\displaystyle\sum_{t=0}^{\infty}(1+i)^{-(t+1)}d_{25+t}}{\displaystyle\sum_{t=0}^{19}(1+i)^{-t}\ l_{25+t}} = \$512.10$$

This problem was solved using a personal computer and the l_x and d_x values from Appendix B.

The symbol for the net annual premium for an m-pay whole life policy of \$1 issued to a person aged x is

$$_mP_x = \frac{A_x}{\ddot{a}_{x:\overline{m|}}} = \frac{\displaystyle\sum_{t=0}^{\infty}(1+i)^{-(t+1)}d_{x+t}}{\displaystyle\sum_{t=0}^{m-1}(1+i)^{-t}\ l_{x+t}} \qquad (11.14)$$

11.18 Find the net annual premium, P, for a 5-year, \$80 000 term insurance policy issued to a female aged 28, if $j_1 = 6\%$.

At age 28:

$$P\ddot{a}_{28:\overline{5|}} = 80\ 000A^1_{28:\overline{5|}}$$

$$P = \frac{80\,000 \sum\limits_{t=0}^{4} (1+i)^{-(t+1)} d_{28+t}}{\sum\limits_{t=0}^{4} (1+i)^{-t} l_{28+t}} = \$101.61$$

The symbol for the net annual premium for an n-year term insurance policy of \$1 issued to a person aged x is

$$P^1_{x:\overline{n}|} = \frac{A^1_{x:\overline{n}|}}{\ddot{a}_{x:\overline{n}|}} \equiv \frac{\sum\limits_{t=0}^{n-1} (1+i)^{-(t+1)} d_{x+t}}{\sum\limits_{t=0}^{n-1} (1+i)^{-t} l_{x+t}} \tag{11.15}$$

11.19 Find the net annual premium, P, for a \$5000, 5-pay, 10-year endowment insurance policy issued to a male at age 42, if $j_1 = 8\%$.

At age 42:

$$P(\ddot{a}_{42:\overline{5}|}) = 5000 A_{42:\overline{10}|} = 5000(A^1_{42:\overline{10}|} + {}_{10}E_{42})$$

$$P = \frac{5000\left[\sum\limits_{t=0}^{9} (1+i)^{-(t+1)} d_{42+t} + (1+i)^{-10} l_{52}\right]}{\sum\limits_{t=0}^{4} (1+i)^{-t} l_{42+t}} = \$551.51$$

The symbol for the net annual premium for an m-pay, n-year endowment insurance policy of \$1 issued to a person aged x is

$$_mP_{x:\overline{n}|} = \frac{A_{x:\overline{n}|}}{\ddot{a}_{x:\overline{m}|}} = \frac{A^1_{x:\overline{n}|} + {}_nE_x}{\ddot{a}_{x:\overline{m}|}} \tag{11.16}$$

If $m = n$, we write $P_{x:\overline{n}|}$ instead of $_nP_{x:\overline{n}|}$.

Supplementary Problems

MORTALITY TABLES (see Appendix B throughout)

11.20 Find the probability that a female now aged 20 will (*a*) attain age 45, (*b*) die before age 60, (*c*) die between ages 50 and 65, (*d*) die at age 70.
Ans. (*a*) 0.9581; (*b*) 0.1239; (*c*) 0.1105; (*d*) 0.0168

11.21 For males, determine (*a*) the age at which the number living is approximately 50% of the number alive at age 20, (*b*) the age at which the probability of dying in the coming year is the least, (*c*) the age by which the population has decreased to approximately one-half its original size. *Ans.* (*a*) 75; (*b*) 10; (*c*) 75

11.22 Bob just entered college and is exactly 18 years old. Find the probability that (*a*) he will live to graduate 4 years hence, (*b*) he will die in his senior year. *Ans.* (*a*) 0.9926; (*b*) 0.0019

11.23 The 1993 graduating class of a certain college contained 200 females aged 21 and 300 aged 22. How many can be expected to survive for the class's 50th reunion? *Ans.* 366

11.24 Obtain directly the relation $l_x = d_x + d_{x+1} + d_{x+2} + \cdots$. *Ans.* l_x counts those who will die after age x.

11.25 Given

$$l_x = 100\ 000 \left(1 - \frac{x}{105}\right) \qquad (x \le 105)$$

determine (a) l_0, (b) l_{50}, (c) d_{10}, (d) $_{10}p_{21}$, (e) $_{15}q_{35}$, (f) the probability that a 21-year old dies between age 45 and age 63. (g) Graph the curve of l_x.
Ans. (a) 100 000; (b) 52 381; (c) 952; (d) 0.8810; (e) 0.2143; (f) 0.2571

11.26 The following values of q_x have been derived from a company's mortality experience:

x	q_x
0	0.010
1	0.005
2	0.003

Given $l_0 = 100\ 000$, calculate l_x and d_x through age 2.

Ans.
$l_0 = 100\ 000 \qquad d_0 = 1000$
$l_1 = \ \ 99\ 000 \qquad d_1 = \ \ 495$
$l_2 = \ \ 98\ 505 \qquad d_2 = \ \ 296$

11.27 Prove that: $_tp_x q_{x+t} = \ _tp_x - \ _{t+1}p_x$

11.28 Given $_np_x = \dfrac{100 - (x + n)}{100 - x}$, find:
(a) $_{20}p_{25}$, (b) $_{10}q_{21}$, (c) the probability that a 30-year-old dies at age 45.
Ans. (a) .7333; (b) .1266; (c) .0143

PURE ENDOWMENTS

11.29 On his 21st birthday Alphonse inherits \$50 000 from an estate. For tax reasons, he uses it to purchase a pure endowment payable if and when he attains age 65. Assuming that he survives, how much will he receive, if $j_1 = 8\%$? *Ans.* \$1 962 865.08

11.30 Find the net single premium to a female for a pure endowment of \$10 000 due at the end of 20 years, if (a) she is aged 45 and $j_1 = 4\%$; (b) she is aged 20 and $j_2 = 9\%$.
Ans. (a) \$3945.33; (b) \$1671.06

11.31 Find the present value at $j_1 = 7\%$ of \$5000 due at the end of 20 years, if (a) the payment is certain to be made; (b) the payment is contingent on the survival of a male now aged 30.
Ans. (a) \$1292.10; (b) \$1209.37

11.32 Given

$$l_x = l_0 \left(1 - \frac{x}{100}\right) \qquad (x \le 100)$$

find the net single premium for a pure endowment of \$1000 due in 20 years, if purchased by a 35-year-old. Assume $j_4 = 10\%$. *Ans.* \$96.03

11.33 A 30-year-old male buys a pure endowment at $j_2 = 10\%$ with \$7500. At age 65 he gets \$$X$; find X. *Ans.* \$298 255.86

11.34 Mr. Pershad works for the Acme Steel Company. The company's pension plan provides each retiring employee with \$100 000 at age 65. The company wants Mr. Pershad to retire early at age 59 and offers him the expected value of his retirement cheque calculated at $j_1 = 6\%$. How much was Mr. Pershad offered? *Ans.* \$62 972.42

LIFE ANNUITIES

11.35 Find the net single premium for a life annuity of $1000 per year issued to a female aged 70, if the first payment is at age 95 and $j_1 = 8\%$. *Ans.* $17.44

11.36 Mr. Smith dies and leaves Mrs. Smith $75 000 as the proceeds of a life insurance policy. Mrs. Smith is now aged 94. Using Appendix B, find the annual income to Mrs. Smith from these proceeds, assuming the first payment is one year hence and $j_1 = 7\%$. *Ans.* $49 453.42

11.37 Find the net single premium for a 5-year temporary life annuity due of $10 000 per year for a male aged 27, given (*a*) $j_1 = 4\%$; (*b*) $j_1 = 9\%$. *Ans.* (*a*) $46 147.17; (*b*) $42 264.96

11.38 At age 95, Mr. Jones takes out a life insurance policy on which he agrees to pay premiums of $70 at the beginning of each year for life. Find the discounted value of these premium payments, if $j_1 = 6\%$. *Ans.* $156.60

11.39 At age 65, Mr. Brown receives $100 000 lump sum from his employer to be used for his retirement. He has the option of buying an annuity certain for 5 years or of buying a 5-year temporary life annuity ($a_{65:\overline{5}|}$). (*a*) Find the annual payment in either case, assuming $j_1 = 9\%$. (*b*) If Mr. Brown dies just before his 68th birthday, how much would his estate receive in either case?
Ans. (*a*) $25 709.25, $27 883.36; (*b*) $70 934.68, $0

11.40 Find the net single premium, for a female aged 80, for a $5000 per annum annuity (first payment one year hence), where the first 15 payments are guaranteed and payments thereafter are made as long as the annuitant survives, if $j_1 = 8\%$. *Ans.* $42 950.47

11.41 At age 95, a male pays $40 000 for a whole life annuity due. What annual payment will this provide, if $j_1 = 6\%$. *Ans.* $17 879.71

11.42 Find the net single premium for a 5-year temporary life annuity of $10 000 per annum for a female now aged 40, if the first payment is made at age 65 and $j_1 = 10\%$. *Ans.* $3186.47

11.43 A female receives $9000 a year for life (first payment at age 95). Find the expected discounted value of these payments at age 65 if $j_1 = 8\%$. *Ans.* $97.83

11.44 A 60-year-old worker is paralyzed in an industrial accident. He sues the company for the present value of his lost income through to age 65. If he was earning $45 000 at the time of the injury and if no salary increase is assumed and if it is assumed that the salary is paid at the end of each year, find the value of the lawsuit if $j_1 = 5\%$. *Ans.* $184 808.55

11.45 Show that

$$(a)\ \ddot{a}_x = 1 + (1+i)^{-1}p_x\ddot{a}_{x+1} \qquad (b)\ \ddot{a}_{x:\overline{n}|} - a_{x:\overline{n}|} = 1 - {}_nE_x$$

11.46 Given

$$l_x = l_0\left(1 - \frac{x}{105}\right) \qquad (x \le 105)$$

(*a*) find an expression for a_x.

(*b*) Calculate $\ddot{a}_{20:\overline{5}|}$ if $j_1 = 9\%$.

$$\textit{Ans.}\ (a)\ a_x = \frac{(105 - x) - (1+i)a_{\overline{105-x}|i}}{(105 - x)i} \qquad (x \le 104);\ (b)\ 4.1485$$

11.47 Prove that, for all x, $a_{x:\overline{n}|} < a_{\overline{n}|i}$ at the common rate of interest i.

$$Ans. \quad a_{x:\overline{n}|} = \sum_{t=1}^{n} \frac{(1+i)^{-t} l_{x+t}}{l_x}; \quad a_{\overline{n}|i} = \sum_{t=1}^{n} (1+i)^{-t} \text{ and } \frac{l_{x+t}}{l_x} < 1$$

LIFE INSURANCE

11.48 Find the net single premium for a whole life insurance policy of $25 000 issued at age 95 (*a*) to a female, (*b*) to a male. Assume $j_1 = 9\%$. *Ans.* (*a*) $20 433.24; (*b*) $20 501.53

11.49 At age 93, Bill inherits $50 000. How much whole life insurance can he purchase if he uses the entire inheritance as a net single premium? Assume $j_1 = 7\%$. *Ans.* $60 595.63

11.50 Determine the net single premium for a one-year term insurance policy for $1000 issued to a female aged 34, if $j_1 = 8\%$. *Ans.* $1.46

11.51 Find the net single premium for a five-year term insurance policy for $25 000 issued to a female aged 27, assuming (*a*) $j_1 = 4\%$, (*b*) $j_1 = 12\%$. *Ans.* (*a*) $144.60; (*b*) $116.51

11.52 Find the net single premium for a 5-year term insurance policy of $75 000 issued to a male now aged 35. Interest is at 8% per annum. *Ans.* $714.88

11.53 Find the net single premium for a 5-year endowment insurance policy of $10 000 issued to a female aged 45, if $j_1 = 9\%$. *Ans.* $6523.52

11.54 Calculate $A^1_{40:\overline{5}|}$ assuming that mortality follows the law

$$l_x = l_0 \left(1 - \frac{x}{100}\right) \qquad (x \le 100)$$

and interest is 9% per annum. *Ans.* 0.06482752

11.55 Show that $A_x = (1+i)^{-1}\ddot{a}_x - a_x$. (*Hint*: Review 11.27.)

11.56 Find the rate of interest if $\ddot{a}_x = 14.257$ and $A_x = 0.193$. *Ans.* 6%

11.57 Calculate $A^1_{40:\overline{10}|}$ for a female if $j_1 = 10\%$ for 5 years and then 7% for the next 5 years. *Ans.* 0.02071

ANNUAL-PREMIUM POLICIES

11.58 It is possible for a female aged 31 to buy a certain insurance policy by paying $600 a year for 10 years certain or X a year for 5 years (but only if the policyholder is alive). Find X, assuming $j_1 = 8\%$. *Ans.* $1011.05

11.59 Find the net annual premium, at $j_1 = 6\%$, for a whole life insurance policy of $25 000 issued to a female aged 95. *Ans.* $9587.51

11.60 Find the net annual premium for a 5-pay, 10-year endowment insurance policy of $5000 issued at age 45 (*a*) to a female, (*b*) to a male. Let $j_1 = 10\%$. *Ans.* (*a*) $477.02; (*b*) $481.66

11.61 Find the net annual premium for a $50 000 term-to-65 policy issued to a female aged 60, if $j_1 = 7\%$. *Ans.* $514.91

11.62 Given $l_x = l_0(\frac{105 - x}{105})$, find $_5P_{30:\overline{20}|}$ if $j_1 = 7\%$. *Ans.* .077315

11.63 How large a whole life insurance policy can a male aged 95 purchase for a net annual premium of $2000, if $j_1 = 9\%$? *Ans.* $5314.88

11.64 Show that

$$P_x = \frac{1}{\ddot{a}_x} - \frac{i}{1+i}$$

The following problems require a personal computer for solving.

11.65 A man directs that his estate of \$600 000 be split so that his wife gets 2/3 and his son 1/3. The money is to be used to buy whole life annuities. If his widow is 63 and his son is 37, find the annual income of each annuity, assuming $j_1 = 7\%$. *Ans.* \$42 480.12; \$16 063.12

11.66 Mr. Smith plans to deposit \$X a year in a retirement fund, starting on his 35th birthday and going until his 64th birthday, inclusive. He wants to be able to retire on \$10 000 year for life, beginning at age 65. Find X if $j_1 = 6\%$. *Ans.* \$898.53

11.67 At age 40, Mr. Jones purchases a policy which will pay his beneficiary \$20 000 if he dies before age 65 and which will pay Mr. Jones a life annuity of \$5000 a year, with the first payment at 65, if he is alive at that time. Find the net annual premium if 20 payments are required and $j_1 = 10\%$. *Ans.* \$418.26

11.68 A special policy has death benefits of \$50 000 up to age 65 and \$10 000 after age 65. Premiums are payable to age 65. Find the net annual premium for this policy if issued to a female aged 37. Assume $j_1 = 9\%$. *Ans.* \$202.43

Review Problems

1. A cash discount of 4% is given if a bill is paid 40 days in advance of its due date. (*a*) What is the highest simple interest rate at which you can afford to borrow money if you want to take advantage of this discount? (*b*) If you can borrow money at a 20% simple interest rate, how much money can you save by paying a bill for $5000 forty days in advance of its due date? *Ans.* (*a*) 37.5%; (*b*) $93.33

2. John borrows $1000 on January 2, 1995, at a simple interest rate of 16%. He pays $350 on April 9, $150 on August 11, and $400 on October 4, 1995. What is the balance due on January 2, 1996, using the United States Rule? *Ans.* $203.07

3. Mrs. Arcularis has two options available in repaying a loan: She can pay $200 at the end of 5 months and $300 at the end of 10 months, or she can pay $X at the end of 3 months and $2X at the end of 6 months. Find X at a simple interest rate of 12%, taking the focal date at the end of 6 months. *Ans.* 161.87

4. Kathy needs $1000 cash for 5 months. If she borrows at 8% bank discount, (*a*) what size of loan should she ask for? (*b*) What simple interest rate will she pay on the loan? *Ans.* (*a*) $1034.48; (*b*) 8.28%

5. A 180-day promissory note for $2000 bears 14% simple interest. After 60 days it is sold to a bank which charges a discount rate of 14%. Find the price paid by the bank. *Ans.* $2040.13

6. A note for $800 is due in 90 days with simple interest at 11%. On the maturity date, the maker of the note paid the interest in full and gave a second note for 60 days without interest and for such an amount that when it was discounted at 9% discount rate on the day it was signed, the proceeds were just sufficient to pay the debt. Find the interest paid on the first note and the face value of the second note. *Ans.* $22, $812.18

7. I.C.U. Optical receives an invoice for $2500 with terms 2/10, *n*/50. (*a*) If the company is to take advantage of the discount, what is the highest simple interest rate at which it can afford to borrow money? (*b*) If money can be borrowed at 10% simple interest, how much does I.C.U. Optical actually save by using the cash discount? *Ans.* (*a*) 18.62%; (*b*) $23.15

8. A student lends his friend $10 for one month. At the end of the month he asks for repayment of the $10 plus purchase of a chocolate bar worth 50¢. What simple interest rate is implied. *Ans.* 60%

9. A taxpayer expects an income tax refund of $380 on May 1. On March 10, a tax discounter offers 85% of the full refund in cash. What rate of simple interest will the tax discounter earn, using exact time and exact interest? *Ans.* 123.87%

10. Today Mr. Mueller borrowed $2400 and arranged to repay the loan with three equal payments of $X at the end of 4, 8, and 12 months, respectively. If the lender charges a simple interest rate of 16%, find X using today as a focal date. *Ans.* $883.96

11. A person borrows $3000 at a 14% simple interest rate. He is to repay the debt with 3 equal payments, the first at the end of 3 months, the second at the end of 7 months, and the third at the end of 12 months. Find the difference between payments resulting from the selection of the focal date at the present time and at the end of 12 months. *Ans.* $2.72

12. On June 1, 1996, Sheila borrows $2000 at 12%. She pays $800 on August 17, 1996; $400 on November 20, 1996; and $500 on February 2, 1997. What is the balance due on April 18, 1997, by (*a*) the Merchant's Rule; (*b*) by the United States Rule? *Ans.* (*a*) $416.56; (*b*) $423.26

13. Mr. A has a note for $1500 dated June 8, 1995. The note is due in 120 days with interest at 12%. (a) If Mr. A discounts the note on August 1, 1995, at a bank charging an interest rate of 15%, what will the proceeds be? (b) What rate of interest will Mr. A realize on his investment? *Ans.* (a) $1518.25; (b) 8.11%

14. How much money will be required on December 31, 1999 to repay a loan of $2000 made December 31, 1996, if $j_4 = 12\%$? *Ans.* $2851.52

15. An investment fund advertises that it will triple your money in 10 years. What rate of interest j_4 is implied? *Ans.* 11.14%

16. A debt of $10 000 with interest at $j_4 = 11\%$ is to be paid off in 8 equal quarterly instalments, the first due today. Find the quarterly payment. *Ans.* $1371.85

17. The cost of living rises 8.7% a year for 5 years. Over that period of time, what would be the increase in value of a $160 000 house (due to inflation only)? *Ans.* $82 810.63

18. What amount of money invested today will grow to $1000 at the end of 5 years, if $j_4 = 8\%$? *Ans.* $672.97

19. At what nominal rate compounded daily will your investment double in 5 years? *Ans.* 13.87%

20. How long will it take $1000 to accumulate to $2500 at $j_{365} = 14\%$? *Ans.* 6 years 199 days

21. At what nominal rate j_{12} will money triple itself in 12 years? *Ans.* 9.19%

22. A trust company offers guaranteed investment certificates paying $j_2 = 16\frac{3}{4}\%$, $j_4 = 16\frac{1}{4}\%$, or $j_{12} = 16\frac{1}{8}\%$. Rate the options from best to worst. *Ans.* j_2, j_{12}, j_4

23. If interest on the outstanding balance of a credit card account is charged at $1\frac{3}{4}\%$ per month, what is the annual effective rate of interest? *Ans.* 23.14%

24. At a certain rate of interest j_2, money will double in value in 8 years. If you invest $1000 at this rate of interest, how much money will you have (a) in 5 years? (b) In 10 years? *Ans.* (a) $1542.21; (b) $2378.41

25. On the birth of their first grandchild, the Smiths bought a $100 savings bond which paid interest at $j_1 = 8\%$. How much money did their grandchild receive upon cashing this bond on her 20th birthday? *Ans.* $466.10

26. Find the total value on June 1, 1994, of $1000 due on December 1, 1989, and $800 due on December 1, 1999, at $j_2 = 11.38\%$. *Ans.* $2080.76

27. The population of Canada in 1981 was 24.3 million people; in 1991 it was 27.3 million people. (a) What was the annual growth rate from 1981 to 1991? (b) At this rate of growth, when will the population reach 30 million people? *Ans.* (a) 1.17%; (b) 1999

28. John buys goods worth $1500. He wants to pay $500 at the end of 3 months, $600 at the end of 6 months, and $300 at the end of 9 months. If the store charges $j_{12} = 21\%$ on the unpaid balance, what down payment will be necessary? *Ans.* $228.04

29. Paul has deposited $1000 in a savings account paying interest at 10% per annum and now finds that his deposits have accumulated to $1610.51. If he had been able to invest the $1000 over the same period in a guaranteed investment certificate paying interest at $j_1 = 13\frac{1}{4}\%$ and had deposited this interest in his savings account, to what sum would his $1000 now have accumulated? *Ans.* $1862.91

30. A note with face value $3000 is due without interest on November 18, 1997. On April 3, 1993, the holder of the note has it discounted at a bank charging $j_4 = 14\%$. What are the proceeds? *Ans.* $1587.17

31. On April 7, 1989, a debt of $4000 was incurred at 10% compounded semiannually. What amount will be required to settle the debt on September 19, 1994? *Ans.* $6814.21

32. A note for $3000 without interest is due on August 15, 1997. On June 11, 1996, the holder of the note has it discounted by a lender who charges $j_{12} = 12\%$. What are the proceeds? *Ans.* $2607.31

33. A man leaves an estate of $50 000 which is invested at $j_{12} = 9\%$. At the time of his death, he has two children, aged 13 and 18. Their shares in the estate when each reaches age 21 are to be equal. How much does each child get? (Ignore any probability of a child's death.) *Ans.* $39 929.39

34. A person deposits $1500 into a mutual fund. If the fund earns 9.8% a year compounded daily for 10 years, what will be the accumulated value of the initial deposit? *Ans.* $3996.16

35. If money is worth $j_1 = 8\%$, what single sum of money payable at the end of 2 years will equitably replace $1000 due today, plus a $2000 debt due at the end of 4 years with interest at $12\frac{1}{2}\%$ per annum compounded semiannually? *Ans.* $3951.33

36. A piece of land can be purchased by paying $50 000 cash or by paying $20 000 now, at the end of 2 years, and at the end of 4 years. To pay cash, the buyer would have to withdraw money from an investment earning interest at $j_2 = 8\%$. Which option is better, and by how much? *Ans.* Cash option by $1709.88

37. A trust company offers guaranteed investment certificates paying either $j_2 = 15.5\%$ or $j_1 = 16\%$. Which option yields the higher annual effective rate of interest? *Ans.* $j_2 = 15.5\%$

38. If an investment grows from $4000 to $6000 in 3 years, what was the rate of growth j_{365}? *Ans.* 13.52%

39. If money is worth $j_4 = 16\%$, find the sum of money due at the end of 15 years equivalent to $1000 due at the end of 6 years. *Ans.* $4103.93

40. Find the amount of interest earned between 5 and 10 years after the date of an investment of $100 if interest is paid at $j_2 = 7\%$. *Ans.* $57.92

41. Calculate the payment due on July 1, 1994, that is equivalent to $1000 due on January 1, 1991, plus $2000 due on March 1, 1996, if $j_4 = 8\%$ prior to July 1, 1994, and $j_{12} = 12\%$ thereafter. *Ans.* $2958.57

42. A city increased in population 4% a year during the period 1975 to 1985. If the population was 40 000 in 1975, what is the estimated population in 1995, assuming the rate of growth remains the same? *Ans.* 87 645

43. At what annual growth rate will the population of a city double in 11 years? *Ans.* 6.50%

44. Find the annual effective rate of interest which is equivalent to $j_4 = 6\%$ for 2 years followed by $j_{12} = 8\%$ for 4 years. *Ans.* 7.57%

45. Gus invests $500 for 4 years. The nominal interest rate remains 8% each year although in the first year it is convertible semiannually, in the second year convertible quarterly, in the third year convertible monthly and in the fourth year convertible daily. (a) What is the accumulated value? (b) How much greater is this value than the corresponding value assuming that the first rate had remained unchanged for the 4 years? *Ans.* (a) $686.76; (b) $2.48

46. $1000 was deposited on January 1, 1989, and $2000 was deposited in an account on July 1, 1991. Interest was paid on the account at $j_4 = 7\%$ from January 1, 1989, to October 1, 1991, and at $j_2 = 5\%$ from October 1, 1991, until April 1, 1993. Find (a) the amount in the account on April 1, 1993, (b) the equivalent nominal interest rate compounded monthly actually earned on the investment over the period. *Ans.* (a) $3494.79; (b) 5.83%

47. What is the difference between the annual effective rate of interest and the annual effective rate of discount corresponding to a nominal rate of interest of 12% compounded monthly? *Ans.* 0.0142743

48. A loan of $10 000, taken on January 1, 1992, is to be repaid on January 1, 1998. The debtor would like to pay $2000 on January 1, 1995, and make equal payments on January 1, 1997 and January 1, 1998. What will the size of these payments be if interest is assumed to be at $j_{12} = 15\frac{1}{4}\%$? *Ans.* $10 017.03

49. Jackie invested $500 in an investment fund. Find the accumulated value of her investment at the end of 6 years, if the interest rate was $j_2 = 12.3\%$ for the first two years, $j_{12} = 11.2\%$ for the next three years and $j_{365} = 10\%$ for the last year. *Ans.* $980.21

50. Calculate the present value of $2000 due in $5\frac{1}{2}$ years if $j_4 = 8\%$ for the first 2 years and $j_2 = 10\%$ thereafter. *Ans.* $1213.12

51. A debt of $5000 was originally due to be paid on July 1, 1993. However the borrower repaid $2000 on March 1, 1990. What amount is now due on July 1, 1993, if money is worth $j_{12} = 12\%$ in 1990, $j_4 = 10\%$ in 1991, $j_1 = 9\%$ in 1992 and $j_2 = 8\%$ in 1993? *Ans.* $2235.61

52. Julie bought $2000 in Canada Savings Bonds which paid interest at $j_1 = 9\%$ with interest accruing at a simple interest rate for each month before November 1 (i.e., for any partial year). How much will she receive for the bonds if they are cashed in 5 years and 3 months after the date of issue? *Ans.* $3146.49

53. An obligation of $2000 is due on February 4, 1995. Find the value of this obligation on December 13, 1992, at 10% compounded quarterly, if simple interest is used for part of an interest conversion period. *Ans.* $1618.81

54. A note for $2000 dated October 6, 1993 is due with compound interest at $j_{12} = 8\%$, two years after the date. On January 16, 1995, the holder of the note has it discounted by a lender who charges $j_4 = 9\%$. Find the proceeds and the compound discount. *Ans.* $2199.79, $145.99

55. Starting on his 36th birthday, Fred deposits $2000 a year into a savings fund. His last deposit is at age 65. Starting 1 month later, he makes monthly withdrawals for 15 years. If $j_4 = 12\%$ throughout, find the size of these monthly withdrawals. *Ans.* $6406.43

56. How long will it take to increase your investment by 50% at rate $14\frac{1}{4}\%$ compounded daily? *Ans.* 1039 days

57. Find the accumulated and the discounted value of an annuity of $500 at the end of each month at $j_{12} = 9\%$ for (a) 10 years, (b) 20 years. *Ans.* (a) $96 757.14, $39 470.85; (b) $333 943.43, $55 572.48

58. At age 65, Mrs. Sahas takes her life savings of $100 000 and buys a 20-year annuity certain having quarterly payments. Find the size of these payments, (a) at 10% compounded quarterly, (b) at 12% compounded quarterly. *Ans.* (a) $2902.60; (b) $3311.17

59. A couple needs a loan of $10 000 to buy a boat. One lender will charge $j_{12} = 18\%$, a second lender offers $j_{12} = 16\%$. What will be the monthly savings in interest using the lower rate, if the monthly payments are to run for 5 years? *Ans.* $10.75

60. Find the accumulated and the discounted value of semiannual payments of $500 at the end of every half-year over 10 years, if interest is $j_2 = 10\%$ for the first four years and $j_2 = 12\%$ for the last 6 years. *Ans.* $18 042.31, $6068.87

61. Charlie wants to accumulate $100 000 by making monthly deposits of $1000 into a fund that accumulates interest at $j_{12} = 12\%$. Find the number of full deposits required and the size of the concluding deposit made 1 month after the last full deposit. *Ans.* 69, $323.66

62. Lisa borrows $10 000 at $j_4 = 18\%$. How many $800 quarterly payments will she make, and what will be the size of the partial concluding payment made 3 months after the last full $800 payment? *Ans.* 18, $627.75

63. On June 1, 1993, Ms. Kaminski purchased furniture for $2200. She paid $400 down and agreed to pay the balance in monthly instalments of $100, plus a smaller final payment; the first payment is due on July 1, 1993. If money earns interest at $j_{12} = 18\%$, when is the final payment made, and what is its amount? *Ans.* $13.85 on April 1, 1995

64. A used car sells for $5000 cash or $1000 down and $800 a month for 6 months. Find the interest rate j_{12} for purchase on the instalment plan. *Ans.* 65.66%

65. Find j_{12} at which deposits of $200 at the end of each month will accumulate to $10 000 in 3 years. *Ans.* 21.55%

66. The XYZ Finance Company charges 10% "interest in advance" and allows the client to repay the loan in 12 equal monthly payments. Thus, for a loan of $6000, they would charge $600 interest in advance and then require 12 monthly repayments of $550 each. What is the corresponding rate of interest convertible monthly? *Ans.* 17.97%

67. A refrigerator is priced at $650. If a customer pays $200 down, the balance plus a carrying charge of $50 can be paid in 12 equal monthly instalments. If the customer pays cash, he can get a discount of 15% off the list price. What is the nominal rate converted monthly, if the refrigerator is bought on time? *Ans.* 70.04%

68. Paul wants to accumulate $5000 by depositing $300 every 3 months into an account paying 8% per annum converted quarterly. He makes the first deposit on July 1, 1993. How many full deposits should he make, and what will be the size and the date of the concluding deposit? *Ans.* 14, $111.97 on January 1, 1997

69. How much a month for 5 years at $j_{12} = 6\%$ would you have to save in order to receive $800 a month for 3 years afterward? *Ans.* $376.91

70. A bank account paying $j_{12} = 5.5\%$ contains $5680 on March 1, 1993. On April 1, 1993, the first of a sequence of monthly withdrawals of $400 is made. (*a*) What is the date of the last withdrawal? (*b*) By what date (first of the month) will the balance again exceed $400? *Ans.* (*a*) May 1, 1994; (*b*) July 1, 2000

71. Jones agrees to pay Smith $800 at the end of each quarter for 5 years, but is unable to do so until the end of the 15th month, when he wins $100 000 in a lottery. Assuming that money is worth $j_4 = 6\%$, what single payment at the end of 15 months liquidates his debt? *Ans.* $14 796.40

72. A deposit of $1000 is made to open an account on March 1, 1990. Monthly deposits of $300 are then made for 5 years, starting April 1, 1990. On April 1, 1995, the first of a sequence of 20 monthly withdrawals of $1000 is made. Find the balance in the account on December 1, 1997, assuming $j_{12} = 12\%$. *Ans.* $11 487.31

73. Ashley deposits $500 into an investment fund each January 1 starting in 1989 and continuing to 1998 inclusive. If the fund pays interest at $j_1 = 10\%$, how much will be in her account on January 1, 2003? *Ans.* $12 833.69

74. Mr. Juneau has deposited $800 at the end of each year into an investment fund for the last 10 years. His investments earned $j_1 = 10\%$ for the first 7 years and $j_1 = 9\%$ for the last 3 years. How much money does he have in his account 10 years after his last deposit if interest rates have remained level at $j_1 = 9\%$? *Ans.* $29 477.01

75. Rebecca has deposited $80 at the end of each month for 7 years. For the first 5 years the deposits earned 12% compounded monthly. After 5 years they earned 9% compounded monthly. Find the value of the annuity after (a) 7 years, (b) 10 years. *Ans.* (a) $9911.94; (b) $12 971.21

76. A debt of $18 000 is to be repaid in annual instalments of $3000 with the first instalment due at the end of the second year and a final instalment of less than $3000. Interest is 14% per year compounded annually. Find the final payment. *Ans.* $386.03

77. A man aged 30 wishes to accumulate a fund for retirement by depositing $1000 at the end of each year for the next 35 years. Starting on his 66th birthday he will make 15 annual withdrawals of equal amount. Find the amount of each withdrawal if $j_1 = 10\%$ throughout. *Ans.* $35 632.60

78. Today Brian purchased a cottage and a boat worth a combined present value of $65 000. To purchase the cottage, Brian agreed to pay $10 000 down and $350 at the end of each month for 20 years. To buy the boat, Brian was also required to pay $1000 down and R at the end of each month for 4 years. Find R if $j_{12} = 12\%$. *Ans.* $584.96

79. Marie deposited $150 at the end of each month into an account earning $j_{12} = 8\%$. She made these deposits for 14 years except that in the fifth year she was unable to make any deposits. Find the value of the account 2 years after the last deposit. *Ans.* $49 702.20

80. An annuity pays $2000 at the end of each year for 5 years and then $1000 at the end of each year for the next 8 years. Find the discounted value of these payments if $j_1 = 10\%$. *Ans.* $10 894.14

81. An office space is renting for $30 000 a year payable in advance. Find the equivalent monthly rental payable in advance if money is worth 12% compounded monthly. *Ans.* $2639.07

82. A farmer borrowed $80 000 to buy some farm equipment. He plans to pay off the loan with interest at $j_1 = 13\frac{3}{4}\%$ in 8 equal annual payments, the first to be made 5 years from now. Find the annual payment. *Ans.* $28 630.72

83. Payments of $100 per quarter are made from March 1, 1993, to September 1, 1998, inclusive. If interest is at $j_4 = 18\%$, find (a) the discounted value on December 1, 1991; (b) the accumulated value on March 1, 2000. *Ans.* (a) $1186.38; (b) $5070.61

84. Joshua deposits $500 a year in a fund paying interest at $j_1 = 11\%$. The first deposit is made January 1, 1988, and the last deposit January 1, 1997. How much money is in the fund on (a) January 1, 1994 after the payment is made? (b) January 1, 2000?
Ans. (a) $4891.64; (b) $11 434.77

85. Diana has an insurance policy whose cash value at age 65 will provide payments of $1500 a year for 15 years, first payment at age 66. If the insurance company pays $j_1 = 9\%$ on its funds, what is the cash value at age 65? *Ans.* $12 091.03

86. Five years from now a company will need $150 000 to replace worn-out equipment. Starting now, what monthly deposits must be made in a fund paying $j_{12} = 9\%$ for 5 years to accumulate this sum? *Ans.* $1973.95

87. Fred needs a loan of $15 000 to buy a new car. The dealer offers him the loan at $j_{12} = 16\%$. Fred can get the loan at his bank at $j_{12} = 15\%$. (a) What will be the monthly savings using the lower rate at his bank if the monthly payments are to run for 3 years? (b) What will be the total interest on the loan at his bank? (c) Assume that Fred takes the loan of $15 000 at his bank and decides to pay $600 at the end of each month. How many months will it take him to repay the loan, assuming a smaller final payment 1 month after the last $600 payment? Find the concluding payment. *Ans.* (a) $7.38; (b) $3719.28; (c) 31, $97.97

88. The Sound Warehouse advertises "no interest for 1 year" option. For example, you can buy a $1200 stereo by paying $100 at the end of each month for 1 year. However, if you pay in full at the time of the purchase, you get a 10% cash discount. What rate of interest j_{12} is the Sound Warehouse actually charging? *Ans.* 19.91%

89. (*a*) Elsa wants to accumulate $20 000 by the end of 6 years by making quarterly deposits in a fund which pays interest at 10% compounded quarterly. Find the size of these deposits. (*b*) After 4 years the bank changes the interest rate to $j_4 = 8\%$. Find the size of the quarterly deposits now required if the $20 000 goal is to be met. (*c*) What is the total interest earned on the fund over 6 years? *Ans.* (*a*) $618.26; (*b*) $694.53; (*c*) $4551.60

90. Instead of paying $550 rent at the beginning of each month for the next 10 years, a couple decides to buy a townhouse. What is the cash equivalent of the 10 years of rent, at $j_{12} = 8\%$? *Ans.* $45 634.03

91. A company sets aside $15 000 at the beginning of each year to accumulate a fund for future expansion. What is the amount in the fund at the end of 5 years, if the fund earns $j_{12} = 5\%$? *Ans.* $87 327.44

92. A bank pays interest at 6% compounded quarterly. Interest dates are March 31, June 30, September 30, and December 31. On April 1, 1992, Richard opened an account with a deposit of $200. He continues to make $200 deposits every 3 months until July 1, 1994, when he makes his last deposit. Find the amount in his account on (*a*) September 30, 1993; (*b*) September 30, 1997. *Ans.* (*a*) $1264.60; (*b*) $2597.66

93. A loan of $10 000, taken on January 1, 1992, is to be repaid by January 1, 1998. The debtor would like to pay $2000 on January 1, 1995, and to make equal payments on January 1, 1997, and January 1, 1998. What will the size of these payments be, if interest is at $j_{12} = 15\frac{1}{4}\%$? *Ans.* $10 017.03

94. George deposits $200 at the beginning of each year in a bank account that earns interest at $j_4 = 6\%$. How much money will be in his bank account at the end of 5 years? *Ans.* $1199.86

95. Julia borrows $10 000. The loan is to be repaid with equal payments at the end of each month for the next 5 years. Find the size of these payments if (*a*) $j_4 = 12\%$, (*b*) $j_1 = 12\%$. *Ans.* (*a*) $221.85; (*b*) $219.36

96. Upon graduation, Scott determines that he has borrowed $8000 from the student loan plan over his three years of college. This loan must be repaid with monthly payments (first payment at the end of the first month) over the next 5 years. If $j_2 = 11\%$, find the monthly payment. *Ans.* $172.97

97. How much must be deposited in a bank account at the end of each quarter for 4 years to accumulate $4000 if (*a*) $j_{12} = 6\%$, (*b*) $j_2 = 6\%$, (*c*) $j_{365} = 6\%$? *Ans.* (*a*) $222.93; (*b*) $223.25; (*c*) $222.87

98. Deposits of $200 are made at the end of each month for 5 years into a bank account where interest is paid $j_4 = 9\%$. Find the accumulated value of the account at the end of 5 years if (*a*) simple interest is paid for part of a period, (*b*) compound interest is paid for part of a period. *Ans.* (*a*) $15 059.01; (*b*) $15 058.46

99. An insurance company pays interest at $j_1 = 9\%$ on money left with them. What would be the cost of an annuity certain paying $250 at the end of each month for 10 years if compound interest is paid for each part of a period? *Ans.* $20 034.80

100. What rate of interest j_4 must be earned for deposits of $100 at the end of each month to accumulate to $2000 in 18 months? *Ans.* 14.86%

101. A couple made monthly deposits of $400 into an account paying interest at 8.25% compounded quarterly. The first deposit was made on August 1, 1993; the last deposit was made on December 1, 1996. (*a*) How much money will be in the account on January 1, 1997? (*b*) How much money will they be able to withdraw monthly, starting on February 1, 1997, and ending February 1, 1998, to pay for their expenses during their planned trip around the world? *Ans.* (*a*) $18 980.79; (*b*) $1530.80

102. Melvin borrows $1000 at 16% compounded semiannually. The loan is to be repaid with ten equal monthly payments, the first due 1 year from today. Determine the size of the payments. *Ans.* $123.49

103. Tania has made deposits of $250 at the end of each month for 2 years into a savings account paying interest at $j_4 = 8\%$. What monthly deposits for the next 12 months will bring the fund up to $10 000? *Ans.* $239.89

104. At age 65 a man takes his life savings of $96 000 and buys a 20-year annuity with monthly payments. Find the size of these payments if interest is at 8% per annum compounded (*a*) daily, (*b*) continuously. *Ans.* (*a*) $804.53; (*b*) $804.58

105. A couple is looking at buying one of two houses. The smaller house would require a $55 000 mortgage, the larger house a $70 000 mortgage. If $j_2 = 10\frac{1}{2}\%$, what difference would there be in their monthly payments between these two houses (assume a 25-year repayment period is used in both cases)? *Ans.* $139.25

106. Mr. and Mrs. Battiston are considering the purchase of a house. They would require a mortgage loan of $35 000 at rate $j_2 = 9\frac{3}{4}\%$. If they can afford to pay $900 monthly, how many full payments will be required and what will be the smaller concluding payment? *Ans.* 46, $654.14

107. Find the accumulated and the discounted value of payments of $100 at the end of each quarter-year for 10 years at $j_{12} = 4\%$. *Ans.* $4892.00; $3281.39

108. Deposits of $100 are made at the end of each month to a savings account paying interest at 8% compounded semiannually. How much money will be accumulated in the account at the end of 3 years if (*a*) simple interest is paid for part of a conversion period, (*b*) compounded interest is paid for part of a conversion period? *Ans.* (*a*) $4046.12; (*b*) $4045.61

109. A company wishes to have $150 000 in a fund at the end of 8 years. What deposit at the end of each month must they make if the fund pays interest at 5% compounded daily? *Ans.* $1273.45

110. Steve buys a new car worth $15 000. He pays $3000 down and agrees to pay $500 at the end of each month as long as necessary. Find the number of full payments and the final payment 1 month later if interest is at $j_1 = 14.2\%$. *Ans.* 28, $38.03

111. If it takes $50 per month for 18 months to repay a loan of $800, (*a*) what nominal rate compounded semiannually is being charged? (*b*) What annual effective rate is being charged? *Ans.* (*a*) 15.74%; (*b*) 16.36%

112. A lot is sold for $8000 down and 6 semiannual payments of $3000, the first due at the end of 2 years. Find the cash value of the lot if money is worth 6% compounded daily. *Ans.* $22 830.71

113. Find the accumulated and discounted value of payments of $100 made at the beginning of each month for 5 years at $j_4 = 14\%$. *Ans.* $8681.11

114. Find the present value of a perpetuity of $362.99 payable every 4 years, the first payment 3 years hence, at the annual effective rate of 4%. *Ans.* $2222.49

115. What annual deposits are needed for 15 years to provide for a perpetuity of $2000 per year, with the first payment in 10 years? Assume interest is at $j_1 = 8\%$. *Ans.* $1461.10

116. On September 1, 1993, a wealthy industrialist gives a university a fund of $100 000 which is invested at 8% compounded daily. If semiannual scholarships are awarded for 20 years from this grant, what is the size of each scholarship if the first one is awarded on (*a*) September 1, 1993, (*b*) September 1, 1995? *Ans.* (*a*) $4912.66; (*b*) $5764.96

117. If the semiannual scholarships in question 116 were awarded indefinitely, what would be the size of the payments for the starting dates listed above? *Ans.* (*a*) $3920.64; (*b*) $4600.83

118. Find the discounted and the accumulated value of a decreasing annuity of 20 payments at the end of each year at $j_1 = 12\%$, if the first payment is $2000, the second $1900, and so on, the last payment being $100. *Ans.* $10 442.13; $100 727.85

119. Find the discounted value of a series of payments that start at $18 000 at the end of year 1 and then increase by $2000 each year forever (that is, $20 000, $22 000, etc.) if interest is at $j_1 = 10\%$. *Ans.* $380 000

120. How much a month for 4 years at 12% compounded continuously would you have to save in order to receive $500 a month for 3 years afterward? *Ans.* $245.36

121. Find the present value of an ordinary annuity paid annually for 25 years if payments are $1000 per year for the first 7 years, $5000 for the following 8 years, and $2000 in the final 10 years. Interest is at $j_{12} = 6\%$ throughout this period. *Ans.* $31 787.13

122. $100 is deposited in an account paying $j_{12} = 12\%$ every month for 20 years. Exactly 6 months after the last deposit the rate changes to $j_4 = 10\%$. On this date the first withdrawal is made in a series of quarterly withdrawals for 15 years. Find the amount of each withdrawal. *Ans.* $3314.61

123. A couple requires a $95 000 mortgage loan at $j_2 = 6\frac{3}{4}\%$ to buy a new house. (*a*) Find the couple's monthly mortgage payment based on the following repayment periods: 30 years, 20 years, and 10 years. (*b*) If the couple can afford to pay up to $850 monthly on their mortgage loan, what repayment period in years should they request? What will be their monthly payment? *Ans.* (*a*) $610.31, $717.10, $1086.31; (*b*) 15 years, $835.77

124. An annuity paying $200 at the end of each year for 20 years is replaced by another annuity paying $X at the end of every 6 months for 12 years. Find X if $j_{12} = 12\%$ in both cases. *Ans.* $115.72

125. Starting on her 36th birthday, Ms. Gagnon deposits $2000 a year into a savings account. Her last deposit is at age 65. Starting 1 month later, she makes monthly withdrawals for 15 years. If $j_4 = 8\%$ throughout, find the size of the monthly withdrawals. *Ans.* $2256.98

126. Danielle and Sue would like to open their own business in 3 years. They estimate they will need $40 000 at that time. How much should they deposit at the end of each month into their savings account if they have $5000 in their account now and the interest is at $j_4 = 7\%$? *Ans.* $848.07

127. Find the discounted value of a series of 20 annual payments of $500 if $j_1 = 5\%$ and we want to allow for an inflation factor of $j_1 = 2\%$. *Ans.* $7479.35

128. Find the discounted value of a series of 15 payments made at the end of each year at $j_1 = 6\%$ if the first payment is $300, the second payment is $600, the third $900, and so on. *Ans.* $20 180.04

129. A man has a job that pays $25 000 a year. Each year he gets a $1000 raise. What is the discounted value of his income for the next 10 years if money is worth $j_1 = 7\%$? (Assume the payments are at the end of each year with a first payment of $25 000.) *Ans.* $203 305.09

130. Mrs. Rider has just retired and is trying to decide between two retirement income options as to where she should place her life savings. Fund A will pay her quarterly payments for 25 years starting at $3000 at the end of the first quarter. Fund A will increase her payments each quarter thereafter using an inflation factor equivalent to 4% per annum compounded quarterly. Fund B pays $4500 at the end of each quarter for 25 years with no inflation factor. Which fund should Mrs. Rider choose if she compares the funds using 6% per annum compounded quarterly? *Ans.* Choose Fund A

131. Find the present value of a perpetuity under which an amount p is paid at the end of the second year, $p + q$ at the end of the fourth year, $p + 2q$ at the end of the sixth year, $p + 3q$ at the end of the eighth year, etc., if interest is at rate i per year. *Ans.* $(p + q/is_{\overline{2}|i})/is_{\overline{2}|i}$

132. Beginning on his son's first birthday and continuing until his son's eighteenth birthday, a father deposits a sum equal to $\$X$ times his son's age in a fund earning interest at rate i per annum effective. Find the amount in the fund just after the last deposit.

Ans. $S = \dfrac{X}{i}(s_{\overline{19}|i} - 19)$

133. A loan is being repaid over 10 years with equal annual payments. Interest is at $j_1 = 10\%$. If the amount of principal repaid in the third payment is $\$100$, find the amount of principal repaid in the 7th payment. *Ans.* $\$146.41$

134. A $\$10\,000$ loan is to be repaid with semiannual payments of $\$2500$ for as long as necessary. If interest is at $j_{12} = 12\%$, construct a complete amortization schedule.

135. A $\$16\,000$ car is purchased by paying $\$1000$ down and then equal monthly payments for 3 years at $j_{12} = 6\%$. (*a*) Find the size of the monthly payment, (*b*) Complete the first three lines of the amortization schedule. *Ans.* (*a*) $\$456.33$

136. A debt of $\$10\,000$ will be amortized by payments at the end of each quarter of a year for 10 years. Interest is at $j_4 = 10\%$. Find the outstanding principal at the end of 6 years. *Ans.* $\$5200.37$

137. On May 1, 1995, the Morins borrow $\$4000$ to be repaid with monthly payments over 3 years at $j_{12} = 9\%$. (*a*) The 12 payments made during 1996 will reduce the principal by how much? (*b*) What was the total interest paid in 1996? *Ans.* (*a*) $\$1281.01$; (*b*) $\$245.39$

138. A family buys a house worth $\$326\,000$. They pay $\$110\,000$ down and then take out a 5-year mortgage for the balance at $j_2 = 10\frac{1}{2}\%$ to be amortized over 20 years. Payments will be made monthly. Find (*a*) the outstanding principal balance at the end of 5 years, and (*b*) the owner's equity at that time. *Ans.* (*a*) $\$194\,596.80$; (*b*) $\$131\,403.20$

139. Land worth $\$80\,000$ is purchased by a down payment of $\$12\,000$ and the balance in equal monthly instalments for 15 years. If interest is at $j_{12} = 9\%$, find the buyer's and seller's equity in the land at the end of 9 years. *Ans.* $\$41\,738.99$, $\$38\,261.01$

140. A debt is being amortized at an annual effective rate of interest of 15% by payments of $\$500$ made at the end of each year for 11 years. Find the outstanding principal just after the 7th payment. *Ans.* $\$1427.49$

141. A loan is to be amortized by semiannual payments of $\$802$, which includes principal and interest at 8% compounded semiannually. What is the original amount of the loan if the outstanding principal is reduced to $\$17\,630$ at the end of 3 years? *Ans.* $\$18\,137.44$

142. A loan is being repaid with semiannual instalments of $\$1000$ for 10 years at 10% per annum convertible semiannually. Find the amount of principal in the sixth instalment. *Ans.* $\$481.02$

143. Jones purchased a cottage, paying $\$10\,000$ down and agreeing to pay $\$500$ at the end of every month for the next ten years. The rate of interest is $j_2 = 12\%$. Jones discharges the remaining indebtedness without penalty by making a single payment at the end of 5 years. Find the extra amount that Jones pays in addition to the regular payment then due. *Ans.* $\$22\,626.02$

144. A 5-year $\$6000$ loan is being amortized with monthly payments at $j_{12} = 18\%$. Just after making the 30th payment, the borrower has the balance refinanced at $j_{12} = 12\%$ with the term of the loan to remain unchanged. What will be the monthly savings in interest? *Ans.* $\$10.60$

145. Mr. Fisher is repaying a loan at $j_{12} = 15\%$ with monthly payments of $\$1500$ over 3 years. Due to temporary unemployment, Mr. Fisher missed making the 13th through the 18th payments inclusive. Find the value of the revised monthly payments needed starting in the 19th month if the loan is still to be repaid at $j_{12} = 15\%$ by the end of the original 3 years. *Ans.* $\$2079.31$

calculate the outstanding principal, what is the outstanding principal after 6 years?
Ans. (a) \$3670.61, \$3670.46; (b) \$297.56; (c) \$21 763.21

169. A couple has a \$150 000, 5-year mortgage at $j_2 = 9\%$ with a 20-year amortization period. After exactly 3 years (36 payments) they could renegotiate a new mortgage at $j_2 = 7\%$. If the bank charges an interest penalty of three times the monthly interest due on the outstanding balance at the time of renegotiation, what will their new monthly payment be? *Ans.* \$1198.27

170. Janet wants to borrow \$10 000 to be repaid over 10 years. From one source, money can be borrowed at $j_1 = 10\%$ and amortized by annual payments. From a second source, money can be borrowed at $j_1 = 9\frac{1}{2}\%$ if only the interest is paid annually and the principal repaid at the end of 10 years. If the second source is used, a sinking fund will be established by annual deposits that accumulate at $j_4 = 8\%$. How much can Janet save annually by using the better plan? *Ans.* Use amortization and save \$4.90 a year

171. A loan is paid off over 19 years with equal monthly payments. The total interest paid over the life of the loan is \$5681.17. Using the sum-of-digits method, determine the amount of interest paid in the 163rd payment. *Ans.* \$14.36

172. The XYZ Mortgage Company lends you \$100 000 at 9% per annum convertible semiannually. The loan is to be repaid by monthly payments at the end of each month for 20 years and the rate is guaranteed for 5 years. (a) Find the monthly payment. (b) Find the total amount of interest paid over the first 5 years. (c) Split the first monthly payments into principal and interest portions (i) based on the true amortization method, and (ii) based on the sum of digits approximation method. (d) If after 5 years of payments interest rates have increased to 11% per annum convertible semiannually, find the new monthly payment at time of mortgage renegotiation exactly 5 years after the original loan agreement based on the true amortization method.
Ans. (a) \$889.19; (b) \$41 870.28; (c) (i) \$152.88, \$736.31; (ii) −\$51.93, \$941.12; (d) \$992.59

173. A loan of \$2000 is being repaid by equal monthly payments for an unspecified length of time. Interest on the loan is $j_{12} = 15\%$. (a) If the amount of principal in the 4th payment is \$40, what amount of the 18th payment will be principal? (b) Find the regular monthly payment. *Ans.* (a) \$47.60; (b) \$63.54

174. On a loan at $j_{12} = 12\%$ with monthly payments, the amount of principal in the 8th payment is \$62. (a) Find the amount of principal in the 14th payment. (b) If there are 48 equal payments in all, find the amount of the loan. *Ans.* (a) \$65.81; (b) \$3540.41

175. A couple purchases a home worth \$150 000 by paying \$30 000 down and taking out a 5-year mortgage for \$120 000 at $j_2 = 10.25\%$. The mortgage will be amortized over 25 years with equal monthly payments. How much of the principal is repaid during the first year? *Ans.* \$1129.42

176. You take out a \$80 000 mortgage at $j_2 = 9\%$ with a 25-year amortization period. (a) Find the monthly payment required, rounded up to the dime. (b) Find the reduced final payment. (c) Find the total interest paid during the 4th year. (d) At the end of 4 years, you pay down an additional \$2500 (no penalty). (i) How much sooner will the mortgage be paid off? (ii) What would be the difference in total payments over the life of the mortgage?
Ans. (a) \$662.40; (b) \$642.45; (c) \$6754.96; (d) (i) 22 months sooner; (ii) \$14 673.89

177. A \$1000 bond with coupons at $j_2 = 10\%$ is redeemable at par in 20 years. It is callable after 10 years at 110 and after 15 years at 105. Find the price to guarantee a yield rate of (a) $j_2 = 8\%$; (b) $j_2 = 12\%$. *Ans.* (a) \$1181.54; (b) \$849.54

178. The Acme Corporation issues \$10 000 000 of 20-year bonds on March 15, 1995, with semiannual coupons at 13%. The contract requires that Acme will set up a sinking fund earning interest at $j_2 = 10\%$ to redeem the bonds at maturity, the first semiannual sinking-fund deposit to be made September 15, 1995. Find (a) the purchase price of the bond issue to yield $j_2 = 11\%$, (b) the necessary sinking-fund deposit. *Ans.* (a) \$11 604 612.50; (b) \$82 781.61

179. Refer to Problem 178. Rather than issuing 20-year bonds, assume that Acme Corporation decided to issue serial bonds on March 15, 1995, to be redeemed as follows: $5 000 000 on March 15, 2001; $3 000 000 on March 15, 2006; $2 000 000 on March 15, 2011. Compute the purchase price of this serial-bond issue to the public on March 15, 1995, to yield $j_2 = 11\%$. *Ans.* $11 106 504.94

180. The XYZ Corporation issues a $1000, 12% bond with coupons payable February 1 and August 1 and redeemable at par on August 1, 2010. (*a*) How much should be paid for this bond on February 1, 1995, to yield $j_2 = 11\%$? (*b*) If the purchase price was $950, and the bond is held to maturity, determine the overall yield, j_2. *Ans.* (*a*) $1073.62; (*b*) 12.76%

181. A $1000 bond with semiannual coupons at 10%, payable January 1 and July 1 each year, matures on July 1, 2000, for $1050. (*a*) Find the price on January 1, 1995, to yield $j_2 = 10\frac{1}{2}\%$. (*b*) Is the bond purchased at a premium or a discount? (*c*) Calculate the entries in the bond amortization schedule for July 1, 1995, and January 1, 1996. *Ans.* (*a*) $1007.98; (*b*) discount; (*c*) $2.92, $3.07

182. The ABC Corporation $2000 bond, paying bond interest at $j_2 = 13\%$, matures at par on September 1, 2008. (*a*) What did a buyer pay for this bond on July 20, 1995, if the market quotation for the bond was 104.75? (*b*) Estimate using linear interpolation, the yield rate for the buyer in (*a*). (*c*) What should the market quotation for this bond be on July 20, 1995, if the desired yield is $j_2 = 7\%$? *Ans.* (*a*) $2194.62; (*b*) 12.28%; (*c*) 151

183. A corporation issues 20-year, par value bonds on February 1, 1995, with coupons at 13% payable February 1 and August 1. Mr. Brown buys a $1000 bond from this issue on February 1, 1995, to yield $j_2 = 14\%$. On August 1, 1999, Mr. Brown sells this bond to Mr. Black, who wants a yield $j_2 = 12\%$. (*a*) Find the original purchase price. (*b*) Find the sale price on August 1, 1999. (*c*) What yield, j_2, did Mr. Brown realize? (*d*) Find the sale price if the transaction instead took place on September 1, 1999.
Ans. (*a*) $933.34; (*b*) $1069.65; (*c*) 16.26%; (*d*) $1080.46

184. Jones invested $10 000 in the Ace Manufacturing Company four years ago. He was to be paid interest on the loan at $j_2 = 11\%$, and the principal ($10 000) was to be returned after 10 years. Now, having just received the eighth interest payment in full, Jones has been informed that Ace has just been declared bankrupt. Jones is offered, as a settlement, 25% of the present value of all monies due him, determined at $j_1 = 13\%$. How much can he expect to receive? *Ans.* $2334.76

185. A $1000 bond has semiannual coupons at $j_2 = 9\%$. The bond matures at par after 20 years, but can be called after 15 years at $1050. (*a*) Find the price to guarantee a yield of $j_2 = 11\%$. (*b*) What maturity date was assumed in answering (*a*)? (*c*) Determine the yield j_2 realized, if the bond is redeemed otherwise than as anticipated in (*a*).
Ans. (*a*) $839.54; (*b*) 20 years; (*c*) 11.40%

186. A bond with face value $10 000 pays $j_2 = 9\%$. An investor buys it for $8000 and sells it 4 years later for $9000. Find the yield rate j_2 earned by the investor over this period using (*a*) the method of averages; (*b*) interpolation. *Ans.* (*a*) 13.53%; (*b*) 13.70%

187. A $1000 bond with coupons at $j_2 = 10\%$ is redeemable at par in n years. It is purchased at a premium of $300. Another $1000 bond with coupons at $j_2 = 8\%$ is also redeemable at par in n years. It is purchased at a premium of $100 on the same yield basis. (*a*) Find the unknown yield rate j_2. (*b*) Determine n. *Ans.* (*a*) 7%; (*b*) 17.5 years

188. A $1000 callable bond pays $j_2 = 10\%$ and matures at par in 20 years. It may be called at $1100 at the end of year 5-9 inclusive. Find the price to yield at least $j_2 = 9\%$. *Ans.* $1092.01

189. A $10 000 bond has semiannual coupons, and matures at par in 15 years. It is bought at a discount to yield $j_2 = 11\%$. The adjustment in book value at the end of 5 years is $25. Find the purchase price. *Ans.* $8881.54

190. Find the price on December 8, 1996, of a $5000 bond maturing at par on July 2, 2006, paying $j_2 = 11\%$. It is priced to yield $j_2 = 12\%$. *Ans.* $4957.62

191. A $5000 callable bond matures on September 1, 2004, at par. It is callable on September 1, 1999, 2001 or 2003 at $5250. Interest on the bond is $j_2 = 13\%$. (*a*) Find the price on September 1, 1995, to yield an investor $j_2 = 11\%$. (*b*) Draw up the bond schedule for the year September 1, 1995, to September 1, 1996. (*c*) If the bond were called on September 1, 2001, what yield j_2 would the investor have earned? Use the method of averages, followed by interpolation. (*d*) On November 1, 1997, the bond is sold to yield $j_2 = 12\%$ till maturity. (i) Find the purchase price. (ii) What is the market quotation?
Ans. (*a*) $5479.63; (*c*) 11.40%, 11.36%; (*d*) (i) $5338.17; (ii) $104\frac{1}{2}$

192. A 20-year bond with a par value of $1000, paying $j_2 = 14\%$, and a maturity value of $1100 is bought on December 20, 1991, to yield $j_2 = 12\%$ to the investor. (*a*) What is the price of the bond? (*b*) How much of the coupon received on June 20, 1995, can be considered as interest income by the purchaser? (*c*) If the bond were sold on July 20, 1995, to yield $j_2 = 9\%$ to the purchaser, how much would the seller receive? (*d*) Assume the bond is bought for the price calculated in (*a*) and sold for the price calculated in (*c*) on July 20, 1995, what rate j_2 has the original investor earned? Use the method of averages.
Ans. (*a*) $1160.19; (*b*) $69.45; (*c*) $1459.66; (*d*) 17.06%

193. A pension fund can earn $j_1 = 11\%$ with investments in government securities. Determine which of the following investments the fund should accept if the initial investment required is $100 000 in each case.

Project	Estimated Cash Flow at End of Year				
	Year 1	Year 2	Year 3	Year 4	Year 5
A	25 000	25 000	25 000	25 000	25 000
B	10 000	30 000	40 000	30 000	10 000
C	−100 000	40 000	50 000	60 000	70 000
D	70 000	50 000	30 000	−	−30 000

Ans. Accept D only

194. Calculate the internal rate of return for a project that costs $100 000 and returns $30 000 at the end of each of the next 4 years. *Ans.* 7.72%

195. Compare the internal rates of return for the alternative projects with initial cost $100 000 and the following cash flows.

Alternative	Estimated Year-End Cash Flow			
	Year 1	Year 2	Year 3	Year 4
(*a*)	60 000	−	−	60 000
(*b*)	−	60 000	60 000	−
(*c*)	−	−	−	120 000
(*d*)	120 000	−	−	−
(*e*)	140 000	−100 000	130 000	−50 000

Ans. (*a*) 7.84%; (*b*) 7.60%; (*c*) 4.66%; (*d*) 20%; (*e*) 18.17%

196. A machine costing $20 000 has, after 10 years of service, a scrap value of $2500. It produces 1000 units of output a year. The annual maintenance cost is $750. How much can be spent to double its output if its scrap value, lifetime, and maintenance costs remain the same? Assume $j_1 = 7\%$. *Ans.* $23 996.82

197. An asset with an initial value of $10 000 has a salvage value of $1000 after 10 years. Find the difference between the depreciation expenses entered in the books in the seventh year as calculated by the sinking-fund method using $j_1 = 9\%$ and by the constant-percentage method. *Ans.* $476.86

198. (*a*) An asset is being depreciated over an expected lifetime of 10 years, after which it will have no salvage value ($S = 0$). If the depreciation expense in the third year is $2000, find the depreciation expense in the eighth year (i) by the straight-line method; (ii) by the sum-of-digits method; (iii) by the sinking-fund method, where $j_1 = 10\%$. (*b*) Find the original value of the asset, C, under each method of (*a*).
Ans. (*a*) (i) $2000, (ii) $750, (iii) $3221.02; (*b*) (i) $20 000, (ii) $13 750, (iii) $26 342.85

199. An $18 000 car will have a scrap value of $2000 six years from now. Prepare a complete depreciation schedule, using (*a*) the straight-line method; (*b*) the constant-percentage method; (*c*) the physical-service method with the actual mileage indicated.

Year	Miles Driven
1	12 500
2	14 700
3	11 800
4	16 200
5	13 800
6	11 000

Ans. (*a*) $R_1 = $2666.66 (etc.); (*b*) $R_1 = $5519.50 (etc.); (*c*) $R_1 = $2500 (etc.)

200. A machine sells for $100 000 and is expected to have a salvage value of $10 000 after 5 years. The maintenance expense of the machine is estimated to be $2000 per year, payable at the end of each year. (*a*) Construct a depreciation schedule using the sinking-fund method and $j_1 = 10\%$. (*b*) Determine the capitalized cost of the asset. *Ans.* (*a*) deposit = $14 741.77; (*b*) $267 417.73

201. A company is considering buying a certain type of machine. Machine 1 costs $6000, will last 15 years, and have a scrap value of $800 at that time; the cost of maintenance is $500 a year. Machine 2 costs $10 000, will last 25 years, and will have scrap value of $1000 at that time; the cost of maintenance is $800 a year. If money is worth $j_1 = 11\%$, which machine should be purchased? *Ans.* Machine 1

202. An asset with an initial value of $100 000 is being depreciated by the sum-of-digits method over a 50-year period. If the depreciation charge in the 31st year is $1000, find the book value of the asset at the end of 41 years. *Ans.* $38 500

203. A machine sells for $80 000 and is expected to have a salvage value of $10 000 after 5 years. The maintenance expense of the machine is estimated to be $2000 per year payable at the end of each year. (*a*) What is the capitalized cost of this asset, assuming $j_1 = 10\%$? (*b*) Construct a depreciation schedule using the constant percentage method. (*c*) Construct a depreciation schedule using the sinking-fund method, with $j_1 = 6\%$. *Ans.* (*a*) $214 658.24; (*b*) $d = 0.340246045$; (*c*) S.F. deposit = $12 417.75

204. A tool is bought for $1000 and has a salvage value of $50 at the end of 8 years. Calculate the difference in book values at the end of 6 years between the sum-of-digits and declining-balance methods of depreciation. *Ans.* $23.43

205. Equipment with a cost of $80 000 has an estimated salvage value of $10 000 at the end of its expected lifetime of 7 years. Find the book value of the equipment at the end of 5 years, using (*a*) the straight-line method; (*b*) the constant-percentage method; (*c*) the sinking-fund method ($j_1 = 6\%$); and (*d*) the sum-of digits method.
Ans. (*a*) $30 000; (*b*) $18 114.47; (*c*) $32 989.74; (*d*) $17 500

206. A card is drawn at random from a standard deck. What is the probability that the card will be (*a*) an ace or a king? (*b*) a red face card (jack, queen, or king)? *Ans.* (*a*) 2/13; (*b*) 3/26

207. On a particular assembly line, 8 parts in 1000 are defective. Assuming statistical independence, determine the probability that (*a*) out of 100 parts, none is defective; (*b*) two parts in succession are defective; (*c*) the first two parts are not defective, but the third is.
Ans. (*a*) 0.4479; (*b*) 0.000064; (*c*) 0.007873

208. Three cards are drawn from a deck of 52, without replacement. What is the probability of a queen followed by a king followed by an ace? *Ans.* $\frac{8}{16\ 575}$

209. Complete the following table.

x	l_x	d_x	p_x	q_x
101	320			
102		80		
103	48			2/3
104			1/4	
105				1

210. A gambling game is played wherein a single card is drawn from a deck of 52. For the cards 2 through 10, one wins the face value of the card (that is, \$2 through \$10). For a jack, queen, or king, one wins \$20; for an ace, one wins \$25. How much would one be expected to place in the pot to make this a fair game? *Ans.* \$10.69

211. Perishable goods are purchased by a retailer for \$10 each and sold for \$15 each. These goods last only one day. Given table below, determined from past experience, how many items of these goods should the retailer purchase to maximize his expected profit? What is his expected profit, given that purchase? [*Hint:* Let P_N be the profit given a purchase of N items. Show that

$$E(P_N) = P(D = 70)(1050 - 10N) + \cdots P(D = N - 1)(5N - 15) + P(D \geq N)(5N)$$

where D denotes the demand.]

Daily Demand	Probability
70	0.05
71	0.35
72	0.40
73	0.15
74	0.05

Ans. 71, \$354.25

212. Miss Jones goes to the Family Trust Company to borrow some money. Family Trust charges $j_1 = 9\frac{1}{2}\%$ for customers with an established credit rating; otherwise they assume a 5% probability of default (in which case they get nothing). For a loan to be repaid in a single instalment at the end of one year, what interest rate should they charge Miss Jones, who has no credit rating? *Ans.* 15.26%

213. An insurance company issues a special retirement savings policy whereby they will pay the policyholder \$10 000 on his 65th birthday if he is then alive. If the probability that a 40-year old male lives to age 65 is 0.810 and if $j_1 = 10\%$, what is a fair price for this policy (ignoring expenses)? *Ans.* \$747.60

214. How much would you lend a person today if he promises to repay \$1000 at the end of each year for 10 years, if there is a 5% chance of default in any year entered, and $j_1 = 14\%$? (If default occurs, no payment will be made beyond the time of default.) *Ans.* \$4192.47

215. A 20-year, $1000 par value bond with semiannual coupons at $j_2 = 11\%$ is issued by a company that is not strong financially. In fact, the probability of bankruptcy is 0.05 in any 6-month period entered, with no value if that happens. Find the purchase price of this bond, for an investor's rate $j_2 = 10\%$. *Ans.* $531.22

216. An insurance policy issued to a female aged 32 can be purchased by paying 5 annual premiums (the first one due now) of $100 each. What net single premium would buy the same policy, if $j_1 = 8\%$? *Ans.* $430.01

217. Mr. Anderson wants to buy a $25 000 whole life insurance policy at age 94. At $j_1 = 10\%$, find (*a*) the net annual premium and (*b*) the net single premium for this policy.
Ans. (*a*) $8217.06; (*b*) $19 583.47

218. The XYZ Insurance Company sells a "life-paid-up-at-65" policy, which offers whole life insurance coverage with premiums payable to age 65. If the net annual premium to a female aged 55 for $10 000 coverage is $200, what is the net annual premium payable for a maximum of 5 years if $j_1 = 7\%$? *Ans.* $336.30

219. To what age does a 65-year old male have a 50% chance of living (based on Appendix B)?
Ans. 79

220. If

$$l_x = l_0 \left(\frac{100 - x}{100} \right) \qquad (x \leq 100)$$

find (*a*) p_{70}, (*b*) $_3p_{40}$, (*c*) $_5q_{20}$. *Ans.* (*a*) 29/30; (*b*) 19/20; (*c*) 1/16

221. Complete the following table.

x	l_x	d_x	p_x	q_x
101	320			
102		80		
103	48			2/3
104			1/4	
105				1

222. Determine the present value to a female at age 28 of a deferred life annuity having 3 payments of $1000 each, the first payment being at age 40, if (*a*) $j_1 = 11\%$, (*b*) $j_1 = 4\%$.
Ans. (*a*) $758.68; (*b*) $1763.70

223. Determine the net single premium for a life annuity due issued to a male aged 97 for $1500 a year, if (*a*) $j_1 = 10\%$, (*b*) $j_1 = 4\%$. *Ans.* (*a*) $2429.21; (*b*) $2496.27

224. Find the net single premium at $j_1 = 9\%$ for a female aged 95 of a whole life annuity of $1000 per year, with the first payment (*a*) at age 96, (*b*) at age 95, (*c*) at age 98.
Ans. (*a*) $1212.34; (*b*) $2212.34; (*c*) $227.33

225. Using a personal computer, find the net annual premium at $j_1 = 4\%$ for a whole life insurance policy of $40 000 issued to a 35-year-old male, with annual premiums payable (*a*) for life, (*b*) for 20 years, (*c*) to age 60. (*d*) What would be the net single premium?
Ans. (*a*) $504.17; (*b*) $718.19; (*c*) $631.98; (*d*) $9872.95

The Number of Each Day of the Year

Day of month	Jan.	Feb.	Mar.	Apr.	May	June	July	Aug.	Sept.	Oct.	Nov.	Dec.	Day of month
1	1	32	60	91	121	152	182	213	244	274	305	335	1
2	2	33	61	92	122	153	183	214	245	275	306	336	2
3	3	34	62	93	123	154	184	215	246	276	307	337	3
4	4	35	63	94	124	155	185	216	247	277	308	338	4
5	5	36	64	95	125	156	186	217	248	278	309	339	5
6	6	37	65	96	126	157	187	218	249	279	310	340	6
7	7	38	66	97	127	158	188	219	250	280	311	341	7
8	8	39	67	98	128	159	189	220	251	281	312	342	8
9	9	40	68	99	129	160	190	221	252	282	313	343	9
10	10	41	69	100	130	161	191	222	253	283	314	344	10
11	11	42	70	101	131	162	192	223	254	284	315	345	11
12	12	43	71	102	132	163	193	224	255	285	316	346	12
13	13	44	72	103	133	164	194	225	256	286	317	347	13
14	14	45	73	104	134	165	195	226	257	287	318	348	14
15	15	46	74	105	135	166	196	227	258	288	319	349	15
16	16	47	75	106	136	167	197	228	259	289	320	350	16
17	17	48	76	107	137	168	198	229	260	290	321	351	17
18	18	49	77	108	138	169	199	230	261	291	322	352	18
19	19	50	78	109	139	170	200	231	262	292	323	353	19
20	20	51	79	110	140	171	201	232	263	293	324	354	20
21	21	52	80	111	141	172	202	233	264	294	325	355	21
22	22	53	81	112	142	173	203	234	265	295	326	356	22
23	23	54	82	113	143	174	204	235	266	296	327	357	23
24	24	55	83	114	144	175	205	236	267	297	328	358	24
25	25	56	84	115	145	176	206	237	268	298	329	359	25
26	26	57	85	116	146	177	207	238	269	299	330	360	26
27	27	58	86	117	147	178	208	239	270	300	331	361	27
28	28	59	87	118	148	179	209	240	271	301	332	362	28
29	29	..	88	119	149	180	210	241	272	302	333	363	29
30	30	..	89	120	150	181	211	242	273	303	334	364	30
31	31	..	90	...	151	...	212	243	...	304	...	365	31

Note: For leap year add 1 to the tabulated number after February 28.

Appendix B

Commissioners Standard Ordinary Mortality Table
(As at 1/1/96)

				Female Lives			
Age	l_x	d_x	$1000q_x$	Age	l_x	d_x	$1000q_x$
0	10000000	28900	2.89	50	9219130	45727	4.96
1	9971100	8675	0.87	51	9173403	48711	5.31
2	9962425	8070	0.81	52	9124692	52011	5.70
3	9954355	7864	0.79	53	9072681	55797	6.15
4	9946491	7659	0.77	54	9016884	59602	6.61
5	9938832	7554	0.76	55	8957282	63507	7.09
6	9931278	7250	0.73	56	8893775	67326	7.57
7	9924028	7145	0.72	57	8826449	70876	8.03
8	9916883	6942	0.70	58	8755573	74160	8.47
9	9909941	6838	0.69	59	8681413	77612	8.94
10	9903103	6734	0.68	60	8603801	81478	9.47
11	9896369	6828	0.69	61	8522323	86331	10.13
12	9889541	7120	0.72	62	8435992	92458	10.96
13	9882421	7412	0.75	63	8343534	100289	12.02
14	9875009	7900	0.80	64	8243245	109223	13.25
15	9867109	8387	0.85	65	8134022	118675	14.59
16	9858722	8873	0.90	66	8015347	128246	16.00
17	9849849	9357	0.95	67	7887101	137472	17.43
18	9840492	9644	0.98	68	7749629	146003	18.84
19	9830848	10027	1.02	69	7603626	154810	20.36
20	9820821	10312	1.05	70	7448816	164693	22.11
21	9810509	10497	1.07	71	7284123	176494	24.23
22	9800012	10682	1.09	72	7107629	190982	26.87
23	9789330	10866	1.11	73	6916647	208260	30.11
24	9778464	11147	1.14	74	6708387	227616	33.93
25	9767317	11330	1.16	75	6480771	247825	38.24
26	9755987	11610	1.19	76	6232946	267830	42.97
27	9744377	11888	1.22	77	5965116	286564	48.04
28	9732489	12263	1.26	78	5678552	303519	53.45
29	9720226	12636	1.30	79	5375033	319008	59.35
30	9707590	13105	1.35	80	5056025	333647	65.99
31	9694485	13572	1.40	81	4722378	347567	73.60
32	9680913	14037	1.45	82	4374811	360484	82.40
33	9666876	14500	1.50	83	4014327	371446	92.53
34	9652376	15251	1.58	84	3642881	378167	103.81
35	9637125	15901	1.65	85	3264714	379033	116.10
36	9621224	16933	1.76	86	2885681	373090	129.29
37	9604291	18152	1.89	87	2512591	360105	143.32
38	9586139	19556	2.04	88	2152486	340480	158.18
39	9566583	21238	2.22	89	1812006	315180	173.94
40	9545345	23100	2.42	90	1496826	285520	190.75
41	9522245	25139	2.64	91	1211306	253005	208.87
42	9497106	27257	2.87	92	958301	219269	228.81
43	9469849	29262	3.09	93	739032	185874	251.51
44	9440587	31343	3.32	94	553158	154503	279.31
45	9409244	33497	3.56	95	398655	126501	317.32
46	9375747	35628	3.80	96	272154	102259	375.74
47	9340119	37827	4.05	97	169895	80695	474.97
48	9302292	40279	4.33	98	89200	58502	655.85
49	9262013	42883	4.63	99	30698	30698	1000.00

Commissioners Standard Ordinary Mortality Table
(As at 1/1/96)

Age	l_x	d_x	$1000q_x$	Age	l_x	d_x	$1000q_x$
0	10000000	41800	4.18	50	8966618	60166	6.71
1	9958200	10655	1.07	51	8906452	65017	7.30
2	9947545	9848	0.99	52	8841435	70378	7.96
3	9937697	9739	0.98	53	8771057	76396	8.71
4	9927958	9432	0.95	54	8694661	83121	9.56
5	9918526	8927	0.90	55	8611540	90163	10.47
6	9909599	8522	0.86	56	8521377	97655	11.46
7	9901077	7921	0.80	57	8423722	105212	12.49
8	9893156	7519	0.76	58	8318510	113049	13.59
9	9885637	7315	0.74	59	8205461	121195	14.77
10	9878322	7211	0.73	60	8084266	129995	16.08
11	9871111	7601	0.77	61	7954271	139518	17.54
12	9863510	8384	0.85	62	7814753	149965	19.19
13	9855126	9757	0.99	63	7664788	161420	21.06
14	9845369	11322	1.15	64	7503368	173628	23.14
15	9834047	13079	1.33	65	7329740	186322	25.42
16	9820968	14830	1.51	66	7143418	198944	27.85
17	9806138	16376	1.67	67	6944474	211390	30.44
18	9789762	17426	1.78	68	6733084	223471	33.19
19	9772336	18177	1.86	69	6509613	235453	36.17
20	9754159	18533	1.90	70	6274160	247892	39.51
21	9735626	18595	1.91	71	6026268	260937	43.30
22	9717031	18365	1.89	72	5765331	274718	47.65
23	9698666	18040	1.86	73	5490613	289026	52.64
24	9680626	17619	1.82	74	5201587	302680	58.19
25	9663007	17104	1.77	75	4898907	314461	64.19
26	9645903	16687	1.73	76	4584446	323341	70.53
27	9629216	16466	1.71	77	4261105	328616	77.12
28	9612750	16342	1.70	78	3932489	329936	83.90
29	9596408	16410	1.71	79	3602553	328012	91.05
30	9579998	16573	1.73	80	3274541	323656	98.84
31	9563425	17023	1.78	81	2950885	317161	107.48
32	9546402	17470	1.83	82	2633724	308804	117.25
33	9528932	18200	1.91	83	2324920	298194	128.26
34	9510732	19021	2.00	84	2026726	284248	140.25
35	9491711	20028	2.11	85	1742478	266512	152.95
36	9471683	21217	2.24	86	1475966	245143	166.09
37	9450466	22681	2.40	87	1230823	220994	179.55
38	9427785	24324	2.58	88	1009829	195170	193.27
39	9403461	26236	2.79	89	814659	168871	207.29
40	9377225	28319	3.02	90	645788	143216	221.77
41	9348906	30758	3.29	91	502572	119100	236.98
42	9318148	33173	3.56	92	383472	97191	253.45
43	9284975	35933	3.87	93	286281	77900	272.11
44	9249042	38753	4.19	94	208381	61660	295.90
45	9210289	41907	4.55	95	146721	48412	329.96
46	9168382	45108	4.92	96	98309	37805	384.55
47	9123274	48536	5.32	97	60504	29054	480.20
48	9074738	52089	5.74	98	31450	20693	657.98
49	9022649	56031	6.21	99	10757	10757	1000.00